P9-DFB-499

LABORATORY GUIDE

NINTH EDITION

Human Physiology

LABORATORY GUIDE

NINTH EDITION

Human Physiology

Stuart Ira Fox

PIERCE COLLEGE

Mc
Graw
Hill

Boston   Burr Ridge, IL   Dubuque, IA   Madison, WI   New York   San Francisco   St. Louis
Bangkok   Bogotá   Caracas   Kuala Lumpur   Lisbon   London   Madrid   Mexico City
Milan   Montreal   New Delhi   Santiago   Seoul   Singapore   Sydney   Taipei   Toronto

McGraw-Hill Higher Education

A Division of The **McGraw-Hill** *Companies*

LABORATORY GUIDE FOR HUMAN PHYSIOLOGY
NINTH EDITION

Published by McGraw-Hill, a business unit of The McGraw-Hill Companies, Inc., 1221
Avenue of the Americas, New York, NY 10020. Copyright © 2002, 1999, 1996 by The
McGraw-Hill Companies, Inc. All rights reserved. No part of this publication may be
reproduced or distributed in any form or by any means, or stored in a database or retrieval
system, without the prior written consent of The McGraw-Hill Companies, Inc., including,
but not limited to, in any network or other electronic storage or transmission, or broadcast
for distance learning.

Some ancillaries, including electronic and print components, may not be available to
customers outside the United States.

This book is printed on recycled, acid-free paper containing 10% postconsumer waste.

1 2 3 4 5 6 7 8 9 0 QPD/QPD 0 9 8 7 6 5 4 3 2 1

ISBN 0–07–027236–0

Publisher: *Colin H. Wheatley*
Senior developmental editor: *Lynne M. Meyers*
Marketing manager: *Michelle Watnick*
Project manager: *Joyce Watters*
Production supervisor: *Sherry L. Kane*
Designer: *K. Wayne Harms*
Cover designer: *Diane Beasley*
Cover illustration: *William Westwood*
Photo research coordinator: *John C. Leland*
Photo research: *LouAnn K. Wilson*
Senior supplement producer: *Stacy A. Patch*
Media technology senior producer: *Barbara R. Block*
Compositor: *Shepherd, Inc.*
Typeface: *10/12 Goudy*
Printer: *Quebecor World Dubuque, IA*

www.mhhe.com

Contents

4
THE ENDOCRINE SYSTEM 145

5
SKELETAL MUSCLES 169

6
BLOOD: GAS TRANSPORT, IMMUNITY, AND CLOTTING FUNCTIONS 197

7
THE CARDIOVASCULAR SYSTEM 233

8

RESPIRATION AND METABOLISM   285

9

RENAL FUNCTION AND HOMEOSTASIS   325

10

DIGESTION AND NUTRITION   349

11

REPRODUCTIVE SYSTEM   385

Appendix 1

Appendix 2

Appendix 3

Preface

The ninth edition, like the previous editions, is a stand-alone human physiology manual that can be used in conjunction with any human physiology textbook. It includes a wide variety of exercises that support most areas covered in a human physiology course, allowing instructors the flexibility to choose those exercises best suited to meet their particular instructional goals. Background information that is needed to understand the principles and significance of each exercise is presented in a concise manner, so that little or no support is needed from the lecture text.

However, lecture and laboratory segments of a human physiology course are most effectively wedded when they cover topics in a similar manner and sequence. Thus, this laboratory guide is best used in conjunction with the textbook *Human Physiology*, seventh edition, by Stuart Ira Fox (McGraw-Hill, © 2002).

The laboratory experiences provided by this guide allow students to become familiar—in an intimate way that cannot be achieved by lecture and text alone—with many fundamental concepts of physiology. In addition to providing hands-on experience in applying physiological concepts, the laboratory sessions allow students to interact with the subject matter, with other students, and with the instructor in a personal, less formal way. Active participation is required to carry out the exercise procedures, collect data, and to complete the laboratory report. Critical thinking is necessary to answer all questions at the end of each exercise.

NEW TO THE NINTH EDITION

UPDATED INFORMATION

The ninth edition is a thorough renovation of the eighth edition. Each exercise has been carefully refined and updated to keep pace with continual changes in laboratory technology, vendor supply sources, and biohazard health concerns. Laboratories that utilize the *Biopac* or *Intelitool* systems for computer-assisted data acquisition will find references and correlations to the use of these systems with the exercises presented in this edition. Similarly, those that use the *A.D.A.M.* interactive physiology programs to supplement their classroom instruction will find correlations to those programs in the exercises of this edition.

The review activities in the laboratory reports at the end of each exercise are thoroughly revised in this edition. They now present questions at three levels: *Test Your Knowledge of Terms and Facts*, *Test Your Understanding of Concepts*, and *Test Your Ability to Analyze and Apply Your Knowledge*. These three levels of questions are consistent between laboratory exercises, and consistent with the Review Activities approach in the textbook *Human Physiology*, seventh edition, by Stuart Ira Fox.

Clinically oriented laboratory exercises that heighten student interest and demonstrate the health applications of physiology have been a hallmark of previous editions and continue to be featured in this latest edition.

We are indebted to our colleagues and students for their suggestions and encouragement in the development of these exercises. Drawing on these recommendations, many of the laboratory procedures have been altered to accommodate both fluctuations in class size and laboratory time constraints. Some alterations were necessary since some of the sources of laboratory supplies and equipment have changed. New sources are indicated for some of the reagents, test strips, or kits required for certain exercises, reflecting changes made by the vendors.

SAFETY

Special effort has been made to address concerns about the safe use and disposal of body fluids. For example, normal and abnormal artificial serum can be used as a substitute for blood in Section 2 (plasma chemistry); artificial saliva is suggested in exercise 10.2 (digestion); and in Section 9 (renal function) both normal and abnormal artificial urine is now available. In the interest of safety, a substitute for the use of benzene (previously required in two exercises) is now provided.

The international symbol for caution is used throughout the laboratory guide to alert the reader when special attention is necessary while preparing for or performing a laboratory exercise. For reference, laboratory safety guidelines appear on the inside front cover.

TECHNOLOGY

Computer-assisted and computer-guided instruction in human physiology laboratories has greatly increased in recent years. Computer programs provide a number of benefits: some experiments that require animal sacrifice can be simulated; data can be analyzed against a data bank and displayed in an appealing and informative manner; class data records can be analyzed; and costs can be reduced by eliminating the use of some of the most expensive equipment.

 This edition continues to reference programs offered by Intelitool, and new to this edition, A.D.A.M. Benjamin/Cummings *InterActive PHYSIOLOGY* Modules (800–755–2326; www.adam.com), and the ***Virtual Physiology Lab CD-ROM*** (ISBN 0–697–37994–9) by McGraw-Hill and Cypris Publishing.

ART PROGRAM

Almost every figure in this edition has been revised or improved, with a few deletions, and many new, exciting figures and tables added. These new figures enhance the pedagogical value and add to the aesthetic appeal of the laboratory manual. Furthermore, the design was reworked, adding icons (such as the ⚖ balance icon for normal values), boxes, and shading to important concepts to enhance visual comprehension by students and to improve overall continuity.

ORGANIZATION OF THE LABORATORY GUIDE

The exercises in this guide are organized in the following manner:

1. Each exercise begins with a list of **materials** needed to perform the exercise, so that it is easier to set up the laboratory. This section is identified by a materials icon.

2. Following the materials section is an overview paragraph describing the **concept** behind the laboratory exercise.

3. Following the concept paragraph is a list of **learning objectives,** to help students guide their learning while performing the exercise.

4. A box providing **textbook correlations** is a new feature of this edition. This section can be used to help integrate the lecture textbook (if *Human Physiology, seventh edition,* by Stuart Ira Fox, is used) with the laboratory material.

5. A brief **introduction** to the exercise presents the essential information for understanding the physiological significance of the exercise. This concisely written section eliminates the need to consult the lecture text.

6. Boxed information, set off as screened insets, provide the **clinical significance** of different aspects of the laboratory exercise. This approach was pioneered by this laboratory manual and the current edition continues that tradition.

7. The **procedure** is stated in the form of easy-to-follow steps. These directions are set off from the textual material through the use of a distinctive typeface, making it easier for students to locate them as they perform the exercise.

8. A **laboratory report** follows each exercise. Students enter data here when appropriate, and answer questions. The questions in the laboratory report begin with the most simple form (objective questions) in most exercises and progress to essay questions. The essay questions are designed to stimulate conceptual learning and to maximize the educational opportunity provided by the laboratory experience.

SUPPLEMENTAL MATERIALS

Instructor's Manual for the Laboratory Guide to accompany Human Physiology, ninth edition, by Laurence G. Thouin, Jr. (ISBN 0–697–34221–2) provides a suggested correlation between the textbook and laboratory manual for *Human Physiology,* introductions, materials needed, approximate completion times, and solutions to the laboratory reports for each exercise, a listing of laboratory supply houses, and commonly used solutions.

Virtual Physiology Lab CD-ROM by McGraw-Hill and Cypris Publishing (ISBN 0–697–37994–9) features ten simulations of the most common and important animal-based experiments. The flexibility of this multimedia tool offers many pre-lab, actual lab, and post-lab options.

Laboratory Atlas of Anatomy and Physiology, second edition, by Douglas Eder et al. (ISBN 0–697–39480–8), is a full-color atlas including histology, skeletal and muscular anatomy, dissections, and reference tables.

EXPERIMENT IN THE VIRTUAL WORLD

With ten simulations of the most common laboratory experiments, *Virtual Physiology Lab* lets you conduct lifelike research—without the animals or the lab.

You can work at your own pace and practice essential techniques over and over. The flexibility of this multimedia tool offers many prelab, actual lab, and postlab options. You can work in a computer lab, at home, or in teams.

Each lab features: Objectives, Foundation, Experiment, Results, and Self-Testing.

Contents
1. Action Potential
2. Synaptic Transmission
3. Frog Muscle
4. Effects of Drugs on the Frog Heart
5. Electrocardiogram
6. Pulmonary Function

7. Respiration and Exercise
8. Digestion of Fat
9. Diffusion, Osmosis, and Tonicity
10. Enzyme Characteristics

1998 CD-ROM for Macintosh and Windows ISBN 0-697-37994-9

To order a copy of the *Virtual Physiology Lab*, check your bookstore or call McGraw-Hill Customer Service at 1-800-338-3987.

ACKNOWLEDGMENTS

The ninth edition was greatly benefited by input from my colleague Dr. Laurence G. Thouin, Jr. His numerous suggestions helped to make the ninth edition more accurate and student friendly. I am also grateful to Dr. Jenine Tanabe (Yuba College) for her help in incorporating the *Biopac* procedures into this edition.

The shaping of the ninth edition was also aided by suggestions from other colleagues and students. Ms. Karen Gebhardt was particularly instrumental in checking laboratory sources for materials and reworking some of the procedures that are new to this edition. I greatly appreciate the support of the editors at McGraw-Hill, Colin Wheatley and Lynne Meyers; their contributions help to make this the best edition yet of the *Laboratory Guide to accompany Human Physiology*.

Table 1.3 Structure and Function of Cellular Components

Component	Structure	Function
Cell (plasma) membrane	Membrane composed of phospholipid and protein molecules	Gives form to cell and controls passage of materials in and out of cell
Cytoplasm	Fluid, jellylike substance between the cell membrane and the nucleus in which organelles are suspended	Serves as matrix substance in which chemical reactions occur
Endoplasmic reticulum	System of interconnected membrane-forming canals with (rough) or without (smooth) attached ribosomes	Smooth endoplasmic reticulum metabolizes nonpolar compounds and stores Ca++ in striated muscle cells; rough endoplasmic reticulum assists in protein synthesis
Ribosomes	Granular particles composed of protein and RNA	Synthesize proteins
Golgi apparatus	Cluster of flattened, membranous sacs	Synthesizes carbohydrates and packages protein and lipid molecules for secretion
Mitochondria	Double-walled membranous sacs with folded inner partitions	Release energy from food molecules and transform energy into usable ATP
Lysosomes	Single-walled membranous sacs	Digest foreign molecules and worn and damaged cells
Peroxisomes	Spherical membranous vesicles	Contain enzymes that produce hydrogen peroxide and use this for various oxidation reactions
Centrosome	Nonmembranous mass of two rodlike centrioles	Helps organize spindle fibers and distribute chromosomes during mitosis
Vacuoles	Membranous sacs	Store and excrete various cytoplasmic substances
Fibrils and microtubules	Thin, rodlike, or hollow tubes of varying lengths	Support cytoplasm and transport materials within the cytoplasm (e.g., cytoskeleton)
Cilia and flagella	Small cytoplasmic projections containing microtubules	Move particles along surface of cell and enable sperm to migrate
Nuclear membrane	Porous, double membrane surrounding nucleus composed of protein and lipid molecules	Supports nucleus and controls passage of materials between nucleus and cytoplasm
Nucleolus	Dense, nonmembranous mass composed of protein and RNA molecules	Forms ribosomes
Chromatin	Fibrous strands composed of DNA molecules and protein	Controls cellular activity for carrying on life processes, such as protein synthesis

Table 1.4 Major Events in Mitosis

Stage	Major Events
Prophase	Chromosomes form from the chromatin material, centrioles migrate to opposite sides of the nucleus, the nucleolus and nuclear membrane disappear, and spindles appear and become associated with centrioles and centromeres.
Metaphase	Duplicated chromosomes align themselves on the equatorial plane of the cell between the centrioles, and spindle fibers become attached to duplicate parts of chromosomes.
Anaphase	Duplicated chromosomes separate, and spindles shorten and pull individual chromosomes toward the centrioles.
Telophase	Chromosomes elongate and form chromatin threads, nucleoli and nuclear membranes appear for each chromosome mass, and spindles disappear.

(a) Mitosis

Interphase
- The chromosomes are in extended form and seen as chromatin in the electron microscope.
- The nucleus is visible.

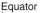

Prophase
- The chromosomes are seen to consist of two chromatids joined by a centromere.
- The centrioles move apart toward opposite poles of the cell.
- Spindle fibers are produced and extend from each centrosome.
- The nuclear membrane starts to disappear.
- The nucleolus is no longer visible.

Metaphase
- The chromosomes line up at the equator of the cell.
- The spindle fibers from each centriole are attached to the centromeres of the chromosomes.
- The nuclear membrane has disappeared.

Anaphase
- The centromeres split, and the sister chromatids separate as each is pulled to an opposite pole.

Telophase
- The chromosomes become longer, thinner, and less distinct.
- New nuclear membranes form.
- The nucleolus reappears.
- Cell division is nearly complete.

Figure 1.3   Cell division. (a) The stages of mitosis. (b) The stages of meiosis. Note that meiosis occurs only in the cells of the gonads that produce the gametes (sperm and ova).

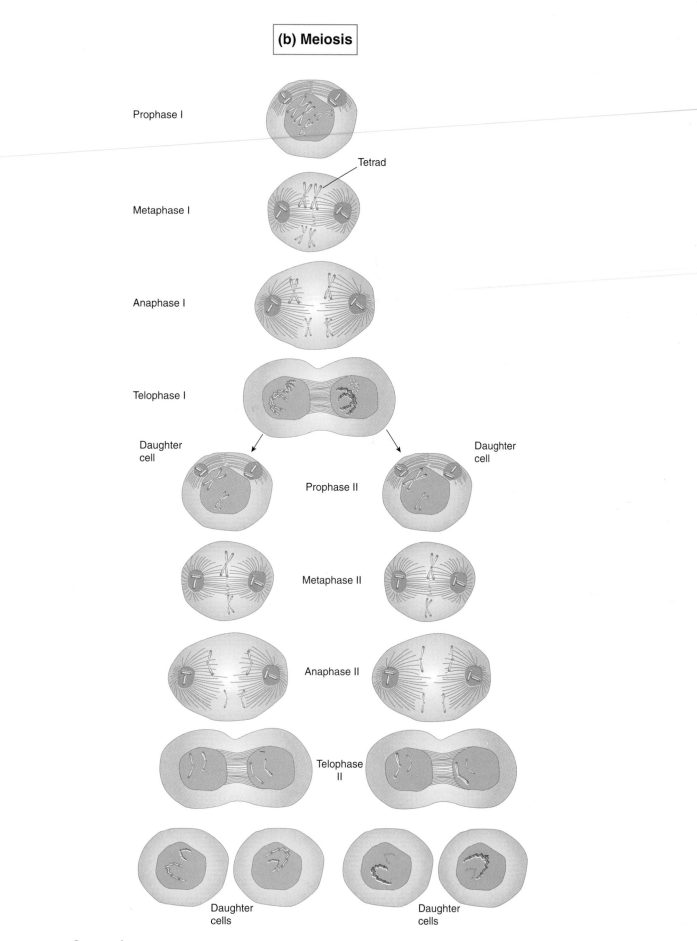

(b) Meiosis

Prophase I

Tetrad

Metaphase I

Anaphase I

Telophase I

Daughter cell

Prophase II

Daughter cell

Metaphase II

Anaphase II

Telophase II

Daughter cells

Daughter cells

Figure 1.3 *Continued*

Microscopic Examination of Tissues and Organs

MATERIALS

1. Compound microscopes
2. Lens paper
3. Prepared microscope slides of tissues

The body is composed of only four primary tissues, and each is specialized for specific functions. Most organs of the body are composed of all four primary tissues, which cooperate in determining the overall structure and function of the organ.

OBJECTIVES

1. Define the terms *tissue* and *organ*.
2. List the distinguishing characteristics of the four primary tissues.
3. Identify and describe the subcategories of the primary tissues.
4. In general terms, correlate the structures of the primary tissues with their function.

Textbook Correlations

Before performing this exercise, you may want to consult the following references in *Human Physiology*, seventh edition, by Stuart I. Fox:

- *The Primary Tissues.* Chapter 1, pp. 8–16.
- *Organs and Systems.* Chapter 1, pp. 17–18.

Those using different physiology textbooks may want to consult the corresponding information in those books.

The trillions of cells that compose the human body have many basic features in common, but they differ considerably in size, structure, and function. Furthermore, cells neither function as isolated units nor are they haphazardly arranged in the body. An aggregation of cells that are similar in structure and that work together to perform a specialized activity is referred to as a tissue. Groups of tissues that are integrated to perform one or more common functions constitute organs. Tissues are categorized into four principal types, or **primary tissues:** (1) *epithelial*, (2) *connective*, (3) *muscular*, and (4) *nervous*.

A. Epithelial Tissue

Epithelial tissue, or *epithelium*, functions to protect, secrete, or absorb. Epithelial membranes cover the outer surface of the body (epidermis of the skin) and the outer surfaces of internal organs; and line the body cavities and the *lumina* (the inner hollow portions) of ducts, vessels, and tubes. All *glands* are derived from epithelial tissue. Epithelial tissues share the following characteristics:

1. The cells are closely joined together and have little intercellular substance (matrix) between them.
2. There is an exposed surface either externally or internally.
3. A *basement membrane* is present to anchor the epithelium to underlying connective tissue.

Epithelial tissues that are composed of a single layer of cells are called *simple*; those composed of more than one layer are known as *stratified*. Epithelial tissues may be further classified by the shape of their surface cells: *squamous* (if the cells are flat), *cuboidal*, or *columnar*. Using these criteria, one can identify the following types of epithelia:

1. **Simple squamous epithelium** (fig. 1.4, *top*). This type is adapted for diffusion, absorption, filtration, and secretion—present in such places as the lining of air sacs, or *alveoli*, within the lungs (where gas exchange occurs); parts of the kidney (where blood is filtered); and the lining, or *endothelium*, of blood vessels (where exchange between blood and tissues occurs).
2. **Stratified squamous epithelium** (fig. 1.4, *middle*). This type is found in areas that receive a lot of wear and tear. The outer cells are sloughed off and replaced by new cells, produced by mitosis in the deepest layers. Stratified squamous epithelium is found in the mouth, esophagus, nasal cavity, and in the openings into the ears, anus, and vagina. A special *keratinized*, or *cornified*, layer of dead surface

Figure 1.4 Squamous and cuboidal epithelial membranes. The structures shown in each photomicrograph are depicted in the accompanying diagrams.

cells is found in the stratified squamous epithelium of the skin (the epidermis).

3. **Simple cuboidal epithelium** (fig. 1.4, *bottom*). This type of epithelium is usually simple and is found lining such structures as small tubules of the kidneys, and the ducts of the salivary glands or of the pancreas.

4. **Simple columnar epithelium** (fig. 1.5, *top*). This simple epithelium of tall columnar cells is found lining the lumen of the gastrointestinal tract, where it is specialized to absorb the products of digestion. It also contains mucus-secreting *goblet cells*.

5. **Simple ciliated columnar epithelium** (fig. 1.5, *upper middle*). These columnar cells support hairlike *cilia* on the exposed surface. These cilia produce wavelike movements that are characteristic along the luminal surface of female uterine tubes and the ductus deferens (vas deferens) of the male.

6. **Pseudostratified ciliated columnar epithelium** (fig. 1.5, *lower middle*). This epithelium is really simple but appears stratified because the nuclei are at different levels. Also characterized by hairlike cilia, this epithelium is found lining the respiratory passages of the trachea and bronchial tubes.

7. **Transitional epithelium** (fig. 1.5, *bottom*). This type is found only in the urinary bladder and ureters, and is uniquely stratified to permit periodic distension (stretching).

PROCEDURE

1. Observe slides of the mesentery, esophagus, skin, pancreas, vas deferens or uterine tube, trachea, and urinary bladder.
2. Identify the type of epithelium in each of the slides.

Simple columnar (e.g., digestive tract)

Lumen of small intestine

Nucleus

Basement membrane

Goblet cell

Cilia

Simple ciliated columnar (e.g., uterine tube)

Lumen of uterine tube

Cilia

Cell membrane

Nucleus

Basement membrane

Pseudostratified ciliated columnar (e.g., lung bronchus)

Cilia
Goblet cell
Nucleus
Basement membrane
Connective tissue

Transitional (e.g., urinary bladder)

Lumen of urinary bladder

Transitional epithelium

Smooth muscle tissue

Figure 1.5   Columnar and transitional epithelial membranes. The structures shown in each photomicrograph are depicted in the accompanying diagrams.

Figure 1.6 Connective tissue proper. The structures shown in each photomicrograph are depicted in the accompanying diagrams.

B. CONNECTIVE TISSUES

Connective tissue is characterized by abundant amounts of extracellular material, or *matrix*. Unlike epithelial tissue, which is composed of tightly packed cells, the cells of connective tissue (which may be of many types) are spread out. The large extracellular spaces in connective tissue provide room for blood vessels and nerves to enter and leave organs.

There are five major types of connective tissues: (1) *mesenchyme*, an undifferentiated tissue found primarily during embryonic development; (2) *connective tissue proper*; (3) *cartilage*; (4) *bone*; and (5) *blood*.

Connective tissue proper (fig. 1.6) refers to a broad category of tissues with a somewhat loose, flexible matrix. This tissue may be *loose* (*areolar*), which serves as a general binding and packaging material in such areas as the skin and the fascia of muscle, or *dense*, as is found in ten-

Hyaline (e.g., larynx)

— Lacuna
— Intercellular matrix
— Chondrocyte

Fibrocartilage (e.g., symphysis pubis)

— Lacuna
— Chondrocyte
— Intercellular matrix
— White fibers

Elastic (e.g., outer ear)

— Lacuna
— Chondrocyte
— Elastic fibers

Figure 1.7 **Different forms of cartilage.** The structures shown in each photomicrograph are depicted in the accompanying diagrams.

dons and ligaments. The degree of denseness relates to the relative proportion of protein fibers to fluid in the matrix. These protein fibers may be made of *collagen*, which gives tensile strength to tendons and ligaments; they may be made of *elastin* (*elastic fibers*), which are prominent in large arteries and the lower respiratory system; or they may be *reticular fibers* providing more delicate structural support to the lymph nodes, liver, spleen, and bone marrow. *Adipose tissue* is a type of connective tissue in which the cells (*adipocytes*) are specialized to store fat.

Cartilage consists of cells (*chondrocytes*) and a semisolid matrix that imparts strength and elasticity to the tissue. The three types of cartilage are shown in figure 1.7. *Hyaline cartilage* has a clear matrix that stains a uniform blue. The most abundant form of cartilage, hyaline cartilage is found on the articular surfaces of bones (commonly called "gristle"), in the trachea, bronchi, nose, and the costal cartilages between the ventral ends of the first ten ribs and the sternum. *Fibrocartilage* matrix is reinforced with collagen fibers to resist compression. It is found in the symphysis pubis, where the two pelvic bones articulate, and between the vertebrae, where it forms intervertebral

discs. *Elastic cartilage* contains abundant elastic fibers for flexibility. It is found in the external ear, portions of the larynx, and in the auditory canal (eustachian tube).

Bone (fig. 1.8) contains mature cells called *osteocytes*, surrounded by an extremely hard matrix impregnated with calcium phosphate. Arranged in concentric layers, the osteocytes surround a *central canal*, containing nerves and blood vessels, and obtain nourishment via small channels in the matrix called *canaliculi*.

Blood (fig. 1.8) is considered a unique type of connective tissue because its extracellular matrix is fluid (*plasma*) that suspends and transports blood cells (*erythrocytes, leukocytes,* and *thrombocytes*) within blood vessels. The composition of blood will be described in more detail in later exercises.

PROCEDURE

1. Observe slides of skin, mesentery, a tendon, the spleen, cartilage, and bone.
2. Identify the types of connective tissue in each slide.

Bone (osseous)

Lamellae

Central canal

Osteocyte within a lacuna

Canaliculi

Erythrocytes (red blood cells)

Centrifuged blood sample

Blood (centrifuged sample)

Plasma

Cells

Leukocytes (white blood cells)

Platelets (thrombocytes)

Figure 1.8 Bone and blood. The structures shown in each photomicrograph are depicted in the accompanying diagrams.

C. MUSCLE TISSUE

Muscles are responsible for heat production, body posture and support, and for a wide variety of movements, including locomotion. Muscle tissues, which are contractile, are composed of muscle cells, or *fibers,* that are elongated in the direction of contraction. The three types of muscle tissues—*smooth, cardiac,* and *skeletal*—are shown in figure 1.9.

Smooth muscle tissue is found in the digestive tract, blood vessels, respiratory passages, and the walls of the urinary and reproductive ducts. Smooth muscle fibers are long and spindle shaped, with a single nucleus near the center. **Cardiac muscle** tissue, which is found in the heart, is characterized by striated fibers that are branched and interconnected by *intercalated discs.* These interconnections allow electrical impulses to pass from one *myocardial* (heart muscle) *cell* to the next.

Skeletal muscle tissue attaches to the skeleton, and is responsible for voluntary movements. Skeletal muscle fibers are long and thin and contain numerous nuclei. Skeletal muscle is under voluntary control, whereas cardiac and smooth muscles are classified as involuntary. This distinction relates to the type of nerves involved (innervation) and not to the characteristics of the muscles themselves. Both skeletal muscle and cardiac muscle cells are categorized as *striated muscle* because they contain cross striations.

PROCEDURE

1. Observe prepared slides of smooth, cardiac, and skeletal muscles.
2. Identify the major distinguishing features of each type of muscle.

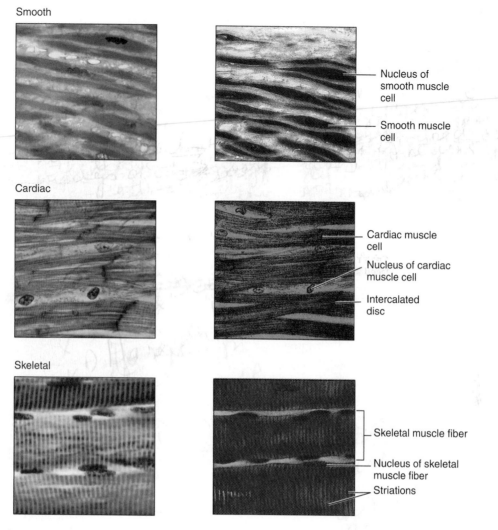

Smooth

Nucleus of
smooth muscle
cell

Smooth muscle
cell

Cardiac

Cardiac muscle
cell

Nucleus of cardiac
muscle cell

Intercalated
disc

Skeletal

Skeletal muscle fiber

Nucleus of skeletal
muscle fiber

Striations

Figure 1.9   Muscle tissue. The structures shown in each photomicrograph are depicted in the accompanying diagrams.

D. Nervous Tissue

Nervous tissue, which forms the nervous system, consists of two major categories of cells. The nerve cell, or **neuron** (fig. 1.10), is the functional unit of the nervous system. The typical neuron has a *cell body* with a nucleus, smaller projections called *dendrites* branching from the cell body, and a single, long, cytoplasmic extension called an *axon,* or *nerve fiber.* The neuron is generally capable of receiving, producing, and conducting electrical impulses. Most neurons release specialized chemicals from the axon endings. A second category of cell found in the nervous system is a **neuroglial cell.** Various types of neuroglia support the neurons both structurally and functionally.

PROCEDURE

1. Observe prepared slides of the spinal cord and the brain.
2. Identify the parts of a neuron.

Some axons of the central nervous system (CNS) and peripheral nervous system (PNS) are surrounded by myelin sheath (are *myelinated*); others lack a myelin sheath (are *unmyelinated*). Neuroglial cells called **Schwann cells** form myelin sheaths in the PNS. When an axon in a peripheral neuron is cut, the Schwann cells form a *regeneration tube* that helps to guide the regenerating axon to its proper destination. Even a severed major nerve may be surgically reconnected, and the function of the nerve largely reestablished, if the surgery is performed before tissue death. Neuroglial cells of the CNS that form myelin sheaths are known as **oligodendrocytes.** In contrast to Schwann cells, oligodendrocytes do not form regeneration tubes. For this and other reasons that are incompletely understood, cut or severely damaged neurons of the brain and spinal cord usually result in permanent damage.

3. Distinguish neurons from neuroglial cells.
4. Without referring to the caption, identify the various tissue types in the photomicrographs in figure 1.10.

E. AN ORGAN: THE SKIN

Organs contain more than one type—usually all four types—of primary tissue. The **skin,** the largest organ of the body, provides an excellent example.

Epithelial tissue is illustrated by the *epidermis* and the *hair follicles* (fig. 1.11). Like all glands, the oily *sebaceous glands* associated with hair follicles and the *sweat glands* are a type of epithelial tissue.

Connective tissue is seen in the *dermis.* Collagen fibers that form dense connective tissue are located in the dermis, whereas adipose connective tissue is embedded in the *hypodermis.*

Muscle tissue is represented by the *arrector pili muscle,* a smooth muscle that attaches to the hair follicle and the matrix of the dermis.

Nerve tissue is featured within skin by the sensory and motor nerves, and by *Meissner's corpuscle* (the oval structure in the dermis near the start of the sensory nerve, fig. 1.11), a sensory structure sensitive to pressure.

PROCEDURE

1. Observe a prepared slide of the skin or scalp.
2. Identify the structures of the skin and try to find all four types of primary tissue.

Figure 1.10 Nervous tissue. Photomicrographs of representative neurons and neuroglia in the CNS.

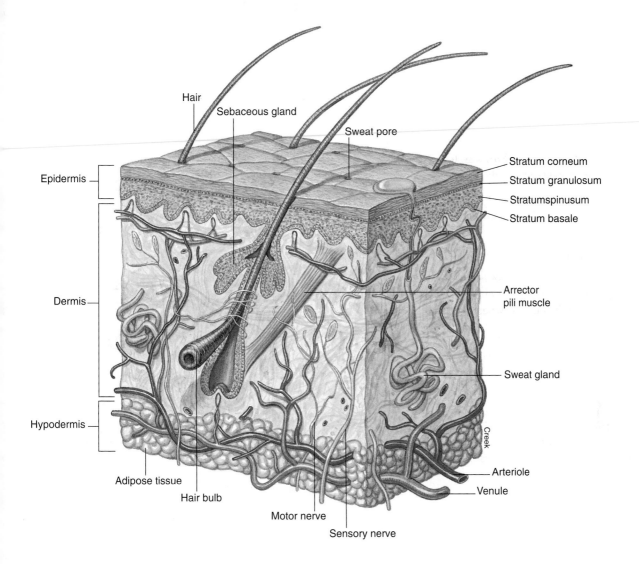

Figure 1.11 Diagram of the skin.

Homeostasis and Negative Feedback

1.3

MATERIALS

1. Watch or clock with a second hand
2. Constant-temperature water bath; thermometer

The regulatory mechanisms of the body help to maintain a state of dynamic constancy of the internal environment known as homeostasis. Most systems of the body maintain homeostasis by operating negative feedback mechanisms that control effectors (muscles and glands).

OBJECTIVES

1. Define the term *homeostasis.*
2. Explain how the negative feedback control of effectors helps to maintain homeostasis.
3. Explain why the internal environment is in a state of dynamic, rather than static, constancy.
4. Define the terms *set point* and *sensitivity.*
5. Explain how a normal range of values for temperature or heart rate is obtained, and discuss the significance of these values.

Textbook Correlations

Before performing this exercise, you may want to consult the following references in *Human Physiology,* seventh edition, by Stuart I. Fox:

- *Negative Feedback Loops.* Chapter 1, pp. 5–8.
- *Feedback Control of Hormone Secretion.* Chapter 1, pp. 8–9.

  Those using different physiology textbooks may want to consult the corresponding information in those books.

Although the structure of the body is functional, the study of body function involves much more than a study of body structure. The extent to which each organ performs the functions endowed by its genetic programming is determined by regulatory mechanisms that coordinate body functions in the service of the entire organism. The primary prerequisite for a healthy organism is the maintenance of **homeostasis,** or constancy of the internal environment.

When homeostasis is disturbed—for example, by an increase or decrease in body temperature from its normal value, or *set point*—a *sensor* detects the change. The sensor then activates an *effector,* which induces changes opposite to those that activated the sensor. Activation of the effector attempts to correct the initial disturbance, so that the initial change and its compensatory reaction result in only slight deviations from the normal value. In this way, temperature and other body parameters are maintained at a relative constancy. Homeostasis is therefore a state of *dynamic,* rather than absolute, constancy (fig. 1.12).

Since a disturbance in homeostasis initiates events that lead to changes in the opposite direction, the cause-and-effect sequence is described as a **negative feedback mechanism** (or a *negative feedback loop*). A constant-temperature water bath, for example, uses negative feedback mechanisms to maintain the temperature at which the bath is set (the **set point**). Deviations from the set point are detected by a thermostat (temperature sensor), which turns on a heating unit (the effector) when the temperature drops below the set point, and turns off the unit when the temperature rises above the set point (fig. 1.12).

By means of the negative feedback control of the heating unit, the water-bath temperature is not allowed to rise or fall too far from the set point. Keep in mind, however, that the temperature of the water is at the set point only briefly. The set point is in fact only the *average value* within a *range* (from the highest to the lowest value) of temperatures. The *sensitivity* of this negative feedback mechanism is measured by the temperature deviation from the set point required to activate the compensatory (negative feedback) response (turning the heater on or off).

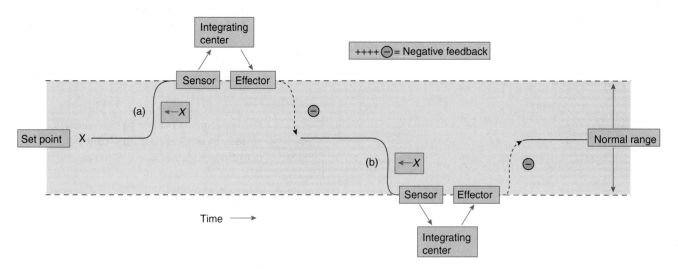

Figure 1.12 Homeostasis is maintained by negative feedback loops. (a) A rise in X (←X), a factor of the internal environment, activates a negative feedback mechanism that produces opposing changes in the body and thereby restores homeostasis. (b) A fall in X (←X) similarly activates a negative feedback mechanism that causes opposing changes in the body and restores homeostasis.

A. NEGATIVE FEEDBACK IN A CONSTANT-TEMPERATURE WATER BATH

In the healthy individual, homeostasis works to maintain a constant internal environment by successfully responding to various forms of physiological stress (such as changes in temperature). In this exercise, "stress" can be imposed by adding cold (or hot) water to the water bath and observing the ability of the negative feedback mechanisms of the water bath to compensate.

PROCEDURE

1. The temperature of the water bath is set by the instructor somewhere between 40°C and 60°C.
2. A red indicator light goes on when the heating unit is activated; it goes off when the heater is turned off. In the spaces provided in the laboratory report, record the temperature of the water when the light first goes on and when the light first goes off.
3. Determine the temperature range, the set point, and the sensitivity of the water bath to deviations from the set point.
4. Record your data in the laboratory report.
5. Add a relatively large volume of cold water (*stress*) to the water bath at time zero. Record the time required for the light to first go on; and the time for the light to go out again, indicating temperature compensation. Record your observations and conclusions in the laboratory report.

B. RESTING PULSE RATE: NEGATIVE FEEDBACK CONTROL AND NORMAL RANGE

Homeostasis—the dynamic constancy of the internal environment—is maintained by negative feedback mechanisms that are far more complex than those involved in maintaining a constant-temperature water bath. In most cases, several effectors, many with antagonistic effects, are involved in maintaining homeostasis. It is as if the temperature of a water bath were determined by the antagonistic actions of both a heater and a cooling system. The cardiac rate (or pulse rate) is largely determined by the antagonistic effects of two different nerves. One of these (a *sympathetic nerve*, described in section 7) stimulates an increase in cardiac rate. A different nerve (a *parasympathetic nerve*) produces inhibitory effects that slow the cardiac rate.

The resting cardiac rate or pulse rate, measured in *beats per minute*, is maintained in a state of dynamic constancy by negative feedback loops initiated by sensors in response to changes in blood pressure and other factors. Therefore, the resting pulse rate is not absolutely constant but instead varies about a set-point value. This exercise will demonstrate that your pulse rate is in a state of dynamic constancy (implying negative feedback controls). From the data you can determine your own pulse-rate set point as the average value of the measurements.

PROCEDURE

1. Gently press your index and middle fingers (not your thumb) against the radial artery in your wrist until you feel a pulse. Alternatively, the carotid

pulse in the neck may be used for these measurements.

2. The pulse rate is usually expressed as beats per minute. However, only the number of beats per 15-second interval (quarter minute) need be measured; multiplying this by four gives the number of beats per minute. Record the number of beats per 15-second interval in the data table provided in the laboratory report.

3. Pause 15 seconds, and then count your pulse during the next 15-second interval. Repeat this procedure over a 5-minute period. Recording your count once every half minute for 5 minutes, a total of 10 measurements (expressed as beats per minute) will be obtained.

4. Using the grid provided in the laboratory report, graph your results by placing a dot at the point corresponding to the pulse rate for each measurement, and then connect the dots.

Normal Values

Students often ask, How do my measurements compare with those of others? and Are my measurements normal? Normal values are those that healthy people have. Since healthy people differ to some degree in their particular values, what is considered normal is usually expressed as a range of values that encompasses the measurements of most healthy people. An estimate of the **normal range** is a statistical determination that is subject to statistical errors and also subject to questions about what is meant by the term *healthy*.

Healthy, in this context, means the absence of cardiovascular disease. Included in the healthy category, however, are endurance-trained athletes, who usually have lower than average cardiac rates, and relatively inactive people, who have higher than average cardiac rates. For this reason, determinations of normal ranges can vary, depending on the relative proportion of each group in the sample tested. A given class of students may therefore

The concept of homeostasis is central to medical diagnostic procedures. Through the measurement of body temperature, blood pressure, concentrations of specific substances in the blood, and many other variables, the clinical examiner samples the internal environment. If a particular measurement deviates significantly from the range of normal values—that is, if that individual is *not* able to maintain homeostasis—the cause of the illness may be traced and proper treatment determined to bring the measurement back within the normal range.

have an average value and a range of values that differ somewhat from those of the general population.

PROCEDURE

1. Each student in the class determines his or her own average cardiac rate (pulse rate) from the previous data either by taking an arithmetic average or simply by observing the average value of the fluctuations in the previously constructed graph. Record your own average in the laboratory report.

2. Record the number of students in the class with average pulse rates in each of the rate categories shown in the laboratory report. Also, calculate the percentage of students in the class who are within each category and record this percentage in the laboratory report.

3. Divide the class into two groups: those who exercise on a regular basis (at least three times a week) and those who do not. Determine the average pulse rate and range of values for each of these groups. Enter this information in the given spaces in the laboratory report.

Laboratory Report 1.3

Name _____

Date _____

Section _____

DATA FROM EXERCISE 1.3

A. Negative Feedback in a Constant-Temperature Water Bath

_____ temperature at which light goes on and heater is activated
_____ temperature at which light and heater go off
_____ temperature range permitted by negative feedback mechanism
_____ set point of constant-temperature water bath
_____ sensitivity of water bath to temperature deviations
_____ time (minutes) required for the light to first go *on* when cold water was added
_____ time (minutes) required for the light to first go *out*

1. How long did the water bath system take to compensate for the stressful change (cold water)?

2. What predictions can you make regarding the response times to stress if either colder or warmer water is added to the water bath in homeostasis (or stable)?

B. Resting Pulse Rate: Negative Feedback Control and Normal Range

Measurement	1	2	3	4	5	6	7	8	9	10
Beats per 15 seconds										
Beats per minute										

1. Your *average* pulse rate: _____ beats/minute.
2. What is the *range* of values in the 10 measurements? _____
3. What is the *sensitivity* of values in the 10 measurements? _____
4. Pulse rate averages of the class:

Pulse Rate (beats per minute)	Number of Students	Percentage of the Total
Over 100 bpm		
90–100		
80–89		
70–79		
60–69		
50–59		
Under 50 bpm		

5. Data for the exercise and nonexercise groups:

	Exercise Group	Nonexercise Group
Range of Pulse Rates		
Average of Pulse Rates		

REVIEW ACTIVITIES FOR EXERCISE 1.3

Test Your Knowledge of Terms and Facts

1. Define the term *homeostatsis*. _____

2. Define the term *set point*. _____

Test Your Understanding of Concepts

3. Explain how negative feedback mechanisms operate to maintain homeostasis. Use the terms *sensor*, *integrating center*, and *effector* in your answer.

4. Draw a flow diagram, illustrating cause and effect with arrows, to show how constant temperature is maintained in a water bath. (*Note:* Flow diagrams are pictorial displays of processes that occur in sequence, using arrows to indicate the possible directions or flow of activity.)

5. Suppose that a constant-temperature water bath contained two antagonistic effectors: a heater and a cooler. Draw a flow diagram to show how this dual system could operate to maintain a constant temperature about some set point.

Test Your Ability to Analyze and Apply Your Knowledge

6. Explain why your graph of pulse rate measurements suggests the presence of negative feedback control mechanisms.

7. Sympathetic nerves to the heart increase the rate of beat, while parasympathetic nerves decrease the rate of beat. Draw a negative feedback loop showing how sympathetic and parasympathetic nerves are affected in someone experiencing a fall in blood pressure (the initial stimulus). (*Note:* the sensor detects the fall in blood pressure.)

8. Why would there be different published values for the normal range of a particular measurement? Do these values have to be continuously updated? Why?

Cell Function and Biochemical Measurements

Physiological control systems maintain homeostasis of the internal chemical environment to which the organ systems are exposed. The concentrations of *glucose, protein,* and *cholesterol* in plasma (the fluid portion of the blood), for example, are maintained within certain limits despite the expected variety in dietary food selections and variations in our eating schedules. This regulation is necessary for health. If plasma glucose levels fall too low, for example, the brain may "starve" and a coma may result. A drop in plasma protein, as another example, may disturb the normal distribution of fluid between the blood and tissues. An abnormal rise in these values, or other abnormal changes in the chemical composition of plasma, can endanger a person's health in various ways.

Abnormal changes in the internal chemical environment, which can contribute to disease processes, are usually themselves the result of diseases that affect cell function. For example, since most plasma proteins are produced by liver cells, diseases of the liver can result in the lowering of plasma protein concentrations. Similarly, abnormal lowering of plasma glucose levels may result from excess secretion of the hormone insulin by certain cells of the pancreas. Thus, homeostasis of the internal chemical environment depends on proper cell function.

All of the molecules found in the body's internal environment, aside from those few obtained directly from food, are produced within the cells. Some molecules remain within the cells; others are secreted into the tissue fluids and blood. Almost all of these molecules are produced by chemical reactions catalyzed by special proteins known as **enzymes.** All enzymes in the body are produced within tissue cells according to information contained in the **DNA** (genes). In this way, the overall metabolism of carbohydrates, lipids, proteins, and other molecules in the cell is regulated largely by genes. Defects in these genes can result in the production of defective enzymes, which result in impaired metabolism. Thus, the study of organ system physiology is intertwined with the study of cell function and biochemistry, as well as with the study of genetics.

Proper cell function also depends upon the integrity of the plasma (cell) membrane. Composed primarily of two semifluid phospholipid layers, cell membranes can regulate the passive transport of molecules moving from higher to lower concentration by diffusion. Special membrane proteins can serve as channels for the passage of larger or more polar molecules, whereas other membrane proteins serve as carriers that require the expenditure of energy to "pump" molecules across the membrane "uphill" from lower to higher concentrations (a process called active transport).

Measurements of Plasma Glucose, Cholesterol, and Protein

MATERIALS

1. Pyrex (or Kimax) test tubes, mechanical pipettors for 40 μL, 50 μL, 100 μL, and 5.0 mL volumes; and corresponding pipettes (0.10 mL and 5.0 mL total volume—see fig. 2.1)
2. Constant-temperature water bath, set at 37° C
3. Colorimeter and cuvettes
4. Glucose kit ("Glucose LiquiColor Test," Stanbio Laboratory, Inc.)
5. Cholesterol kit ("Cholesterol Liquicolor Test," Stanbio Laboratory, Inc.)
6. Total Protein Standard (10g/dL from Stanbio Laboratory, Inc.); the following concentrations: 2, 4, 6, 8 g/dL can be prepared by dilution.
7. Biuret reagent. To a 1.0-L volumetric flask, add 45 g of sodium potassium tartrate and 15 g of $CuSO_4 \cdot 5 H_2O$. Fill 2/3s full with 0.2N NaOH and shake to dissolve. Add 5 g of potassium iodide and fill to 1.0 L volume with 0.2N NaOH.
8. Serum (Artificial "Normal" and "Abnormal Control" sera can be purchased from Stanbio Laboratory, Inc.)

The concentrations of glucose, protein, and cholesterol in plasma (or serum) can be measured using colorimetric techniques in the laboratory. Abnormal concentrations of these molecules are associated with specific disease states.

OBJECTIVES

1. Describe how Beer's law can be used to determine the concentration of molecules in solution.
2. Use the formula method and graphic method to determine the concentration of molecules in plasma (serum) samples.
3. Explain the physiological roles of glucose, protein, and cholesterol in the blood.
4. Explain why abnormal measurements of plasma glucose, protein, and cholesterol are clinically significant.

Textbook Correlations

Before performing this exercise, you may want to consult the following references in *Human Physiology*, seventh edition, by Stuart I. Fox:

- *Carbohydrates and Lipids.* Chapter 2, pp. 31–37.
- *Proteins.* Chapter 2, pp. 38–42.
- *Exchange of Fluid between Capillaries and Tissues.* Chapter 14, pp. 414–416.

Those using different physiology textbooks may want to consult the corresponding information in those books.

(a) (b)

Figure 2.1 Automatic devices for dispensing fluids.
(a) Device to dispense milliliters (such as 5.0 mL) of reagent.
(b) An automatic microliter pipettor (Eppendorf) for dispensing 100 μL (0.10 mL) of solution, or similar volumes.

Table 2.1   Examples of Monomers and Polymers

Monomer	Examples	Polymer	Examples
Monosaccharides	glucose, fructose	Polysaccharides	starch, glycogen
Amino acids	glycine, phenylalanine	Proteins	hemoglobin, albumin
Fatty acids and glycerol		Triglycerides	fats, oils
Ribonucleotides and deoxyribonucleotides		Nucleic acids	DNA and RNA

*O*rganic molecules found in the body contain the atoms carbon (C), hydrogen (H), and oxygen (O) in various ratios, and some of these molecules also contain the atoms nitrogen (N), phosphorus (P), and sulfur (S). Many organic molecules are very large. They consist of smaller repeating subunits that are chemically bonded to each other. The term *monomer* refers to the individual subunits; the term *polymer* refers to the long chain formed from these repeating subunits.

When two monomers are bonded together, a molecule of water (HOH) is released. This reaction is called **condensation,** or **dehydration synthesis.**

$$A—OH + HO—B \rightarrow A—B + HOH$$

The new molecule (**A—B**) formed from the two monomers (**A** and **B**) is called a *dimer*. This dimer may participate in a condensation reaction with a third monomer to form a *trimer*. The stepwise addition of new monomers to the growing chain by condensation reactions will result in the elongation of the chain and the formation of the full polymer. Examples of monomers and polymers are given in table 2.1.

When the chemical bond between monomers is broken, a molecule of water is consumed. This **hydrolysis reaction** is the reverse of a condensation reaction.

$$A—B + HOH \rightarrow A—OH + B—OH$$

Ingested foods are usually polymers—mainly proteins, carbohydrates, and triglycerides. In the stomach and small intestine, these polymers are hydrolyzed (in the process of *digestion*) into their respective monomers: amino acids, monosaccharides, fatty acids, and glycerol. These monomers are then moved across the wall of the small intestine into the blood of the capillaries (a process called *absorption*). The vascular system transports them primarily to the liver and then to all the other tissues of the body.

Once inside the cells of the body, the monomers can be either hydrolyzed into smaller molecules, by a process that yields energy for the cell, or condensed to form new, larger polymers in the cytoplasm. Some of these new polymers are released into the blood (e.g., hormones and the plasma proteins), whereas others remain inside the cell and contribute to its structure and function. In turn, some of the new polymers of the cell can eventually be hydrolyzed to form new monomers, which may be used by the cell or released into the blood for use by other cells in the body.

In the healthy person, the concentrations of the different classes of monomers and polymers in the blood plasma are held remarkably constant and vary only within narrow limits. When the concentration of one of these molecules in the blood deviates from the normal range, specific compensatory mechanisms are activated that bring the concentration back to normal (negative feedback). Homeostasis is thus maintained.

When the concentration of any of the monomers or polymers in the blood remains consistently above or below normal, the health of the person may be threatened. Abnormal concentrations of different molecules in the blood are characteristic of different diseases and aid in their diagnosis. The disease *diabetes mellitus*, for example, is characterized by a high blood glucose concentration. Therefore, accurate measurement of the concentrations of different molecules in the blood is extremely important in physiology and clinical laboratories.

THE COLORIMETER

The colorimeter is a device used in physiology and clinical laboratories to measure the concentration of a substance in a solution. This is accomplished by the application of **Beer's law,** which states that the concentration of a substance in a solution is directly proportional to the amount of light absorbed (*Absorbance, A*) by the solution and inversely proportional to the logarithm of the amount of light transmitted (*Percent Transmittance, %T*) by the solution.

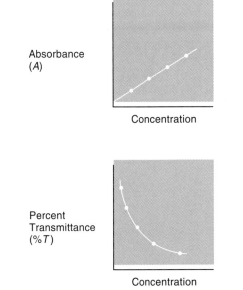

35

Beer's law will apply only if the incident light (the light entering the solution) is monochromatic—that is, light composed of a single wavelength. White light is a mixture of many different wavelengths between 380 and 750 nanometers (nm), or millimicrons (mμ). The rods and cones within the eyes respond to the light waves, and the brain interprets these different wavelengths as different colors.

Violet	380–435 nanometers (nm)
Blue	436–480 nanometers (nm)
Green	481–580 nanometers (nm)
Yellow	581–595 nanometers (nm)
Orange	596–610 nanometers (nm)
Red	611–750 nanometers (nm)

By means of a prism or diffraction grating, the colorimeter can separate white light into its component wavelengths. The operator of this device can select incident light of any wavelength by simply turning the appropriate dial to that wavelength. This light enters a specific tube, the *cuvette,* which contains the test solution. A given fraction of the incident light is absorbed by the solution and the remainder of the light passes through the cuvette. The transmitted light generates an electric current by means of a photoelectric cell, and the amount of this current is registered on a galvanometer scale.

The colorimeter scale indicates the percent transmittance (%). Since the amount of light that goes into the solution and the amount of light that leaves the solution are known, a ratio of the two indicates the light absorbance (A) of that solution. The colorimeter also includes an absorbance scale. In the following exercises, the absorbance scale will be used rather than the percent transmittance scale because absorbance and concentration are directly proportional to each other. This relationship can be described in a simple formula, where 1 and 2 represent different solutions:

$$\frac{\text{Concentration}_1}{\text{Absorbance}_1} = \frac{\text{Concentration}_2}{\text{Absorbance}_2}$$

One solution might be a sample of plasma whose concentration (e.g., of glucose) is unknown. The second solution might be a *standard,* which contains a known concentration of the test substance (such as glucose). When the absorbances of both solutions are recorded from the colorimeter, the concentration of the test substance in plasma (i.e., the unknown) can easily be calculated:

$$C_x = C_{std}\frac{A_x}{A_{std}}$$

where

x = the unknown plasma
std = the standard solution
A = the absorbance value
C = the concentration

Suppose there are four standards. Standard 1 has a concentration of 30 mg per 100 mL (or mg per deciliter, dL). Standards 2, 3, and 4 have concentrations of 50 mg/dL, 60 mg/dL, and 70 mg/dL, respectively. Since standard 3 has twice the concentration of standard 1, it should (according to Beer's law) have twice the absorbance. The second standard (at 50 mg/dL), similarly, should have an absorbance value midway between that of the first and the fourth standard, since its concentration is midway between 30 and 70 mg/dL. Experimental errors, however, make this unlikely. Therefore, it is necessary to average the answers obtained for the unknown concentration when different standards are used. This can be done either arithmetically by applying the previous formula, or by means of the graph below.

A graph plotting the four standard data points, including a straight line of "best fit" drawn closest to these points, is called a **standard curve.**

Standard	Concentration (mg/dL)	Absorbance
1	30 mg/100 mL	0.25
2	50 mg/100 mL	0.38
3	60 mg/100 mL	0.41
4	70 mg/100 mL	0.57
Unknown	**??**	**0.35**

Now suppose that a solution of unknown concentration has an absorbance of 0.35. The standard curve graph can be used to determine its concentration.

Standardizing the Colorimeter

The following procedure is intended specifically for the Spectronic 20 (Bausch & Lomb) colorimeter (fig. 2.2). Although the general procedure is similar for all colorimeters, specific details may vary between different models.

Figure 2.2 A typical colorimeter (spectrophotometer).
(a) Cuvette holder; (b) power switch/zero control;
(c) 100% transmittance control; and (d) wavelength dial.

PROCEDURE

1. Turn on the colorimeter by rotating knob *b* in figure 2.2 to the right.
2. Set the wavelength dial (*d* in fig. 2.2) so that the correct wavelength in nanometers (provided in each exercise) is lined up with the indicator in the window adjacent to this dial.
3. When there is no cuvette in the cuvette holder (*a* in fig. 2.2), the light source is blocked. The pointer should read zero transmittance or infinite absorbance at the left end of the scale. Turn knob *b* to align the pointer with the *left* end of the scale.
4. Place the cuvette, which contains all the reagents *except* the test solution (e.g., glucose), into the cuvette holder. This tube is called the **blank** because it has a concentration of test substance equal to zero. It should therefore have an absorbance of zero (or a transmittance of 100%) that is read at the right end of the scale. Turn knob *c* to align the pointer with the *right* end of the scale (fig. 2.2).
5. Repeat steps 3 and 4 to confirm settings.
6. Place the other cuvettes, which contain the standard solutions and the unknown, in the cuvette holder. Close the hatch and read the absorbance value of each solution.

Note: *Before placing each cuvette in the chamber, wipe it with a lint-free, soft paper towel. If the cuvette has a white indicator line, place the cuvette so that this line is even with the line at the front of the cuvette holder.*

A. CARBOHYDRATES: MEASUREMENT OF PLASMA GLUCOSE CONCENTRATION

The monomers of the carbohydrates are the **monosaccharides,** or simple sugars. The general formula for these molecules is $C_nH_{2n}O_n$, where n can be any number. *Glucose,* for example, has the formula $C_6H_{12}O_6$. The monosaccharide *fructose* has the same formula but differs from glucose in the arrangement of the atoms (glucose and fructose are *isomers*).

Two monosaccharides can join together by means of a dehydration synthesis (condensation) reaction to form a **disaccharide.** *Sucrose* (common table sugar), for example, is a disaccharide of glucose and fructose, whereas *maltose* is a disaccharide of two glucose subunits.

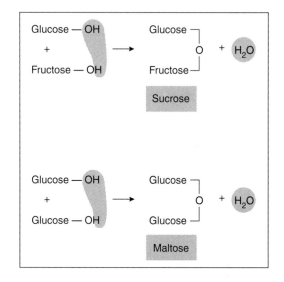

The continued addition of glucose subunits to maltose will result in the production of a long, branched chain of repeating glucose subunits, forming the **polysaccharide** *glycogen* (or animal starch). This polysaccharide is formed inside muscle and liver cells and serves as an efficient storage form of glucose. When the blood glucose level drops below normal, the liver cells can hydrolyze stored glycogen and release glucose into the blood. Conversely, when the blood sugar level rises above normal, the liver cells can take glucose from the blood and store it as glycogen for later use. In this way, the equilibrium between blood glucose and liver glycogen helps to maintain constancy (*homeostasis*) of the blood sugar level. This process is regulated by hormones that include epinephrine (adrenaline), insulin, hydrocortisone, and glucagon.

The most important regulator of the blood glucose level is the hormone *insulin,* produced by the islets of Langerhans in the pancreas. This hormone stimulates the transport of blood glucose into the cells of the body; hence, it lowers the blood glucose level. An elevated blood glucose level, *hyperglycemia,* results from insufficient insulin secretion. This disease is called **diabetes mellitus.** Low blood glucose, *hypoglycemia,* is clinically rare, but can result from the excessive secretion (or injection) of insulin. In addition, slight hypoglycemia may be associated with arthritis, renal disease, and the late stages of pregnancy.

PROCEDURE
Measurement of Plasma Glucose Concentration

1. Obtain three test tubes, and label them
   U (unknown), *S* (standard), *B* (blank).
2. Using a mechanical pipettor, pipette 5.0 mL of the
   glucose reagent into *each* tube.
3. Use a microliter pipettor to add 40 μL (0.04 mL) of
   the following solutions into each of the indicated
   test tubes to avoid contamination. Use different
   pipette tips for adding each solution.

Tube	Serum	Standard	Water	Reagent
Unknown (*U*)	40 μL	—	—	5.0 mL
Standard (*S*) (100 mg/dL)	—	40 μL	—	5.0 mL
Blank (*B*)	—	—	40 μL	5.0 mL

Note: *All tubes must contain equal volumes (5.04 mL) of
solution, (1 deciliter (dL) = 100 milliliters (mL).*

4. Gently tap each tube to mix the contents and allow
   the tubes to stand at room temperature for 10
   minutes.
5. Set the monochromator (wavelength) dial at
   500 nm and standardize the colorimeter, using
   solution *B* as the blank.
6. Record the absorbance values of solutions *U* and *S*
   in the chart in the laboratory report.
7. Using Beer's law formula, calculate the
   concentration of glucose in the unknown plasma
   sample and enter the value in the laboratory report.
8. Using graph paper that follows this exercise, draw a
   graph of absorbance versus glucose concentration
   (mg/dL). Plot the standard (*S*) absorbance value
   and draw a standard curve. Then determine the
   unknown glucose concentration from the graph.

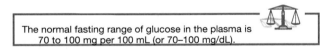
The normal fasting range of glucose in the plasma is
70 to 100 mg per 100 mL (or 70–100 mg/dL).

B. LIPIDS: MEASUREMENT OF PLASMA CHOLESTEROL CONCENTRATION

The lipids are an extremely diverse family of molecules
that share the common property of being soluble (dissolv-
able) in organic solvents such as benzene, ether, chloro-
form, and carbon tetrachloride, but *are not soluble in water
or plasma.* The lipids found in blood can be classified as
free fatty acids (FFA), also known as nonesterified fatty
acids (NEFA); **triglycerides** (or neutral fats); **phospholipids;** and **steroids.** As found in carbohydrate molecules,
carbon, hydrogen, and oxygen form the basic structure of
lipids; however, these elements are not present in the
same predictable ratio.

Fatty acids are long chains, ranging from sixteen to
twenty-four carbons in length. When adjacent carbons
are linked by single bonds, the fatty acid is said to be *saturated;* when adjacent carbons are linked by double bonds,
the fatty acid is said to be *unsaturated.*

   Saturated: $-CH_2-CH_2-CH_2-CH_2-CH_2-$
   Unsaturated: $=CH-CH_2-CH=CH-CH_2-$

The triglycerides consist of three fatty acids bonded
to a molecule of the alcohol *glycerol.* Triglycerides with
few sites of unsaturation and that are solid at room temperature are called *fats,* whereas those with many sites of
unsaturation and that are liquid at room temperature are
called *oils.*

Like the triglycerides, the phospholipids consist of
two fatty acids bonded to a glycerol molecule. However,
as their name implies, each phospholipid also contains
the element phosphorus (in the form of phosphate,
PO_4^{3-}) bonded to the third position on the glycerol molecule. Phospholipids are important components of cell
membranes.

The steroids are characterized by a structure consisting of four rings. One of the most important steroids in
the body is cholesterol.

Cholesterol

There is evidence that high blood cholesterol, together with other risk factors,
such as hypertension and cigarette
smoking, contributes to *atherosclerosis.*
In atherosclerosis, deposits of cholesterol and other lipids, calcium salts, and
smooth muscle cells build up in the walls of arteries and
reduce blood flow. These deposits—called *atheromas*—
also serve as sites for the production of *thrombi* (blood
clots), which further occlude blood flow. The reduction
in blood flow through the artery may result in heart disease or cerebrovascular accident (stroke). It is generally
believed that blood cholesterol levels, and the risk of
atherosclerosis, may be significantly lowered by a diet
low in cholesterol and saturated fats.

PROCEDURE

Measurement of Plasma Cholesterol Concentration

1. Obtain three test tubes, and label them U (unknown), S (standard), B (blank).
2. Using a mechanical pipettor, pipette 5.0 mL of the cholesterol reagent into each tube.
3. Use a microliter pipettor to add 50 μL (0.05 mL) of the following solutions into each of the indicated test tubes. Use different pipette tips for adding each solution to avoid contamination.

Tube	Serum	Standard	Water	Reagent
Unknown (*U*)	50 μL	—	—	5.0 mL
Standard (*S*) (200 mg/dL)	—	50 μL	—	5.0 mL
Blank (*B*)	—	—	50 μL	5.0 mL

Note: *All tubes must contain equal volumes (5.05 mL) of solution, (1 deciliter (dL) = 100 milliliters (mL).)*

4. Gently tap each tube to mix the contents and allow the tubes to stand at room temperature for 10 minutes.
5. Transfer the solutions to three cuvettes. Standardize the spectrophotometer at 500 nm, using solution B as the blank.
6. Record the absorbance values of solutions U and S in the laboratory report.
7. Using Beer's law formula, calculate the concentration of cholesterol in the unknown plasma sample. Enter this value in the laboratory report.
8. Using graph paper that follows this exercise, draw a graph of absorbance versus cholesterol concentration (mg/dL). Plot the standard (S) absorbance value and draw a standard curve. Then determine the unknown cholesterol concentration from the graph.

$$C_{plasma} = \frac{A_{plasma}}{A_{standard}} \times C_{standard}$$

Normal values for plasma cholesterol are 130–250 mg/dL.

C. Proteins: Measurement of Plasma Protein Concentration

Proteins are long chains of amino acids bonded to one another by condensation reactions. Each amino acid has an amino (—NH₂) end and a carboxylic acid (—COOH) end, as shown by the general formula:

When amino acids bond (through a **peptide bond**) to form a protein, one end of the protein will have a free amino group and the other end will have a free carboxyl group.

Carboxyl end Peptide bond Amino end

There are more than twenty-two different amino acids in nature, each differing from the others with respect to the combination of atoms in the R *group*, sometimes known as the *functional group*. The amino acid **glycine,** for example, has a hydrogen atom (H) in the R position, whereas the amino acid **alanine** has a methyl group (CH_3) in the R position.

Glycine Alanine

An abnormally low concentration of total blood protein (*hypoproteinemia*) may be due to an inadequate production of protein by the liver caused by liver disease (such as cirrhosis or hepatitis), or to the loss of protein in the urine (*albuminuria*) caused by kidney diseases. Hypoproteinemia decreases the colloid osmotic pressure of the blood and may lead to accumulation of excess fluid in the tissue spaces, a condition called **edema.**

An abnormally high concentration of total plasma protein (*hyperproteinemia*) may be due to dehydration or to an increased production of the plasma proteins. An increased production of gamma globulins (antibodies), for example, is characteristic of pneumonia and many other infections, and of parasitic diseases such as malaria.

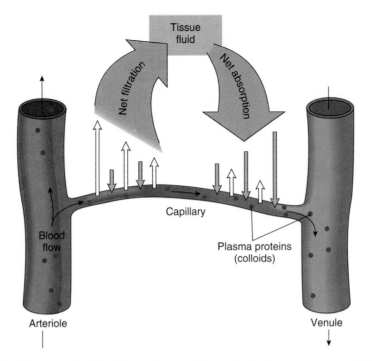

Figure 2.3 **Circulation of fluid between the blood plasma in a capillary and the tissues.** Arrows pointing away from the capillary indicate the force exerted by the blood pressure, whereas arrows pointing toward the capillary indicate the force exerted by the colloid osmotic pressure of the plasma proteins. The arrow labeled "blood flow" indicates the direction of flow along the capillary from arteriole to venule.

Proteins in the plasma serve a variety of functions. Some proteins may be active as enzymes, hormones, or carrier molecules (transporting lipids, iron, or steroid hormones in the blood), while others have an immune function (antibodies). The **plasma proteins** are classified according to their behavior during biochemical separation procedures. These classes include the *albumins,* the *alpha* and *beta globulins* (synthesized mainly in the liver from amino acids absorbed by the intestine), and *gamma globulins* (antibodies produced by the lymphoid tissue).

In addition to the separate functions of the different plasma proteins, the *total* concentration of proteins in the plasma is physiologically important. The plasma proteins exert an osmotic pressure, the **colloid osmotic** (or **oncotic) pressure,** which pulls fluid from the tissue spaces into the capillary blood. This force compensates for the continuous filtration of fluid from the capillaries into the tissue spaces produced by the *hydrostatic pressure* of the blood (fig. 2.3).

PROCEDURE

Measurement of Total Plasma Protein Concentration

1. Obtain seven test tubes, and label them 1–7.
2. Using a mechanical pipettor, pipette 5.0 mL of biuret reagent into each tube.
3. Use a microliter pipettor to add 50 μL (0.05 mL) of the following solutions into each of the indicated test tubes. Use different pipette tips for adding each solution to avoid contamination.

	Tube Number		Serum	Standard	Water	Reagent
1	Blank (B)	50 μL	—	—	5.0 μL	
2	Standard (S)	(2.0 g/dL)	—	50 μL	—	5.0 mL
3		(4.0 g/dL)	—	50 μL	—	5.0 mL
4		(6.0 g/dL)	—	50 μL	—	5.0 mL
5		(8.0 g/dL)	—	50 μL	—	5.0 mL
6		(10.0 g/dL)	—	50 μL	—	5.0 mL
7	Unknown (U)		—	—	50 μL	5.0 mL

Note: *All tubes must contain equal volumes (5.05 mL) of solution. (The expression g/dL is equivalent to g per 100 mL or g%.)*

4. Gently tap each tube to mix the contents, and allow the tubes to stand at room temperature for at least 5 minutes.
5. Transfer the solutions to seven cuvettes. Standardize the spectrophotometer at 550 nm, using solution *B* as the blank.
6. Record the absorbance values of the unknown (*U*) and the five standard solutions (*S* tubes 2–6) in the laboratory report.

7. Using Beer's law formula, calculate the concentration of total protein in the unknown plasma sample. Enter this value in the laboratory report.
8. Using graph paper that follows this exercise, draw a graph of absorbance versus total protein concentration (g/dL). Plot the standard (*S* tubes 2–6) absorbance values and draw a standard curve. Then determine the unknown total protein concentration from the graph.

The normal fasting total protein level is 6.0–8.0 g/dL.

Laboratory Report 2.1

DATA FROM EXERCISE 2.1

A. CARBOHYDRATES: Measurement of Plasma Glucose Concentration

1. Record the absorbance values of solutions U and S.
   Absorbance of solution U: _____
   Absorbance of solution S: _____

2. Enter the unknown serum glucose concentration derived from the equation for Beer's law in the space below.
   _____ mg/dL
   How does this value for the unknown compare with that derived from the graph? Explain.

3. Is the glucose concentration of the serum sample normal or abnormal? Explain.

B. LIPIDS: Measurement of Plasma Cholesterol Concentration

1. Record the absorbance values of solutions U (unknown) and S (standard) in the spaces below:
   Absorbance of solution U: _____
   Absorbance of solution S: _____

2. Enter the unknown serum cholesterol concentration derived from the equation for Beer's law
   _____ mg/dL
   How does this value for the unknown compare with that derived from the graph? Explain.

3. Is the cholesterol concentration of the serum sample normal or abnormal? Explain.

C. PROTEINS: Measurement of Plasma Protein Concentration

1. Record your absorbance values in this data table.

Tube Number	Protein Concentration (g/dL)	Absorbance
1	0 (blank)	0
2	2.0	
3	4.0	
4	6.0	
5	8.0	
6	10.0	
7	**Unknown** (serum sample)	

2. Use the graph paper at the end of this exercise to plot a standard curve.
3. Derive an estimated protein concentration for the unknown serum sample.
   Unknown protein concentration from the curve: _____ g/dL.
4. Use the equation for Beer's law to determine the protein concentration of the unknown serum.
   Unknown protein concentration from Beer's equation: _____ g/dL .
5. How does this value for the unknown compare with that derived from the graph? Explain.

6. Is the total protein concentration of the unknown serum sample normal or abnormal? Explain: What clinical condition(s) might have caused an abnormal result?

REVIEW ACTIVITIES FOR EXERCISE 2.1

Test Your Knowledge of Terms and Facts

1. The concentration of a solution is _____ proportional to its absorbance.
2. The above relationship is described by _____law.
3. Hyperglycemia is characteristic of the disease _____ .
4. Hyperglycemia may be caused by a deficiency of the hormone _____ .
5. Cholesterol belongs to the general category of molecules known as _____and to the specific category of molecules known as _____ .
6. High blood cholesterol, along with other risk factors, is a contributing factor in the disease_____ .
7. Most of the plasma proteins are produced by the _____ (organ).
8. Low plasma protein concentration is described clinically as _____ , and can produce a physical condition called _____ .
9. The colloid osmotic pressure of the blood is related to the plasma concentration of _____ .
10. "All fats are lipids, but not all lipids are fats." Explain.

11. Describe the functions of the plasma proteins. Where do these proteins originate?

12. What does the blank tube contain, and what is its function in a colorimetric assay?

Test Your Ability to Analyze and Apply Your Knowledge

13. Do any of the proteins in your plasma come from food proteins? Does the starch (glycogen) in your liver come from food starch? Explain your answers.

14. Why do you draw a linear (straight line) graph of absorbance vs. concentration even though your experimental values deviated slightly from a straight line? Why must your line intersect the origin of the graph (zero concentration equals zero absorbance)? Explain.

Thin-Layer Chromatography of Amino Acids

MATERIALS

1. Silica gel plates (F-254 rapid, adhered to plastic or glass); capillary tubes; chromatography (or hair) dryers; rulers

Note: *Chromatography paper can be used as an alternative*

2. Developing chambers or oven set at about 60° C
3. Amino acid solutions: arginine, cysteine, aspartic acid, phenylalanine—1.0 mg/mL of each dissolved in 0.1N HCl:isopropyl alcohol (9:1); "unknown" solution of amino acids, containing two of these amino acids in the same solution
4. Developing solvent: 20 mL of 17% NH_4OH (dilute concentrated NH_4OH with an equal amount of water), 40 mL of ethyl acetate, and 40 mL of methanol per developing chamber
5. Ninhydrin spray

Amino acids, the subunits of protein structure and function, are divisible into approximately twenty chemically unique molecules. The distinctive properties of each amino acid provide the basis for its separation from the others and its identification. This information can be clinically useful in the diagnosis of genetic diseases that involve amino acid metabolism.

OBJECTIVES

1. Using the general formula for amino acids, explain how one amino acid differs from another.
2. Explain how thin-layer chromatography can separate different amino acids that are present together in a single solution.
3. Explain what the R_f value signifies. Calculate the R_f values for different spots and use this information to identify unknown amino acids.

Textbook Correlations

Before performing this exercise, you may want to consult the following references in *Human Physiology*, seventh edition, by Stuart I. Fox:

- *Proteins.* Chapter 2, pp. 38–42.
- *Inborn Errors of Metabolism.* Chapter 4, pp. 89–91.

Those using different physiology textbooks may want to consult the corresponding information in those books.

In this exercise, you will attempt to identify two unknown amino acids that are present in the same solution. To accomplish this task, you must (1) *separate* the two amino acids, and (2) *identify* these amino acids by comparing their behavior with that of known amino acids.

Since each amino acid has a chemically different R group, each will dissolve in a given solvent to a different degree. These differences will be used to separate and identify the amino acids on a **thin-layer plate.**

The thin-layer plate consists of a thin layer of porous material (in this procedure, silica gel) that is coated on one side of a plastic, glass, or aluminum plate. The solutions of amino acids are applied to different spots on the plate (a procedure called *spotting*) and allowed to dry. The plate is then placed on edge in a solvent bath with the spots above the solvent.

As the solvent creeps up the plate by capillary action, it will wash the amino acids off their original spots (or *origins*) and carry them upward toward the other end of the plate. Since the solubility of each amino acid is different, the ability of the solvent to dissolve and carry a particular amino acid will vary with the properties of that amino acid. The most soluble amino acids in the given solvent will wash and carry farther than those that are less soluble. Since the process is halted shortly before the solvent front reaches the top of the plate, some amino acids will have migrated farther from the origin than others.

If this chromatography were repeated using the same amino acids and the same solvent, the final pattern (*chromatogram*) would be the same as that obtained previously. In other words, the distance that a given amino acid migrates in a given solvent, relative to the *solvent front*, can be used as an identifying characteristic of that amino acid. A numerical value (the R_f value) can be assigned to this characteristic by calculating the distance the amino acid traveled relative to that traveled by the solvent front, as follows:

$$R_f = \frac{\text{distance from origin to spot}}{\text{distance from origin to solvent front}}$$

An unknown amino acid can be identified by comparing its R_f value in a given solvent with the R_f values of known amino acids in the same solvent.

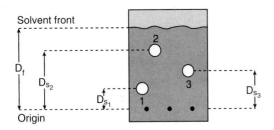

PROCEDURE

1. Fill a capillary tube with amino acid solution 1 (arginine). Gently touch it to the first spot on the plate 1 1/2 inches from the bottom and 1/2 inch from the left-hand edge. Dry the spot with a hair dryer. Repeat the spotting and drying procedure until you have made *five* applications of the *same* amino acid to the *same* spot.

An abnormally high concentration of certain amino acids or their metabolites in the blood frequently results in deterioration of the central nervous system and mental retardation. These abnormal concentrations are usually due to defective enzymes that are involved in the degradation of the amino acids.

In the disease **phenylketonuria (PKU),** for example, the enzyme that converts the amino acid phenylalanine to tyrosine is defective, leaving phenylalanine and its other metabolites to accumulate in the body. This condition, which affects one baby in every 10,000 to 20,000, results in severe mental retardation. Other diseases of similar etiology (cause) include *homocystinuria, alkaptonuria,* and *maple syrup disease* (the name refers to a characteristic odor of the urine). The defective enzymes are synthesized by defective genes; hence, the diseases of amino acid metabolism that have this etiology are referred to as **inborn errors of metabolism** (see exercise 2.5).

Note: (1) Be careful to apply the amino acid solution to exactly the same spot each time (use a spotting guide, or make X's lightly with pencil before spotting); (2) dry the spot thoroughly between applications; and (3) spot gently so that you do not gouge out the silica gel.

2. Repeat the spotting procedure for amino acid 2 (cysteine), using a new capillary tube and applying the spot about 1/2 inch to the right of the first amino acid.
3. Repeat the spotting procedure for amino acids 3 (aspartic acid) and 4 (phenylalanine), applying them about 1/2 inch to the right of the preceding spot.
4. Repeat the spotting procedure with the solution containing two unknown amino acids, applying it about 1/2 inch to the right of amino acid 4.
5. Carefully place the thin-layer plate in a chromatography developing chamber that has been previously filled with solvent. Cover the chamber and allow the solvent to migrate up the plate for *1 hour.*
6. Remove the plate from the developing chamber, dry it, and then spray it with ninhydrin in a well-ventilated area.

Note: Since amino acids are colorless, it is necessary to react them with ninhydrin, a reagent that combines with the amino acids to produce a blue-colored complex.

7. Heat the plate in an oven set at approximately 60°C for *10–15* minutes.
8. Remove the plate and measure the distance, in centimeters, from the origin to the solvent front and from the origin to the center of each amino acid spot. Record these values and calculate R_f values. Enter your data in the laboratory report.

Laboratory Report 2.2

Name _____

Date _____

Section _____

DATA FROM EXERCISE 2.2

Record your data in this table and calculate the R_f value for each spot.

Amino Acid	D_s	D_f	R_1
Arginine			
Cysteine			
Aspartic acid			
Phenylalanine			
Unknown 1			
Unknown 2			

The unknown solution contained the two amino acids _____ and _____ .

REVIEW ACTIVITIES FOR EXERCISE 2.2

Test Your Knowledge of Terms and Facts

1. Which part of an amino acid distinguishes it and grants it chemical specificity? _____
   _____

2. Inherited defects in the ability to convert one amino acid into another are in a class of disorders called _____
   _____ .

3. Define the term R_f. _____

Test Your Understanding of Concepts

4. Why do different amino acids have different R_f values?

5. Suppose an amino acid is 8 cm from the origin, and the solvent front is 12 cm from the origin. What is the R_f value for this amino acid?

6. What is the maximum R_f value that a spot can have? Explain.

Test Your Ability to Analyze and Apply Your Knowledge

7. Suppose amino acid "A" has a higher R_f value than amino acid "B" in a solvent system "1" and the order is reverse in solvent system "2." Further, suppose that solvent system "1" has a higher ratio of methanol to ethyl acetate (and is thus more polar) than solvent system "2." What can you conclude about the structure of amino acid "A" compared to the structure of amino acid "B"?

Electrophoresis of Serum Proteins

MATERIALS

1. Plastic troughs, buffer chamber, sample applicator, power supply, Sepharose strips (the Sepra Tek System, Gelman), forceps
2. High-resolution buffer, Ponceau S stain (Gelman), 5% acetic acid (v/v)
3. Sterile lancets, 70% ethanol, disposal receptacle for all blood-contaminated objects (Alternatively, test tubes containing previously prepared serum samples may be used.)
4. Unheparinized capillary tubes and microhematocrit centrifuge

Plasma contains classes of proteins that differ in structure and function. These classes can be separated from one another and identified by electrophoresis.

OBJECTIVES

1. Explain what is meant by the term *amphoteric* and describe how amphoteric molecules can be separated by electrophoresis.
2. Identify the different bands of serum proteins in an electrophoresis pattern.
3. Describe the origin and functions of the different classes of plasma proteins.

📖 Textbook Correlations

Before performing this exercise, you may want to consult the following references in *Human Physiology*, seventh edition, by Stuart I. Fox:

- *Proteins*, Chapter 2, pp. 38–42.
- *Plasma Proteins*, Chapter 13, p. 368.
- *Antibodies*, Chapter 15, pp. 455–457.

Those using different physiology textbooks may want to consult the corresponding information in those books.

The unique structure and physiological role of each type of protein is determined by the specific number, type, and sequence of its component amino acids. Proteins differ in size, range in shape from elliptical (globular) to fibrous, and contain different numbers of positive and negative charges.

Under acidic conditions, the amino group ($-NH_2$) of an amino acid tends to gain an H^+ and become positively charged ($-NH_3^+$), whereas under basic conditions, the carboxyl group ($-COOH$) loses an H^+ and becomes negatively charged ($-COO^-$). Since amino acids can have either polarity, depending on the pH, they are said to be **amphoteric.**

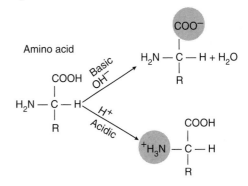

When an amino acid is electrically neutral, its amphoteric nature can be shown by the *zwitterion formula,* in which neutrality is indicated by a balance between positive and negative charges.

Amino acid (zwitterion formula)

In addition to the amino and carboxyl ends of a protein, the functional groups (R groups) of many amino acids have either an amino-containing functional group (such as lysine or arginine) or a carboxylic acid–containing functional group (such as aspartic acid or glutamic acid). Since each type of protein has a characteristic ratio of these two types of amino acids, each protein will have a characteristic net charge at a given pH.

Anode + ———————————————————— — Cathode

A α_1 α_2 β γ

Figure 2.4 **Electrophoresis pattern of normal serum.** The bands include albumin (A) and the α_1, α_2, β, and γ globulins.

In many instances, the diagnosis of a disease can be aided by an analysis of the electrophoresis pattern made by migrating plasma proteins. Although direct observation is useful, more reliable information can be gained by a quantitative measurement of the proteins in each band. These measurements are made by a *densitometer,* a device that optically scans the electrophoresis pattern and graphically records the absorbance of different regions of the strip.

Diseases that can be accurately diagnosed through electrophoresis include acute inflammations (elevated alpha-2 proteins), viral hepatitis (change in gamma globulin and albumin), and cirrhosis of the liver (broad gamma globulin band). In addition, electrophoresis is valuable in the diagnosis of nephrotic syndrome, malignant tumors, and many other diseases.

At a pH of 8.8 (slightly basic), each of the variety of proteins found in plasma will have a different degree of net negative charge. When plasma proteins are placed in an electric field, each protein will migrate away from the negative pole (cathode) and toward the positive pole (anode) at different rates. The rates at which they travel will also be influenced by their size and shape. This technique, known as **electrophoresis,** can be used to separate and identify the different classes of plasma proteins.

There are two main types of plasma proteins: albumin and the globulins (see section C, exercise 2.1). The latter type is composed of four primary subclasses: *alpha-1* (α_1), *alpha-2* (α_2), beta (β), and *gamma* (γ) globulins (fig. 2.4).

The fluid part of the blood as it circulates in the vessels is **plasma.** When blood clots, a soluble protein in the plasma (*fibrinogen*) is converted into an insoluble threadlike protein called *fibrin*. The strands of fibrin intertwine to form the meshwork of the blood clot. **Serum,** which is the fluid that remains after the clot has formed, does not contain fibrinogen and is incapable of further clotting.

Albumins are the most abundant of the serum proteins, serving as carrier molecules for hormones, lipids, and bile pigment; and they are responsible for most of the colloid osmotic pressure exerted by the blood. The *alpha* and *beta globulins* serve a variety of functions and, like albumin and fibrinogen, are synthesized by the liver. The *gamma globulins* are **antibodies,** which are produced by white blood cells known as lymphocytes.

PROCEDURE

1. Float a strip of cellulose acetate in buffer for 1 minute; then immerse it in the buffer for *10 minutes.*
2. Using forceps, remove the cellulose acetate strip from the buffer and blot it with filter paper that has been premoistened with buffer.
3. Raise the tension latch of the frame assembly, placing the movable support bridge in the vertical position (figs. 2.5 and 2.6a).*
4. Center the cellulose acetate strip on the support bridges and fasten it with the membrane clamps. Tension the membrane by releasing the latch (fig. 2.6b,c).
5. Place the membrane frame assembly in the chamber, bringing the strip ends into contact with the buffer, and position the cover on the chamber.
6. Cleanse the tip of a finger with 70% ethanol and puncture it with a sterile lancet.

Note: *Extreme caution must be exercised when handling blood to guard against contracting infectious agents. Handle only your own blood and discard all objects containing blood into the receptacles provided by the instructor.*

7. Quickly fill an *unheparinized* capillary tube with blood, seal one end, and immediately centrifuge for 3 minutes. (Do *not* use heparinized capillary tubes—the anticoagulant heparin is a protein and will interfere with the test.)
8. Break the capillary tube at the junction of the packed red blood cells and the pale yellow serum. Place a drop of serum on the sample well of the applicator block by lightly touching it with the capillary tube. Similarly, a sample of serum from six different students can be placed in each of six wells (fig. 2.6d).

---

*For Gelman Sepra Tek System.

Support bridge

Membrane clamp

Membrane frame assembly

Buffer chamber

Figure 2.5   An electrophoresis system. A membrane frame assembly and buffer chamber for the Gelman Sepra Tek electrophoresis system.

9. Fill a new capillary tube with the serum provided by the instructor and place it on sample wells 7 and 8.

10. Position the applicator on the applicator block, and load it by depressing the button for 4 seconds (fig. 2.6*e*).

11. Position the loaded applicator on the chamber cover. Depress the button for 4 seconds to place the serum on the cellulose acetate strip (fig. 2.6*f*).

12. Connect the chamber to the power supply and electrophorese for *20 minutes* at 200 V (fig. 2.6*g,h*).

Note: *During this time, clean the applicator by placing it on filter paper moistened with the buffer. Rinse with tap water and distilled water.*

13. When the voltage is off, open the chamber and remove the membrane frame assembly. Raise the tension latch and strip clamps and remove the cellulose acetate strip with forceps.

14. Float the strip on Ponceau S stain for *1 minute*; then immerse it completely in the stain for *10 minutes*.

15. Remove the strip with forceps and rinse it in two successive baths of 5% acetic acid. Tape the cellulose acetate strip in the space provided (or draw a facsimile) and label the bands.

Note: *The albumin band (nearest the anode) is the darkest band; the gamma globulin band (nearest the cathode) is the widest band.*

(a)

(b)

(c)

(d)

(e)

(f)

(g)

(h)

Figure 2.6 Procedure for performing electrophoresis of serum proteins. Steps a–h are described in the text.

DATA FROM EXERCISE 2.3

1. In the space below, tape your electrophoresis strip (or draw a facsimile).

2. Label the protein bands in the strip above.

REVIEW ACTIVITIES FOR EXERCISE 2.3

Test Your Knowledge of Terms and Facts

1. Describe the property of an amphoteric molecule. _____
   _____

2. The amphoteric nature of an amino acid that is electrically neutral can best be shown by its
   _____ formula. Draw one here.

3. Name two amino acids that have
   (a) an extra amino group: _____ and
       _____ .
   (b) an extra carboxylic acid group: _____ and
       _____ .

4. Proteins have a net _____ charge in acidic solutions and a net _____ charge in basic solutions.
5. The soluble protein in plasma that aids in the formation of blood clots is _____ .
6. The insoluble protein that aids in the formation of blood clots is _____ .
7. The most abundant class of plasma proteins is the _____ .
8. Albumin and many other plasma proteins are made by the_____ .
9. The class of plasma proteins containing antibodies is the _____ .

10. What is the difference between serum and plasma? Explain.

11. What factors contribute to the distance that a protein will migrate during electrophoresis?

Test Your Ability to Analyze and Apply Your Knowledge

12. Why is the albumin band the darkest of the bands? What does the width of the gamma globulin band reveal about the structure of antibodies?

13. How might liver disease affect the density of the bands? Which bands would be the most affected? What effects might these changes in plasma proteins have on the body?

Measurements of Enzyme Activity

MATERIALS

1. Beakers, rusty nails, chicken liver
2. Hydrogen peroxide
3. Test tubes, mechanical pipettors, automatic microliter pipettes
4. Constant-temperature water bath set at 37° C
5. Cuvettes and spectrophotometer (colorimeter)
6. Serum or reconstituted normal and abnormal serum (available, for example, from Bio-Analytic Laboratories, Inc., or Stanbio Laboratory, Inc.).
7. Alkaline Phosphatase Test and Lactate Dehydrogenase Test (available, for example, from Bio-Analytic Laboratories, Inc., or Sigma)

Textbook Correlations*

Before performing this exercise, you may want to consult the following references in *Human Physiology,* seventh edition, by Stuart I. Fox:

- *Enzymes as Catalysts.* Chapter 4, p. 84.
- *Effects of Temperature and pH.* Chapter 4, p. 87.
- *Cofactors and Coenzymes.* Chapter 4, p. 88.

Those using different physiology textbooks may want to consult the corresponding information in those books.

The presence of a specific enzyme can be detected by the reaction it catalyzes, and the enzyme concentration can be measured by the amount of product it forms in a given period of time. Enzyme activity is affected by pH, temperature, and the availability of substrates and coenzymes.

OBJECTIVES

1. Describe the lock-and-key model of enzyme activity and use this model to explain enzyme specificity.
2. Describe the effects of pH and temperature on enzyme activity.
3. Describe how enzyme concentration is measured.

Enzymes are biological **catalysts;** that is, enzymes are substances that increase the rate of chemical reaction without changing the nature of the reaction and without being altered by the reaction. The catalytic process occurs in two stages: (1) the reactants (hereafter referred to as the *substrates* of the enzyme) bond in a specific manner to the enzyme, forming an *enzyme-substrate complex,* and (2) the enzyme-substrate complex dissociates into the free, unaltered enzyme and *products.*

All enzymes are proteins (although not all proteins are enzymes). The polypeptide chain of each enzyme bends and folds in a unique way to produce a characteristic three-dimensional structure. The substrates interact with a specific part of this structure, the **active site,** which is complementary in shape to the substrate molecules. This is most easily visualized by the *lock-and-key model* of enzyme action (fig. 2.7).

The shape of the active site is determined by the amino acid sequence of the protein; this shape is different for different enzymes. Enzymes are *relatively specific,* interacting only with specific substrates and catalyzing selective reactions.

The relative specificity of an enzyme can often be deduced from its name. Thus, the enzyme *lactate dehydrogenase* removes a hydrogen from lactic acid (the suffix *-ase* denotes an enzyme), whereas *phosphatase* enzymes hydrolyze the phosphate group from a wide variety of organic compounds. These guidelines do not apply to enzymes that were discovered before a systematic terminology was developed, such as the digestive enzymes *pepsin* and *trypsin* that hydrolyze peptide bonds.

Since enzymatic activity is dependent on the delicate bending and folding of polypeptide chains, changes in pH and temperature, which affect the three-dimensional structure of proteins, also affect enzymatic activity.

---

*See Appendix 3 for correlations to the *Virtual Physiology Laboratory* CD-ROM by McGraw-Hill and Cypris Publishing, Inc.

$$\boxed{\begin{array}{ccc} A + B & \xrightarrow{\text{Enzyme}} & C + D \\ \text{(Reactants)} & & \text{(Products)} \end{array}}$$

Substrate A

Active sites

Substrate B

(a) Enzyme and substrates (A and B)

(b) Enzyme–substrate complex

Product C

Product D

(c) Enzyme (unchanged) and reaction products (C and D)

Figure 2.7 The lock-and-key model of enzyme activity. (a) Substrates A and B fit into the active sites of the enzyme, forming (b) an enzyme-substrate complex. This complex then dissociates (c) to yield free enzyme and the products of the reaction.

Although most enzymes display maximum activity at pH 7 and at 40°C to 45°C, the **pH optimum** and **temperature optimum** of one enzyme may be significantly different from those of another enzyme. Enzymatic activity diminishes when the pH or temperature varies from the optimal value for that enzyme.

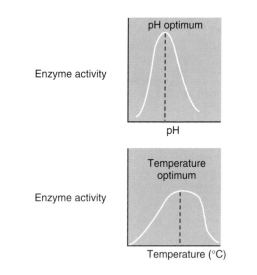

The activity of many enzymes is absolutely dependent on the presence of specific, smaller, nonprotein molecules called *cofactors* or *coenzymes*. Cofactors are specific inorganic ions (e.g., Mg^{2+}) required for enzyme activity, and coenzymes are organic compounds derived from the water-soluble vitamins that play a similar role.

To assay (measure) the enzyme activity in an unknown sample, the appropriate cofactors or coenzymes must be provided, and the pH and temperature must be standardized. In addition, the concentration of substrate should be very large relative to the concentration of enzyme, so that the availability of substrate does not limit the rate of the reaction. Under these conditions, the amount of product doubles when the concentration of enzyme doubles or when the reaction time doubles. If the reaction time is held constant, the amount of product formed is *directly proportional* to the enzyme concentration.

The concentration of the enzyme in a sample is usually measured in **units of activity,** determined under conditions of specified pH, temperature, reaction time, and other controlled factors. The *international* unit is a measure of the reaction rate and may be determined by either the quantity of substrate consumed or the quantity of product formed in a given time interval. Specifically, one **international unit (IU)** is defined as *the quantity of enzyme required to convert one micromole of substrate per minute into products*, under specified conditions of pH, temperature, and other controlled conditions. For example, the normal concentration of the enzyme alkaline phosphatase in serum ranges from 9 to 35 international units per liter (IU/L).

In the following exercises, you will assay the activity of three enzymes found in serum. These enzymes are not normally active in serum but are enzymes released into the blood from ruptured cells of damaged tissues. These enzymatic assays are often valuable in the diagnosis of certain diseases.

A. Catalase in Liver

Catalase is an enzyme that converts hydrogen peroxide to water and oxygen gas. Found in many tissues, catalase is one of the most rapidly acting enzymes in the body.

$$2H_2O_2 \xrightarrow{\text{Catalase}} 2H_2O + O_2$$

Substrate → Products

PROCEDURE

1. Fill two small 100-mL beakers to the halfway mark with *hydrogen peroxide*.
2. Immerse a rusty nail or similar object in the solution in one beaker and then gently stir. Observe the effect of an *inorganic* catalyst and record your observations in the laboratory report.
3. Mince a fresh chicken liver with scissors and add it to the hydrogen peroxide in the second beaker. Stir the solution and record your observations in the laboratory report.

B. Measurement of Alkaline Phosphatase in Serum

Plasma contains a number of enzymes that are released from damaged tissue cells. Assays of these enzymes can thus be useful in clinical diagnosis. In this exercise, you will determine the concentration of plasma **alkaline phosphatase** by measuring the increase in the absorbance of a solution caused by the accumulation of the product of the enzymatic reaction over time. The increase in absorbance of the solution with time is due to the formation of the yellow product, p-nitrophenol, which occurs when alkaline phosphatase hydrolyzes the phosphate from the substrate p-nitrophenylphosphate at a high pH.

 There are two enzymes in serum that display phosphatase activity (remove phosphate from organic compounds). One of these has a pH optimum of 4.9 (*acid phosphatase*); the other has a pH optimum of 9.8 (*alkaline phosphatase*). Abnormally high levels of serum acid phosphatase activity may be noted in patients with cancer of the prostate, whereas elevated alkaline phosphatase activity is primarily associated with various liver and bone diseases.

PROCEDURE

1. Obtain three cuvettes. Label one *U* (for unknown serum), one *S* (for standard reference), and one *B* (for the water blank).

2. Pipette 0.5 mL of substrate reagent into each cuvette and place in a 37° C water bath. Let the solutions sit in the water bath for approximately *5 minutes*.
3. Pipette 50 μL (0.05 mL) of the unknown serum to the cuvette marked *U*. Wait *30 seconds*, and pipette 50 μL of the enzyme standard (containing 25 IU/L) to the cuvette marked *S*. Wait *30 seconds* and pipette 50 μL of distilled water to the cuvette marked *B*. Be sure to return the cuvettes to the water bath immediately.
4. Ten minutes after having added the serum to cuvette *U*, add 2.5 mL of color developer to this cuvette, cap, and mix. Repeat this procedure 30 seconds later with cuvette *S*, and then with cuvette *B*. Note that each cuvette, in this way, will have incubated for exactly 10 minutes before you terminated the reaction by adding color developer.
5. Set the colorimeter at a wavelength of 405 nm. Standardize the colorimeter using the blank cuvette (*B*) and read the absorbance values of cuvettes *U* and *S*. Record these values in your laboratory report.
6. Calculate the concentration of alkaline phosphatase in the unknown serum sample (cuvette *U*) using the Beer's law formula below and enter your value in the laboratory report.

Example

Suppose the enzyme reference standard (25 IU/L) had an absorbance of 0.28, and the unknown serum had an absorbance of 0.34. The concentration of the unknown could be calculated as follows:

$$\frac{0.34}{0.28} \times 25 \text{ IU/L} = 30 \text{ IU/L}$$

The normal range for *alkaline phosphatase* concentrations in plasma is 9–35 IU/L.

C. Measurement of Lactate Dehydrogenase in Serum

An enzyme of great importance in physiology and medicine is **lactate dehydrogenase (LDH).** In skeletal muscles during heavy exercise, this enzyme catalyzes the conversion of pyruvic acid to lactic acid. In the process of this reaction, the coenzyme NADH + H$^+$ is oxidized to NAD:

$$\text{NADH + H}^+ \qquad \text{NAD}$$
$$\text{Pyruvic acid} \xrightarrow{\text{LDH (in skeletal muscle)}} \text{Lactic acid}$$

The same enzyme also catalyzes the reverse reaction, whereby lactic acid is converted into pyruvic acid.

This reverse reaction occurs in normal heart tissue. It also occurs in the liver, which can utilize lactic acid for energy production (cell respiration) or in the formation of glucose. This reaction is accompanied by the reduction of NAD to NADH + H⁺ as follows:

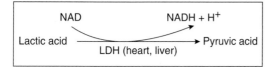

Lactic acid → Pyruvic acid, LDH (heart, liver), NAD → NADH + H⁺

Serum **lactate dehydrogenase (LDH)** concentrations are elevated following a myocardial infarction (MI or "heart attack"). Measurements of these levels, in conjunction with other tests, can help diagnose heart attacks. Serum LDH levels are also elevated in some renal (kidney) diseases, cirrhosis of the liver, and hepatitis.

PROCEDURE

1. Obtain three cuvettes and label them U (for the unknown serum), S (for the reference standard), and B (for the water blank).
2. Pipette 1.0 mL of enzyme reagent to each cuvette, and place the cuvettes in a 37° C water bath for approximately 5 minutes.
3. Pipette 25 µL (0.025 mL) of the unknown serum to cuvette U. *Thirty seconds later*, pipette 25 µL of

reference standard to cuvette S. (The LDH concentration will be printed in the literature accompanying the standard.) *Thirty seconds* later, pipette 25 µL of distilled water into cuvette B. Be sure to return the cuvettes to the water bath following each step.

4. After cuvette U has incubated exactly *10 minutes*, pipette 2.0 mL of acid diluent into the cuvette, cap, and mix. This will stop the reaction. At *30-second* intervals, repeat this procedure with cuvette S and cuvette B, so that all three cuvettes will have incubated exactly 10 minutes.
5. Set the colorimeter at a wavelength of 500 nm, standardize with the water blank (B), and read the absorbance values of cuvettes U and S. Record these values in your laboratory report.
6. Calculate the lactate dehydrogenase concentration of the unknown serum sample (in cuvette U) using the Beer's law formula and enter this value in your laboratory report.

Example
Suppose the reference standard had an LDH concentration of 530 IU/L and an absorbance of 0.82, while the unknown serum had an absorbance of 0.22. The concentration of the unknown could be calculated as follows:

$$\frac{0.22}{0.82} \times 530 \text{ IU/L} = 142 \text{ IU/L}$$

The normal LDH concentration for healthy adult males is 90–221 IU/L; and for healthy adult females it is 89–187 IU/L.

Laboratory Report 2.4

Name _____

Date _____

Section _____

DATA FROM EXERCISE 2.4

A. Catalase in Liver

1. Describe what occurred when the *rusty nail* was placed in the beaker of hydrogen peroxide. Write a simple equation that might describe this reaction.

2. Describe what occurred when the chicken liver was placed in the beaker of hydrogen peroxide. Explain what occurred and write a simple equation that might describe this reaction.

B. Measurement of Alkaline Phosphatase in Serum

1. Record your absorbance values for cuvettes U and S in the spaces below:
   Absorbance of U: _____
   Absorbance of S: _____
   Calculate the alkaline phosphatase concentration of the unknown serum: _____ IU/L.
2. Was the value you obtained in the normal range? What might an abnormally high value indicate?

C. Measurement of Lactate Dehydrogenase in Serum

1. Record your absorbance values for cuvettes U and S in the space below:
   Absorbance of U: _____
   Absorbance of S: _____
   Calculate the lactate dehydrogenase concentration of the unknown serum: _____ IU/L.

2. Was the value you obtained in the normal range? What might an abnormally high value indicate?

REVIEW ACTIVITIES FOR EXERCISE 2.4

Test Your Knowledge of Terms and Facts

1. Define an *international unit (IU)* of enzyme activity. _____
   _____

2. Define the *pH optimum* of an enzyme. _____
   _____

3. Define the *temperature optimum* of an enzyme. _____
   _____

4. What are cofactors and coenzymes? _____
   _____

5. Judging from its name, describe the activity of the following enzymes:
   (a) phosphatase _____
   _____
   (b) glycogen synthetase _____
   _____
   (c) lactate dehydrogenase _____
   _____
   (d) DNA polymerase _____
   _____

Test Your Understanding of Concepts

6. What must you add to a sample of plasma in order to measure the activity of a particular enzyme in plasma? How might you measure the activity of a different enzyme in the same tube of plasma?

7. What happens to the structure of an enzyme when the pH and temperature are changed from the optima for that enzyme?

8. Describe the reactions catalyzed by lactate dehydrogenase in the skeletal muscles, liver, and heart. What is the metabolic significance of these reactions? What happens to NAD and NADH in these reactions?

Test Your Ability to Analyze and Apply Your Knowledge

9. Why were you so careful to have each tube incubate for exactly 10 minutes? What might have occurred if you allowed the tubes to incubate overnight before finally stopping the reaction? Would this be an accurate test? Explain.

10. Pancreatic amylase is an enzyme normally secreted by the pancreas into the small intestine, where it catalyzes the hydrolysis of starch. Should it be in the blood? What might explain its presence in the blood? Would it have any activity as it circulates in the blood plasma? Explain.

Genetic Control
of Metabolism

2.5

MATERIALS

1. Test tubes, Pasteur pipettes (droppers), urine collection cups
2. Phenistix (Ames Laboratories), silver nitrate (3 g per 100 mL), 10% ammonium hydroxide, 40% sodium hydroxide, lead acetate (saturated)

Since a different gene codes for the production of each enzyme, a defective gene can result in a specific metabolic disorder. These inborn errors of metabolism can be detected by tests for specific enzyme products.

OBJECTIVES

1. Define the terms *genotype, phenotype, transcription,* and *translation.*
2. Describe how genes regulate metabolic pathways.
3. Describe how inborn errors of metabolism are produced.
4. Explain the etiology (cause) of phenylketonuria (PKU).

Before performing this exercise, you may want to consult the following references in *Human Physiology*, seventh edition, by Stuart I. Fox:

- *Cell Nucleus and Gene Expression.* Chapter 3, pp. 60–65.
- *Metabolic Pathways.* Chapter 4, pp. 89–91.

Those using different physiology textbooks may want to consult the corresponding information in those books.

The physical appearance of an individual **(phenotype)** is largely determined by the individual's genetic endowment **(genotype).** The control of the phenotype by the genotype is achieved by means of the genetic regulation of cellular metabolism.

All of the chemical reactions of cellular metabolism are catalyzed by specific enzymes, and the information for the synthesis of each specific enzyme is coded by a specific gene. That is, one gene makes one enzyme, which is capable of catalyzing one type of reaction (changing A to B, for example).

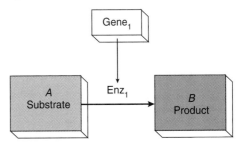

The product of this reaction may become the substrate of a second enzyme, made by a second gene, which converts B into a new product, C. A third enzyme, made by a third gene, may then convert C into D. Thus, a **metabolic pathway** is formed, where the product of one enzyme becomes the substrate of the next. In a metabolic pathway, some initial substrate, A, is converted into a final product (such as D) through a number of *intermediates* (B and C, for example).

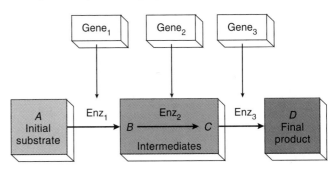

Genes contain the information for the synthesis of *all* proteins, not only those with enzymatic activity. The genetic code is based on the sequence of **DNA** components known as **nucleotide bases** (*adenine, guanine, cytosine,*

thymine). Thus, the sequence of these bases is different in different genes.

The sequence of bases on the DNA that composes one gene is used as a template for the synthesis of one **RNA** molecule. The RNA molecule consists of a linear sequence of nucleotide bases (*uracil*, cytosine, guanine, and adenine), which is precisely complementary to the sequence of bases on the region of the DNA (gene) on which it was made. This complementarity is ensured by the fact that only a specific base on the RNA can bind to a specific base on the DNA.

A part of the genetic code is thus transcribed by the synthesis of a specific RNA molecule—a process called **transcription.** This RNA contains a part of the genetic message and is called *messenger RNA (mRNA)*.

Transcription

The messenger RNA, in association with ribosomes, forms the template for **protein synthesis,** where the sequence of amino acids in the protein is specified by the sequence of bases in the mRNA. The genetic code is based on the fact that a sequence of three mRNA bases (a *base triplet*) can bind only to a specific amino acid through an intermediate compound called *transfer RNA (tRNA)*.

Translation — New Protein Assembly

Base triplets

In this way, the sequence of bases in the mRNA determines the sequence of amino acids in the protein. This process is called **translation.** The sequence of amino acids, in turn, determines how the protein will fold (i.e., its three-dimensional structure). The three-dimensional structure of a protein directly determines its function.

INBORN ERRORS OF METABOLISM

When a gene is missing or defective, the enzyme that it makes will also be missing or defective. This will result in a hereditary metabolic disorder in which there will be a decrease in the intermediates formed *after* the step normally catalyzed by that enzyme. The intermediates formed *before* this step will accumulate in the blood and body tissues and will be excreted in the urine.

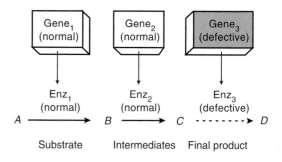

Thus, intermediate D decreases, while intermediates A, B, and C increase.

There are a number of genetic defects associated with the metabolism of the amino acids *phenylalanine* and *tyrosine*. The phenotypic (expressed) effects of these **inborn errors of metabolism** depend on which enzymes are defective and therefore on which intermediate products either are absent or accumulate abnormally in the body tissues.

Probably the best known phenotypic effect of this group of hereditary disorders is the lack of *melanin* pigment characteristic of **albinism. Phenylketonuria (PKU)** is the most clinically serious disorder of this group, since the accumulation of phenylpyruvic acid can affect the developing central nervous system and produce mental retardation. The odious effects of PKU can be avoided only by placing the affected child on a special low phenylalanine diet.

Defective genes cannot be replaced with "good" genes, although the first such treatment has recently been reported. At present, therefore, treatment of those with genetic disorders depends upon early diagnosis and preventive measures. Newborn babies are routinely tested for PKU; and those with the enzyme defect can be placed on low phenylalanine diets. Unfortunately, many other inborn errors of metabolism cannot be treated by dietary restrictions. The only way to prevent some of these diseases is through genetic counseling for prospective parents who are carriers of the disease.

When there is a defective enzyme in a metabolic pathway, the molecule that is the substrate of that enzyme accumulates in particular tissues of the body. Such inborn errors of metabolism are not restricted to pathways of amino acid metabolism. In *Tay-Sachs disease,* for example, there is a defect in the enzyme (hexosaminidase A) that breaks down a complex type of lipid known as a ganglioside, resulting in lipid accumulation in the brain and retina. This disease, which occurs primarily in families of Eastern European Jewish origin (specifically, Ashkenazic Jews), is invariably fatal. Inborn errors also occur in carbohydrate metabolism. In *glycogen storage disease,* for example, the enzyme that breaks down glycogen may be defective, resulting in the excessive accumulation of glycogen that causes liver damage.

A. Phenylketonuria

A person with PKU excretes large amounts of phenylpyruvic acid because of the inability to convert the amino acid phenylalanine into the amino acid tyrosine.

PROCEDURE

1. Dip the test end of a Phenistix strip into a sample of urine.

⚠️ **Note:** Use care when handling all body fluids, including urine. Clean spills and dispose of urine containers properly, as described by your instructor.

2. Compare the color of the strip with the color chart provided.

⚠️ **Note:** Federal law now mandates the Guthrie test using a spot of blood from newborn babies for diagnostic screening for PKU. Therefore, this procedural kit may not be readily available.

B. Alkaptonuria

A person with **alkaptonuria** excretes large amounts of homogentisic acid, which reacts with silver to form a black precipitate. This condition does not have immediate adverse effects on health, but may cause the development of a characteristic type of joint degeneration later in life.

PROCEDURE

1. Add 10 drops of urine to a test tube containing 5 drops of 3% silver nitrate ($AgNO_3$).
2. Add 5 drops of 10% ammonium hydroxide (NH_4OH) to the tube and mix.
3. A positive test for homogentisic acid is indicated by the presence of a black precipitate.

C. Cystinuria

The renal tubules can normally reabsorb all of the amino acids filtered in the glomerulus. People with the recessive trait known as **cystinuria,** however, have an impaired ability to reabsorb the amino acid *cystine* and the related amino acids lysine, arginine, and ornithine.

Cystine is the least soluble amino acid and may precipitate in the urinary tract to form stones. This condition accounts for 1% of the cases of renal stones in the United States. (About 10% are uric acid stones, and the remainder are formed from calcium salts, primarily calcium oxalate.)

PROCEDURE

1. Add 1.0 mL of 40% NaOH to a test tube containing 5.0 mL of urine. Allow the tube to cool.
2. Add 3.0 mL of lead acetate. A brown-to-black precipitate indicates the presence of cystine in the urine.

Laboratory Report 2.5

Name _____

Date _____

Section _____

REVIEW ACTIVITIES FOR EXERCISE 2.5

Test Your Knowledge of Terms and Facts

1. Define the following terms:
   (a) *genetic transcription* _____
   (b) *genetic translation* _____
2. In one sentence, distinguish between the terms *genotype* and *phenotype* _____
   _____

Test Your Understanding of Concepts

3. Using letters *A*, *B*, *C*, *D*, and *E* for molecules, illustrate how these would interrelate in an unbranched metabolic pathway. Then, illustrate a metabolic pathway in which *C* branches to form *D* and *E*. Describe how your two pathways would be affected by an inborn error of metabolism between *C* and *D*.

4. Explain the etiology of phenylketonuria (PKU).

5. Propose reasons why it would be difficult to correct an inborn error of metabolism. Explain why a simple pill or even an injection probably wouldn't be effective.

6. Speculate as to how the information obtained from the sequencing of the human genome might be used to someday eliminate inborn errors of metabolism.

The Nervous System and Sensory Physiology

Despite changes in the external temperature, the availability of foods, the presence of toxic and threatening agents, and other influences, the internal environment of the body remains remarkably constant. The science of physiology is largely a study of the regulatory mechanisms that maintain this internal constancy (*homeostasis*). This regulation is largely a function of the nervous system and endocrine system.

To maintain homeostasis, an organism must be able to recognize specific environmental factors and make appropriate responses. At its simplest level, recognition is achieved through the stimulation of specific types of **sensory receptors,** such as cutaneous (skin) receptors. A given receptor will usually be responsive to only one *modality* (specific type) of stimulus. The rod and cone photoreceptors of the eye, for example, are stimulated by light. Each receptor transduces the particular environmental stimulus into electrical nerve impulses that then go to the specific part of the brain where the sensation is identified. This principle is known as the *law of specific nerve energies.*

The appropriate response to the environmental stimulus occurs through the neural activation of **effector organs,** which are *muscles* and *glands.* The voluntary activation of skeletal muscles by somatic motor nerves emerging from the spinal cord may produce a simple reflex action or involve more complex nervous system interaction. Autonomic motor nerves are involuntary and stimulate cardiac muscles, smooth muscles, exocrine glands (such as sweat glands, gastric glands), and some endocrine glands (the adrenal medulla, for example). Many other endocrine glands are indirectly regulated by the hypothalamus through its control of the anterior pituitary, thus wedding the endocrine system to the nervous system. Interposed between these *afferent* (sensory) and *efferent* (motor) pathways are millions of association neurons within the brain and spinal cord that integrate these activities while promoting learning and memory. The brain is the ultimate interpretation center for all sensations, including touch, pain, vision, hearing, equilibrium, and taste.

Exercise 3.1 Recording the Nerve Action Potential

Exercise 3.2 Electroencephalogram (EEG)

Exercise 3.3 Reflex Arc

Exercise 3.4 Cutaneous Receptors and Referred Pain

Exercise 3.5 Eyes and Vision

Exercise 3.6 Ears: Cochlea and Hearing

Exercise 3.7 Ears: Vestibular Apparatus—Balance and Equilibrium

Exercise 3.8 Taste Perception

Recording the Nerve Action Potential

MATERIALS

1. Frogs
2. Dissecting equipment and trays, glass probes, thread
3. Oscilloscope and nerve chamber
4. Frog Ringer's solution (see Materials, exercise 5.1)

The potential difference across axon membranes undergoes changes during the production of action potentials. As the axons of a nerve produce action potentials, the surface of the nerve at this region has a difference in potential in relation to the surface of an unstimulated region. The polarity of this potential difference and its magnitude in millivolts can be seen in an oscilloscope.

OBJECTIVES

1. Describe the resting membrane potential and the distribution of Na^+ and K^+ across the axon membrane.
2. Describe the events that occur during the production of an action potential.
3. Demonstrate the recording of action potentials in the sciatic nerve of a frog and explain how this recording is produced.

Textbook Correlations*

Before performing this exercise, you may want to consult the following references in *Human Physiology,* seventh edition, by Stuart I. Fox:

- *Action Potentials.* Chapter 7, pp. 160–163.
- *Conduction of Nerve Impulses.* Chapter 7, pp. 163–164.

Those using different physiology textbooks may want to consult the corresponding information in those books.

A neuron (nerve cell) consists of three regions that are specialized for different functions: (1) the *dendrites* receive input from sensory receptors or from other neurons; (2) the *cell body* contains the nucleus and serves as the metabolic center of the cell; and (3) the *axon* conducts the nerve impulse to other neurons or to effector organs (fig. 3.1). A bundle of axons that leaves the brain or spinal cord is known as a **peripheral nerve.**

MEMBRANE POTENTIALS

If one lead of a voltmeter is placed on the surface of an axon and the other lead is placed inside the cytoplasm, a *potential difference* (or voltage) will be measured across the axon membrane. The inside of the cell is 60 to 80 millivolts negative (–80 mV) with respect to the outside. (The surface of the axon is positive with respect to the cytoplasm.) This **resting membrane potential** is maintained by the unequal distribution of ions on the two sides of the membrane. Na^+ is present in higher concentrations outside the cell than inside, whereas K^+ is more concentrated inside the cell. These differences are maintained, in part, by active transport processes (the sodium-potassium pump).

When a neuron is appropriately stimulated, the barriers to Na^+ are lifted and the positively charged Na^+ is allowed to diffuse into the cell along its concentration gradient. The flow of positive ions into the cell first eliminates the potential difference across the membrane and then continues until the polarity is actually reversed and the inside of the cell is positive with respect to the outside (about +15 mV). This phase is called **depolarization.** At this point, the barriers to K^+ are lifted, and the flow of this positively charged ion out of the cell along its concentration gradient helps to reestablish the resting potential (**repolarization**). During the time of repolarization, the diffusion of Na^+ into the cell is blocked and the neuron is refractory (not capable of responding) to further stimulation; it is in a *refractory period.* The momentary

---

*See Appendix 3 for correlations to the A.D.A.M. InterActive PHYSIOLOGY Modules.

 See Appendix 3 for correlations to the *Virtual Physiology Laboratory CD-ROM* by McGraw-Hill and Cypris Publishing, Inc.

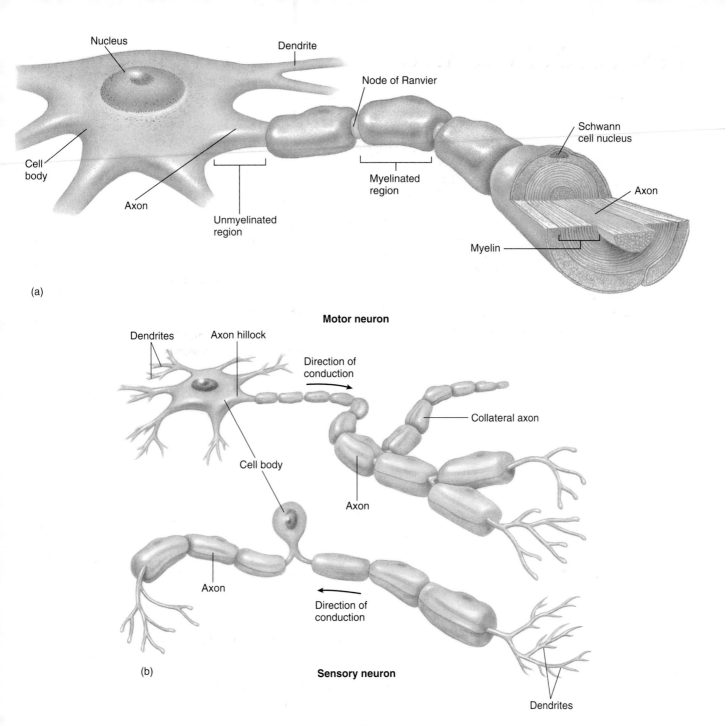

(a)

Motor neuron

(b)

Sensory neuron

Figure 3.1 Neuron structure. (a) The components of a neuron. (b) A motor and a sensory neuron, showing the pathways of conduction (arrows).

reversal and reestablishment of the resting potential is known as the **action potential** (fig. 3.2).

In this exercise, a frog's sciatic nerve will be dissected and placed on two pairs of electrodes. One pair of electrodes (the stimulating electrodes) will deliver a measured pulse of electricity to one point on the nerve; the other pair (the recording electrodes) will be connected to a cathode-ray tube of an oscilloscope that is adjusted to sweep an electron beam horizontally across a screen when the nerve is unstimulated. The delivery of a small pulse of electricity through the stimulating electrodes produces a small vertical deflection at the beginning of the horizontal sweep. This initial vertical deflection (the *stimulus artifact*) increases in amplitude (height) as the strength of the stimulating voltage is increased (fig. 3.3).

When the stimulating voltage reaches a sufficient level (the *threshold potential*), the region of the nerve next

Figure 3.2 **The action potential of a single nerve fiber.** (a) This is an all-or-none change in the membrane potential in response to a depolarizing stimulus above a threshold value. The action potential of a single nerve fiber is observed when a recording electrode is inserted into the cytoplasm of a neuron. Compare this recording with that of figure 3.3. (b) Recordings from a single sensory fiber from a frog nerve stimulated by stretching the gastrocnemius muscle. As more weight (in grams) is used to stretch the muscle, the frequency of action potentials produced by the sensory neuron increases.

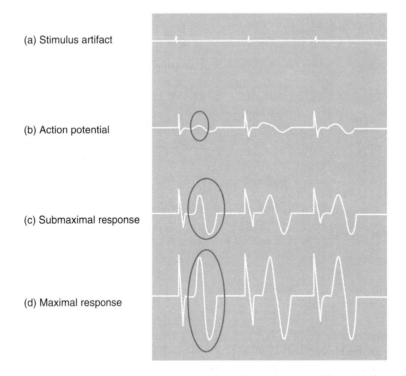

Figure 3.3 **Recording from a frog sciatic nerve.** As the stimulus voltage is increased from (a) through (d), the amplitude of the stimulus artifact and nerve action potential (circled areas) are also increased, but only to a maximum value, shown in (d). This effect is due to the fact that the sciatic nerve contains hundreds of nerve fibers. The number of individual nerve fibers stimulated to produce all-or-none action potentials increases with increasing stimulus intensity because some fibers are located closer to the stimulating electrodes than others.

depolarization wave has not yet reached this second electrode). The potential difference between these two recording electrodes produces a vertical deflection of the electron beam a few milliseconds (msec) after the stimulus artifact. This second event is the action potential of the nerve.

The electrical activity of the nerve can be intensified by increasing either the frequency or the strength of stimulation. Increasing the frequency of stimulation will increase the number of impulses conducted by the nerve in a given time, up to a maximum amount. This maximum (impulses about 2 msec apart) is due to the refractory period of the nerve and ensures that the action potentials will remain separate events even at high frequencies of stimulation. Similarly, increasing the strength of each stimulus will increase the amplitude of each action potential, up to a maximum level. This is because the action potential recorded from a nerve is the sum of the action potentials of all stimulated axons in that nerve. As the strength (voltage) of the stimulus is increased, the number of axons that are depolarized increases, increasing the amplitude of the overall nerve response (fig. 3.3).

When impulses are recorded from individual axons, the amplitude of the action potentials does not increase with increasing strength of stimulation. A neuron either does not "fire" (to any subthreshold stimulus) or it "fires" maximally (to any suprathreshold stimulus). This is the **all-or-none law** of nerve physiology. The stronger the stimulus, the greater the number (not the size) of action potentials carried by a nerve fiber in a given time. The strength of a stimulus, therefore, is coded in the nervous system by the *frequency* (not the amplitude) of action potentials (fig. 3.2b).

USE OF THE OSCILLOSCOPE

Electrical activity in nerves is frequently observed by using an oscilloscope. In an oscilloscope, electrons from a cathode-ray "gun" are sprayed across a fluorescent screen, producing a line of light. Changes in the potential difference between the two recording electrodes cause this line to deflect. Movement of the line upward or downward is proportional to the incoming voltage from the nerve preparation. The oscilloscope can thus function as a fast-responding voltmeter, displaying voltage changes with time as vertical line deflections (fig. 3.3).

The electron beam is made to sweep from left to right across the screen at a particular rate. The image on the screen is a plot of voltage (*y* axis) against time (*x* axis). If the sweep of the electron beam is triggered by a stimulus to the nerve preparation, the electrical response of the nerve will always appear at the same time after the sweep has begun and at the same location on the screen. Since the phosphor in the screen continues to emit light long after it has been struck with electrons, an action potential produced in response to a second stimulus will be superimposed on the one produced in response to the previous stimulus. Therefore, an observer will see an

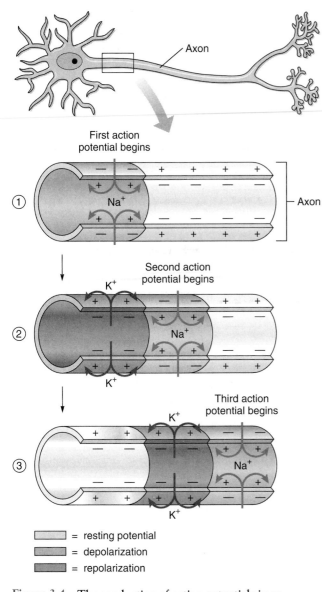

Figure 3.4 **The conduction of action potentials in an unmyelinated axon.** Each action potential "injects" positive charges that spread to adjacent regions. The region that has just produced an action potential is refractory. The next region, not having been stimulated previously, is partially depolarized. As a result, its voltage-regulated Na⁺ gates open and the process is repeated. Successive segments of the axon thereby regenerate, or "conduct," the action potential.

to the stimulating electrodes becomes depolarized (the outside of the nerve becomes negative with respect to the inside). By the creation of "minicircuits," this region depolarizes the adjacent region of the nerve, while it (the region nearest the stimulating electrodes) is being repolarized (fig. 3.4). In this manner, a wave of "surface negativity" is conducted from the stimulating electrodes toward the two recording electrodes. When this wave reaches the first recording electrode, this electrode becomes electrically negative with respect to the second recording electrode (since both are on the surface of the nerve and the

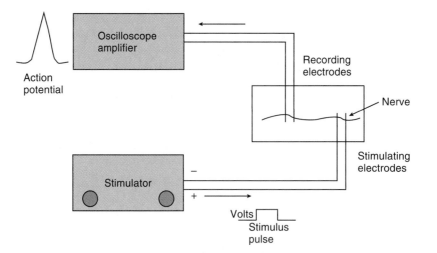

Figure 3.5 Setup for recording from an isolated nerve. A nerve is laid across a pair of stimulating electrodes and a pair of recording electrodes. A stimulator delivers a square-wave pulse of a given voltage and duration to the nerve via the stimulating electrodes. This can be seen at the recording electrodes as the stimulus artifact (not shown). The action potential observed after the stimulus artifact represents the response of the nerve to the stimulus.

apparently stable image of an action potential, even though each action potential lasts for only about 3 msec.

Vertical deflections of the electron beam are produced when a **potential difference,** or voltage, develops between the two *recording electrodes* that touch the nerve some distance away from a pair of *stimulating electrodes* (fig. 3.5). The first action potential, produced near the stimulating electrodes, results from a depolarizing current whose voltage, duration, and frequency (number of "shocks" per second) can be varied by the operator adjusting the oscilloscope stimulator module. Conducted by the nerve, the action potential is recreated in the region of the nerve in contact with the recording electrodes, and a vertical line deflection is observed on the oscilloscope.

 Multiple sclerosis (MS) is a neurological disease characterized by progressive degeneration of the myelin sheaths of neurons in multiple areas of the CNS, resulting in interruption of the normal conduction of impulses and extensive loss of functions. Initially, lesions form on the myelin sheaths and soon develop into hardened *scleroses,* or scars, effectively interfering with the formation of action potentials along the affected axons. Because myelin degeneration is widespread, MS has a wider variety of symptoms than most neurological diseases. This multiplicity of symptoms combined with a characteristic pattern of remission and relapse makes MS a disease that is exceedingly difficult to diagnose and treat.

PROCEDURE

1. Decapitate or double-pith a frog (see section 5 for this procedure), skin its legs and the lower portion of its back, and place it in a prone position in a dissecting tray (fig. 3.6).
2. Make a 1-inch incision on both sides of its spine from the anal region toward the head. Lift the portion of spine free from surrounding muscle, and excise it to expose the right and left sciatic nerves, which run lateral and parallel to the spine.

Note: Be careful not to touch the nerve with your fingers or metal tools; and not to let the nerve touch the surface of the frog's skin or cut muscle. Keep the nerve moist with Ringer's solution.

3. Lift one of the sciatic nerves with a glass probe, tie it with a few inches of thread, and cut the nerve closer to the head, beyond the tie.
4. Separate the large posterior muscles of the thigh to expose the distal portion of the sciatic nerve, and lift it with a glass probe.

Note: The sciatic nerve is easily identified because it runs in the same connective tissue sheath as the sciatic artery. Tie the nerve with a few inches of thread, and cut the nerve distally beyond the tie.

5. Lift the two ends of the nerve with the two lengths of thread (fig. 3.6) and carefully free it from attached muscle and fascia.
6. Lay the nerve across the stimulating and recording electrodes of the nerve chamber, always keeping the nerve moist with Ringer's solution.

Figure 3.6 Exposing the sciatic nerve of a frog. Two glass probes are used to handle the nerve. The two ends have been tied with thread just beyond where they will be cut.

7. Connect the *stimulating* electrodes to the stimulator and the *recording* electrodes to the preamplifier of the oscilloscope.
8. Adjust the horizontal sweep; set the stimulus frequency, duration, and amplitude at their *lowest* values.
9. Slowly increase the stimulus strength until the *threshold* voltage is obtained (the lowest stimulus that produces an observed action potential). Threshold stimulus: _____ V

Note: *The first deflection on the screen is the stimulus artifact; the deflection to the right of the stimulus artifact is the action potential.*

10. Slowly increase the *strength* of the stimulating voltage until the action potential is at its maximum amplitude. Stimulus producing maximum response: _____ V
11. With the stimulating voltage set at a slightly suprathreshold level, gradually increase the *frequency* of stimulation and note this effect on the amplitude of the action potential.

Name _____

Date _____

Section _____

REVIEW ACTIVITIES FOR EXERCISE 3.1

Test Your Knowledge of Terms and Facts

1. The parts of a neuron that receive input from other neurons or from sensory receptors are the _____.
2. The voltage across the membrane of a particular axon may be –70mV. This is known as its _____.
3. During the action potential, the membrane polarity momentarily reverses.
   (a) This process is called _____.
   (b) It is caused by the diffusion of _____ into the axon.
4. The number of action potentials produced per unit time is known at its_____;
   this is proportional to the _____ of the stimulus.
5. What is the all-or-none law regarding action potentials? _____

Test Your Understanding of Concepts

6. Both recording electrodes in this exercise were placed on the surface of the nerve—that is, both were extracellular. What must be done to record the resting membrane potential of an axon?

7. Describe the permeability properties of the axon membrane when the nerve is at rest, and the permeability properties during the production of an action potential. Describe the direction and nature of the ion movements during the production of an action potential.

8. How is the strength of a sensory stimulus coded by a single axon? What additional way could the strength of a stimulus be coded by a collection of axons (a nerve) in the intact nervous system?

9. In this exercise, the amplitude of the action potential produced by the frog sciatic nerve increased when the stimulating voltage increased. How is this possible if the action potentials of individual axons obey the all-or-none law?

Electroencephalogram (EEG)

MATERIALS

1. Oscilloscope and EEG selector box (Phipps and Bird), or physiograph and high-gain coupler (Narco), or electroencephalograph recorder (Lafayette Instrument Company)
2. EEG electrodes and surface electrode
3. Long ECG elastic band and ECG electrolyte gel
4. Alternatively, the *BIOPAC* equipment can be used for visualizing the EEG on a computer (*Biopac* lessons 3 and 4), available in their basic, advanced, and ultimate systems.

Chemical neurotransmitters released by presynaptic nerve fibers produce excitatory or inhibitory postsynaptic potentials. These postsynaptic potentials account for most of the electrical activity of the brain and contribute to the electroencephalogram recorded by surface electrodes positioned over the brain.

OBJECTIVES

1. Describe the structure of a chemical synapse.
2. Describe EPSPs and IPSPs; and explain their significance.
3. Demonstrate the recording of an electroencephalogram; and explain how an EEG is produced.

Textbook Correlations*

Before performing this exercise, you may want to consult the following references in *Human Physiology*, seventh edition, by Stuart I. Fox:

- *The Synapse.* Chapter 7, pp. 164–168.
- *Acetylcholine as a Neurotransmitter.* Chapter 7, pp. 169–174.
- *Cerebral Cortex.* Chapter 8, pp. 191–195.

Those using different physiology textbooks may want to consult the corresponding information in those books.

The basic unit of neural integration is the **synapse,** the functional connection between the axon of one neuron and the dendrites or cell body (occasionally even the axon) of another neuron. A train of action potentials travels down the axon of the first neuron and stimulates the release of packets (*vesicles*) of chemical transmitter substances (fig. 3.7). These chemical transmitters, such as **acetylcholine (ACh),** diffuse across the small space separating the two neurons (the *synaptic cleft*) and reach the membrane of the second neuron (the *postsynaptic membrane*).

The interaction of these chemical transmitters with their specific receptors may stimulate a depolarization in the postsynaptic membrane. Unlike the all-or-none action potential, this **excitatory postsynaptic potential,** or **EPSP,** is a *graded* response—the larger the number of transmitter vesicles released, the greater the depolarization. When the EPSP reaches a critical level of depolarization (the threshold) it generates an action potential, which can then be conducted down the axon of the second neuron to the next synapse. Chemical transmitters released by other neurons that synapse with the second cell may produce the opposite response—a *hyperpolarization* of the postsynaptic membrane (the inside of the cell becomes even more negative with respect to the outside). This is called an **inhibitory postsynaptic potential,** or **IPSP.** The production of action potentials by the second cell, as well as their frequency, will be determined by the algebraic sum of these EPSPs and IPSPs produced by the convergence (fig. 3.8) of multiple neurons on the second cell.

These electrical activities—action potentials, EPSPs and IPSPs—are not equal in two different regions of the brain at the same time. Therefore, an extracellular potential difference (which fluctuates between 50 and 100 millionths of a volt) can be measured by placing two

---

*See Appendix 3 for correlations to the A.D.A.M. *InterActive PHYSIOLOGY Modules*.

 See Appendix 3 for correlations to the *Virtual Physiology Laboratory CD-ROM* by McGraw-Hill and Cypris Publishing, Inc.

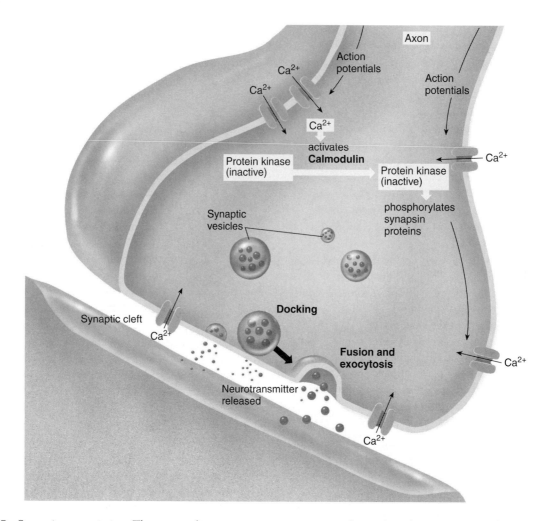

Figure 3.7 Synaptic transmission. The axon ending contains neurotransmitter chemicals within synaptic vesicles. Fusion of the synaptic vesicle with the axon membrane is followed by exocytosis of the neurotransmitter chemicals into the synaptic cleft, where they can diffuse to the postsynaptic membrane. After binding to specific receptor proteins, the neurotransmitters stimulate (or inhibit) the postsynaptic cell. Note that the release of neurotransmitter is stimulated by events set in motion by the entry of Ca^{2+}, which occurs in response to action potentials.

electrode leads on the scalp over two different regions of the brain. The recording of these "brain waves" is known as an **electroencephalogram (EEG).**

In actual practice, nineteen electrodes are placed at various standard positions on the scalp, with each pair of electrodes connected to a different recording pen. The record obtained reflects periodic waxing and waning of synchronous neuronal activity, producing complex wave-forms that are characteristic of the regions of the brain sampled and the state of the subject (fig. 3.9).

There are four characteristic types of EEG wave patterns. **Alpha waves,** consisting of rhythmic oscillations with a frequency of 8 to 12 cycles per second (cps), were the first patterns to be characterized (fig. 3.9). These waves, which can be seen with a single pair of electrodes, are produced by the visual association areas of the parietal and occipital lobes and predominate when the subject is relaxed and awake but has eyes closed. Alpha waves can be suppressed by opening the eyes or by doing mental

arithmetic, and are normally absent in a significant number of people. The alpha rhythm of children under the age of 8 occurs at a lower frequency (4–7 cps).

Beta waves (13–25 cps) are strongest from the frontal lobes and reflect the evoked activity produced by visual stimuli and mental activity; they are enhanced by barbiturate drugs. **Theta waves** (5–8 cps) are emitted from the temporal and occipital lobes and are common in newborn infants. In adults, theta wave recordings generally indicate severe emotional stress and can be a forewarning of a nervous breakdown. **Delta waves** (1–5 cps) seem to be emitted from the cerebral cortex. Delta waves are seen in awake infants. Common in adults during deep sleep, the presence of delta waves in an awake adult indicates brain damage. During a *petit mal epileptic seizure,* the EEG pattern may show regular spikes and waves at 3 cps. Abnormal patterns not observed in the "resting" EEG can often be revealed by stimulation of the subject through hyperventilation or flashing of a strobe light at different frequencies.

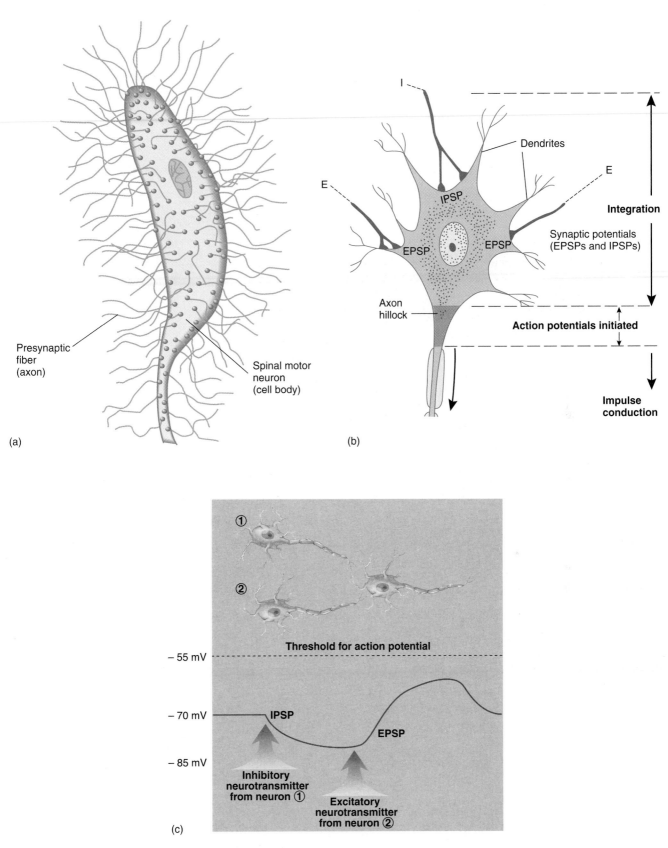

(a)

(b)

I

Dendrites

E

E

IPSP

EPSP

EPSP

Axon
hillock

Integration

Synaptic potentials
(EPSPs and IPSPs)

Action potentials initiated

**Impulse
conduction**

Presynaptic
fiber
(axon)

Spinal motor
neuron
(cell body)

① ②

Threshold for action potential

− 55 mV

− 70 mV

IPSP

EPSP

− 85 mV

**Inhibitory
neurotransmitter
from neuron** ①

**Excitatory
neurotransmitter
from neuron** ②

(c)

Figure 3.8 Synaptic integration. (a) Many presynaptic axons can converge on a single neuron, making synapses at the dendrites and cell body. (b) A diagram illustrating the integration of EPSPs and IPSPs prior to the axon hillock, where the first action potentials are initiated (E = excitatory input; I = inhibitory input, producing EPSPs and IPSPs, respectively). (c) The summation of an IPSP and an EPSP on a postsynaptic cell membrane.

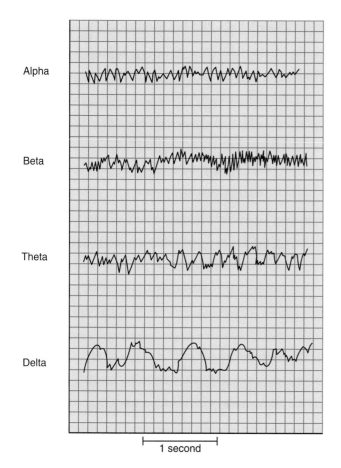

Figure 3.9 Electroencephalograph (EEG) rhythms.

Alpha

Beta

Theta

Delta

├─────── 1 second ───────┤

Red lead

White lead

Black lead
(ground)

Figure 3.10 **Biopac EEG electrode placement.** Be sure that the EEG leads are in firm contact with the scalp.

Use of the *electroencephalograph* may help to diagnose a number of brain lesions, including epilepsy, intracranial infections, and encephalitis. It has been discovered that meditation, as performed by Zen monks and yogis, results in the production of slow alpha waves (7 cps) of increased amplitude and regularity (even with the eyes partially open), and that this change is associated with a decrease in metabolism and sympathetic nerve activity. Many people, through biofeedback techniques, attempt to enhance their ability to produce alpha rhythms so they can relax and lower their sympathetic nerve effects.

PROCEDURE

1. If an **oscilloscope** will be used,
    (a) connect the EEG selector box to the preamplifier of the oscilloscope, and adjust the horizontal sweep and sensitivity;
    (b) plug the lead for the forehead into the EEG selector box outlet labeled *L. Frontal* and the lead for the earlobe into the outlet for the *ground*.

2. If a **physiograph** will be used,
    (a) insert a *high-gain coupler* into the physiograph;
    (b) set the *gain* on ×100, the *sensitivity* on 2 or 5, and the *time constant* on 0.03 or 0.3;
    (c) connect two *EEG electrodes* and one *surface electrode* to the high-gain coupler.
3. If an **electroencephalograph** will be used,
    (a) insert the cable from the electrode box/lead selector into the EEG amplifier within the recorder;
    (b) insert the electrodes into the selector box, and follow the instructions for the particular equipment model used.
4. Obtain a long elastic ECG strap and tie it around the forehead (with the knot at the back of the head) to form a snug headband.
    (a) If an **oscilloscope** will be used for recording, dab a little electrolyte gel onto the single electrode from the left frontal outlet and place it under the headband in the middle of the forehead.
    (b) If a **physiograph** will be used for recording, dab electrolyte gel onto both EEG electrodes and place them under the headband on the right and left sides of the forehead.
5. Dab a little electrolyte gel onto the ground electrode (the surface electrode for the physiograph), and with your fingers, press this electrode against the skin behind the ear.
6. If the ***Biopac*** system is used (fig 3.10), follow the procedural steps provided on the computer screen. You can complete their lab exercise 3 and 4 once the electrodes are in place.

Note: Alpha, beta, delta, and theta waves will be recorded and stored in your student folder.

7. With the subject in a relaxed position (no muscular movements), with his or her eyes closed, observe the EEG pattern, checking particularly for the presence of alpha waves. Since many people do not produce alpha waves in this situation, test a number of subjects.

Note: Interference from room electricity at 60 cps sometimes occurs. This will appear as regular, fast, low-amplitude waves usually superimposed on the slower, more irregular brain waves of larger amplitude.

8. Observe the effect of opening the eyes, doing mental arithmetic, and hyperventilating on the production of alpha waves.

Laboratory Report 3.2

REVIEW ACTIVITIES FOR EXERCISE 3.2

Text Your Knowledge of Terms and Facts

1. A chemical released by an axon is known as a _____.
2. Binding of the above-named chemical to its receptor in the postsynaptic membrane may produce a depolarization called a(n) _____.
3. Different chemicals released by axons may produce a hyperpolarization of the postsynaptic membrane; such a hyperpolarization is called a(n) _____.
4. Action potentials are all-or-none; synaptic potentials, by contrast, are _____.

Match the following:

____5. alpha rhythm   (a) common in awake, tense subjects
____6. beta rhythm   (b) observed in some relaxed subjects
____7. theta rhythm   (c) observed in deep sleep
____8. delta rhythm   (d) seen in some children

Test Your Understanding of Concepts

9. Describe where an EPSP is produced, and explain how it is produced. What is its significance?

10. Compare the properties of an EPSP with those of an action potential.

11. What is the significance of IPSPs? How are they produced?

12. Propose a rationale by which the observation of a person's brain waves might be used to help that person lower his or her blood pressure.

13. Which type of neural activity, action potentials or synaptic potentials, produces the EEG? Explain your answer. Speculate as to why the EEG may be influenced by hyperventilation.

Reflex Arc

MATERIALS

1. Rubber mallets
2. Blunt probes

In a reflex, specific sensory stimuli evoke characteristic motor responses very rapidly because few synapses are involved. Since a specific simple reflex arc occurs at a specific spinal cord segment and involves particular nerves, tests for simple reflex arcs are very useful in diagnosing neurological disorders.

OBJECTIVES

1. Describe the neurological pathways involved in a simple reflex arc.
2. Describe the structure and function of muscle spindles.
3. Demonstrate muscle stretch reflexes and explain the clinical significance of these tests.
4. Demonstrate a Babinski reflex (Babinski's sign) and explain the clinical significance of this test.

Textbook Correlations*

Before performing this exercise, you may want to consult the following references in *Human Physiology*, seventh edition, by Stuart I. Fox:

- *Spinal Cord Tracts.* Chapter 8, pp. 206–209.
- *Cranial and Spinal Nerves.* Chapter 8, pp. 209–212.
- *Neural Control of Skeletal Muscles.* Chapter 12, pp. 347–354.

Those using different physiology textbooks may want to consult the corresponding information in those books.

The speed of a motor response to an environmental stimulus depends, in part, on the number of synapses to be crossed between the *afferent* flow of impulses and the activation of *efferent* nerves. A **reflex** is a relatively simple motor response that is made without the involvement of large numbers of association neurons.

The simplest reflex requires only one synapse between the sensory and motor neurons (the **knee-jerk,** or **patellar reflex,** for example). Impulses traveling on the sensory axons enter the CNS in the *dorsal root* of the peripheral nerve, make a single synapse with a motor neuron (an alpha motoneuron) within the central gray matter, and then leave the CNS in the *ventral root* of the spinal nerve (fig. 3.11). In more complicated reflexes, the sensory impulses may travel longitudinally and transversely within the gray matter, stimulating other motor neurons. This may lead to the contraction of other flexor muscles on the same side (*ipsilateral* muscles) and the contraction of extensor muscles on the opposite side (*contralateral* muscles), while inhibiting the contraction of antagonistic muscles (ipsilateral extensors and contralateral flexors).

A. TESTS FOR SPINAL NERVE STRETCH REFLEXES

In this exercise, a number of reflex arcs will be tested that are initiated by distinctive *stretch receptors* within muscles. These receptors, called **muscle spindles,** are embedded within the connective tissue of the muscle and consist of specialized thin muscle fibers (*intrafusal fibers*) that are innervated by sensory neurons (fig. 3.11). The intrafusal fibers are arranged in parallel with the normal muscle cells (*extrafusal fibers*), so that stretch of the muscle also places tension on the intrafusal fibers.

---

*See Appendix 3 for correlations to the A.D.A.M. *InterActive* PHYSIOLOGY *Modules*.

 See Appendix 3 for correlations to the *Intelitool Physiology Laboratory Exercises*.

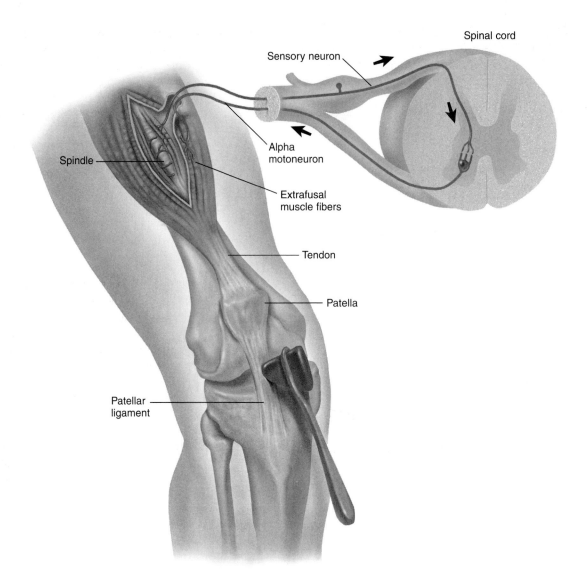

Figure 3.11   The knee-jerk reflex. This is an example of a monosynaptic stretch reflex.

Located within the spindles, the intrafusal fibers respond to the tension by causing the stimulation (depolarization) of the sensory neuron. The sensory neuron arising from the intrafusal fiber synapses with the motor neuron in the spinal cord that, in turn, innervates the extrafusal fibers. The resultant contraction of the extrafusal fibers of the muscle releases tension on the intrafusal fibers and decreases stimulation of the stretch receptors. In a typical clinical examination, this reflex is elicited by striking the muscle tendon with a rubber mallet, creating a momentary stretch.

When the extrafusal muscle fibers contract during the stretch reflex, they produce a short, rapid movement of the limb (the jerk). This is very obvious for the knee-jerk reflex, but can be quite subtle for the biceps- and triceps-jerk reflexes. Use of the *flexicomp* allows the limb movement to be seen as a tracing on the computer screen.

PROCEDURE FOR KNEE-JERK REFLEX (FIG. 3.12A)—TESTS FEMORAL NERVE

1. Allow the subject to sit comfortably with his or her legs free.
2. Strike the ligament portion of the patellar tendon just below the patella (kneecap), and observe the resulting contraction of the quadriceps muscles and extension of the lower leg.

PROCEDURE FOR ANKLE-JERK REFLEX (FIG. 3.12B) — TESTS MEDIAL POPLITEAL NERVE

1. Have the subject kneel on a chair with his or her back to you, and with feet (shoes and socks off) projecting over the edge.
2. Strike the Achilles (calcaneal) tendon at the level of the ankle and observe the resulting plantar extension of the foot.

PROCEDURE FOR BICEPS-JERK REFLEX (FIG. 3.12C) — TESTS MUSCULOCUTANEOUS NERVE

1. With the subject's arm relaxed but fully extended on the desk, gently press his or her biceps tendon in the antecubital fossa with your thumb or forefinger and strike this *finger* with the mallet.
2. If this procedure is performed correctly, the biceps muscle will twitch but usually will not contract strongly enough to produce arm movement.

PROCEDURE FOR TRICEPS-JERK REFLEX (FIG. 3.12D) — TESTS RADIAL NERVE

1. Have the subject lie on his or her back with the elbow bent, so that the arm lies loosely across the abdomen.
2. Strike the triceps tendon about 2 inches above the elbow. If there is no response, repeat this procedure, striking to either side of the original point.
3. If this procedure is correctly performed, the triceps muscle will twitch but usually will not contract strongly enough to produce arm movement.

B. A Cutaneous Reflex: The Plantar Reflex and Babinski's Sign

The **plantar reflex** is elicited by cutaneous (skin) receptors of the foot and is one of the most important neurological tests. In normal individuals, proper stimulation of these receptors located in the sole of the foot results in the flexion (downward movement) of the great toe, while the other toes flex and come together. The normal plantar reflex requires the uninterrupted conduction of nerve impulses along the *pyramidal motor tracts*, which descend directly from the cerebral cortex to motor neurons lower in the spinal cord. Damage anywhere along the pyramidal motor tracts produces a *Babinski reflex,* or **Babinski's sign,** to this stimulation, in which the great toe extends (moves upward) and the other toes fan laterally, as shown in figure 3.12*e*). Infants exhibit Babinski's sign normally because neural control is not yet fully developed.

Tests for simple muscle reflexes are basic to any physical examination when motor nerve or spinal damage is suspected. If there is spinal cord damage, these easily performed tests can help locate the level of the spinal cord that is damaged: motor nerves that exit the spinal cord above the damaged level may not be affected, whereas nerves that originate at or below the damaged level may not be able to produce their normal reflexes. In this regard, the Babinski test is particularly useful, because damage to the pyramidal motor (corticospinal) tract at any level may be detected by a positive Babinski reflex.

PROCEDURE

1. Have the subject lie on his or her back with knees slightly bent, and with the thigh rotated so that the lateral (outer) side of the foot is resting on the couch.
2. Applying firm (but not painful) pressure, draw the tip of a blunt probe along the lateral border of the sole, starting at the heel and ending at the base of the big toe (fig. 3.12*e*). Observe the response of the toes to this procedure.

(a) Knee (patellar) reflex

(b) Ankle (Achilles) reflex

(c) Biceps reflex

(d) Triceps reflex

(e) Plantar reflex

Figure 3.12   Some reflexes of clinical importance.

Cutaneous Receptors and Referred Pain

MATERIALS

1. Thin bristles, cold and warm metal rods
2. Calipers, cold and warm water baths

Specialized sensory organs and free nerve endings in the skin provide four modalities of cutaneous sensation. The modality and location of each sensation is determined by the specific sensory pathway in the brain; the acuteness of sensation depends on the density of the cutaneous receptors. Damage to an internal organ may elicit a perception of pain in a somatic location, such as a limb or a region of the body wall. This type of pain, called referred pain, is very important in clinical diagnosis.

OBJECTIVES

1. Describe the punctate distribution of cutaneous receptors.
2. Describe the structures of cutaneous receptors and the modality of sensations they mediate.
3. Determine the two-point touch threshold in different areas of the skin and explain the physiological significance of the differences obtained.
4. Define and demonstrate sensory adaptation and explain its significance.
5. Define the term *referred pain* and explain how this pain is produced.
6. Demonstrate the referred pain produced by striking the ulnar nerve with a mallet, and explain the clinical significance of other referred pains in the body.

Textbook Correlations

Before performing this exercise, you may want to consult the following references in *Human Physiology*, seventh edition, by Stuart I, Fox:

• *Cutaneous Sensations.* Chapter 10, pp. 242–245.

Those using different physiology textbooks may want to consult the corresponding information in those books.

Sensory receptors in the skin transduce mechanical or chemical stimuli into nerve impulses (action potentials). These impulses are conducted by sensory neurons into the central nervous system, and ultimately to the region of the brain that interprets these impulses as a particular sensation. Different receptors are most sensitive to different stimuli, and thus activate different sensory neurons. It is the brain's interpretation of action potentials along particular sensory neurons that creates the sensory perception.

A. MAPPING THE TEMPERATURE AND TOUCH RECEPTORS OF THE SKIN

Four independent modalities of cutaneous sensation have traditionally been recognized: *warmth, cold, touch,* and *pain.* (Pressure is excluded because it is mediated by receptors deep in the dermis, and the sensations of itch and tickle are usually excluded because of their mysterious origin.) Mapping of these sensations, such as temperature and touch, on the surface of the skin has revealed that the receptors are not generalized throughout the skin but are clustered at different points (have a *punctate distribution*).

Since the punctate distribution is different for each of the four sensory modalities, physiologists initially believed that each sensation was mediated by a different sensory receptor—a view that was supported by the histological identification of different cutaneous receptors (table 3.1 and fig. 3.13). Excision of areas of the skin according to different sensory maps, however, failed to reveal a different distribution of receptors, and more recent experiments have suggested that the four sensations may arise from the brain's analysis of complex patterns of sensory (afferent) impulses.

Table 3.1 Cutaneous Receptors

Receptor	Structure	Sensation	Location
Free nerve endings	Unmyelinated dendrites of sensory neurons	Light touch; hot; cold; nociception (pain)	Around hair follicles; throughout skin
Merkel's discs	Expanded dendritic endings	Sustained touch and pressure	Base of epidermis (stratum basale)
Ruffini corpuscle (endings)	Enlarged dendritic endings within open, elongated capsule	Sustained pressure	Deep in dermis and hypodermis
Meissner's corpuscles	Dendrites encapsulated in connective tissue	Changes in texture; slow vibrations	Upper dermis (papillary layer)
Pacinian corpuscles	Dendrites encapsulated by concentric lamellae of connective tissue structures	Deep pressure; fast vibrations	Deep in dermis

Merkels's discs

Root hair plexus

Pacinian corpuscle

Meissner's corpuscle

Free nerve ending

Ruffini endings

Figure 3.13   A diagram of the skin showing cutaneous receptors.

PROCEDURE

1. With a ballpoint pen, draw a square (2 cm per side) on the ventral surface of the subject's forearm. Alternatively, a square ink stamp may be used.
2. With the subject's eyes closed, gently touch a dry, ice-cold, metal rod to different points in the square. Mark the points of cold sensation with a dark dot.
3. With the subject's eyes closed, gently touch a dry, warm, metal rod (heated to about 45° C in a water bath) to different points in the square. Mark the points of warm sensation with an open circle.
4. Gently touch a thin bristle to different areas of the square and indicate the points of touch sensation with small x's.
5. Reproduce this map in your laboratory report.

b hand 5   lip 3
palm 10
nech f tip 3

bach 20

Figure 3.14 Motor and sensory areas of the cerebral cortex. (a) Motor areas that control skeletal muscles, and (b) sensory areas that receive somatesthetic sensations.

B. THE TWO-POINT THRESHOLD IN TOUCH PERCEPTION

The density of touch receptors in some parts of the body is greater than in other parts. Therefore, the areas of the **sensory cortex** (*postcentral gyrus* of the central fissure, fig. 3.14b) that correspond to different regions of the body are of different sizes. Those areas of the body that have the largest density of touch receptors also receive the greatest motor innervation; the areas of the **motor cortex** (*precentral gyrus* of the central fissure, fig. 3.14a) that serve these regions are correspondingly larger than other areas. Therefore, a map of the sensory and motor areas of the brain reveals that large areas are devoted to the touch perception and motor activity of the face (particularly the tongue and lips) and hands, whereas relatively small areas are devoted to the trunk, hips, and legs.

The density of touch receptors is measured by the **two-point threshold test.** The two points of a pair of adjustable calipers are simultaneously placed on the subject's skin with equal pressure, and the subject is asked whether two separate points of contact are felt. If the answer is yes, the points of the caliper are brought closer together, and the test is repeated until only one point of contact is felt. The minimum distance at which two points of contact can be felt is the two-point threshold.

Sensory information from the cutaneous receptors projects to the **postcentral gyrus** of the cerebral cortex. Therefore, direct electrical stimulation of the postcentral gyrus produces the same sensations as those felt when the cutaneous receptors are stimulated. Much of this information has been gained by the electrical stimulation of the brain of awake patients undergoing brain surgery; the surgeon must often map the areas of the brain in order to locate the site of the lesion and avoid damage to healthy tissue. Since the cutaneous receptors are more densely arranged in the face, tongue, and hands than on the back and thighs, larger areas of the brain are involved in analyzing information from the former areas than from the latter. Consequently, the areas of the brain map representing the face, tongue, and hands are larger than those representing the back and thighs. The map is also upside down, with the feet represented near the superior surface and the head represented more inferiorly and laterally in the cortex (fig. 3.14).

PROCEDURE

1. Starting with the calipers wide apart and the subject's eyes closed, determine the two-point threshold on the back of the hand. (Randomly alternate the two-point touch with one-point contacts, so that the subject cannot anticipate you.)
2. Repeat this procedure with the palm of the hand, fingertip, and back of the neck.
3. Write the minimum distance (in mm) in the data table provided in your laboratory report.

C. ADAPTATION OF TEMPERATURE RECEPTORS

Many of our sense receptors respond strongly to acute changes in our environment and then stop responding when these stimuli become constant. This phenomenon is known as **sensory adaptation.** Our sense of smell, for example, quickly adapts to the odors of the laboratory; and our touch receptors soon cease to inform us of our clothing, until these stimuli change. Sensations of pain, by contrast, adapt little if at all.

When one hand is placed in warm water and another in cold water, the strength of stimulation gradually diminishes until both types of temperature receptors have adapted to their *new* environmental temperature. If the two hands are then placed in water at an intermediate temperature, the hand that was in the cold water will feel warm, and the hand that was in the warm water will feel cold. The "baseline," or "zero," of the receptors has obviously changed. The sensations of temperature are therefore not absolute but relative to the baseline previously established by sensory adaptation.

PROCEDURE

1. Place one hand in warm water (about 40°C) and the other in cold water, and leave them in the water for a minute or two (remove them if the water becomes too uncomfortable).
2. Now place both hands in lukewarm water (about 22°C); and record your observations and conclusions regarding your sensations in the laboratory report.

D. REFERRED PAIN

Receptor organs are *sensory transducers*, changing environmental stimuli into afferent nerve impulses (action potentials). Since the action potentials in one nerve are the same as those in another, the perception of the sensation is determined entirely by the area of the brain stimulated, which is different for each sensory nerve. Although a given sensory nerve is normally stimulated by a specific receptor, trauma to the nerve along the afferent pathway may also evoke action potentials, and this will be interpreted by the brain as the normal sensation. An example of traumatic stimulation would be seeing flashes of light or "stars" when punched in the eye.

Amputees frequently report feelings of pain in their missing limbs as if they were still there; this is part of the **phantom limb phenomenon.** The source of nerve stimulation is trauma to the cut nerve fibers, yet the brain perceives the pain as coming from the amputated region of the body that had originally produced the action potentials along these nerves. This is a **referred pain** because the source of nerve stimulation is different from the perceived location of the stimulus.

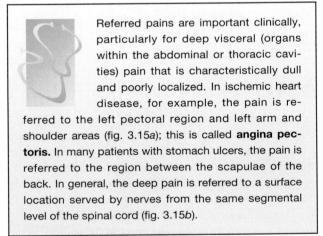

Referred pains are important clinically, particularly for deep visceral (organs within the abdominal or thoracic cavities) pain that is characteristically dull and poorly localized. In ischemic heart disease, for example, the pain is referred to the left pectoral region and left arm and shoulder areas (fig. 3.15a); this is called **angina pectoris.** In many patients with stomach ulcers, the pain is referred to the region between the scapulae of the back. In general, the deep pain is referred to a surface location served by nerves from the same segmental level of the spinal cord (fig. 3.15b).

PROCEDURE

1. Gently tap the ulnar nerve where it crosses the median epicondyle of the elbow.
2. Describe the locations where you perceive tingling or pain.

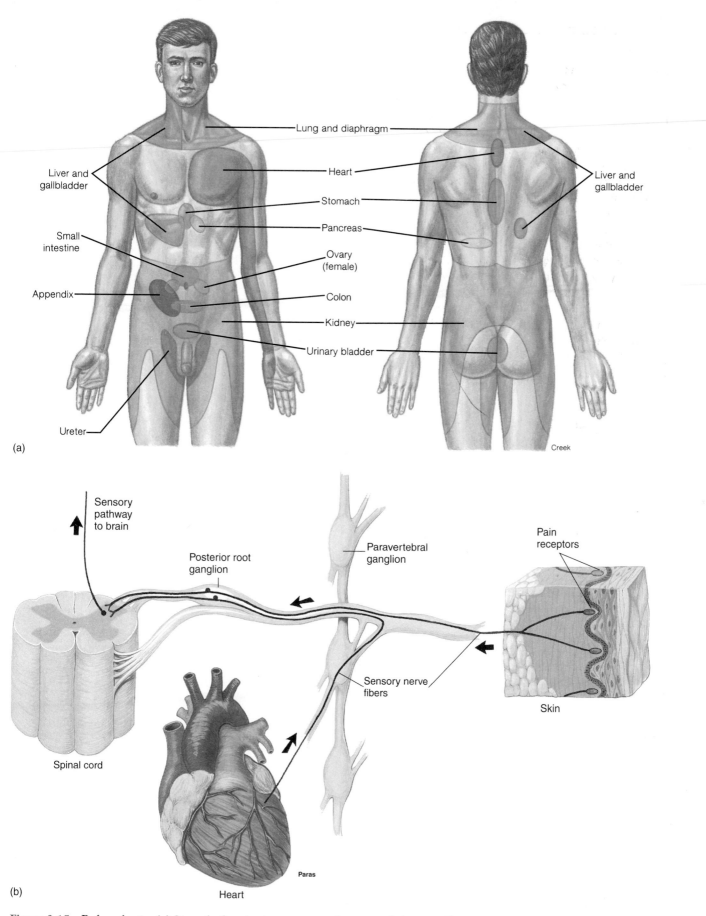

(a)

(b)

Figure 3.15 Referred pain. (a) Sites of referred pain are perceived cutaneously but actually originate from specific visceral organs. (b) One explanation why pain from a visceral organ, such as the heart, might be perceived over a particular area of skin. In this scenario, sensations from the two regions share a common nerve pathway.

A. Right Eye $\frac{H}{2}$ - D - S

$\frac{20}{20}$

Left Eye

$\frac{20}{15}$

B. Right side - astigmatism

Eyes and Vision

MATERIALS

1. Snellen eye chart and astigmatism chart
2. Wire screen and meter stick
3. Ophthalmoscope
4. Lamp
5. Red, blue, and yellow squares on larger sheets of black paper or cardboard

The elastic properties of the eye lens allow its refractive power to be varied so that the image of an object from almost any distance can be focused properly on the retina. Photoreceptors—rods and cones—are located in the retina. The refractive abilities of the eye and the functions of its inner structures are routinely tested in eye examinations.

OBJECTIVES

1. Describe the structure of the eye and the functions of its component parts.
2. Test for visual acuity and accommodation; describe common refractive problems.
3. Identify the extrinsic eye muscles and describe their functions.
4. Describe the optic disc and fovea centralis; explain their significance.
5. Demonstrate the presence of a blind spot; and explain why light focused on this spot cannot be seen.

Textbook Correlations*

Before performing this exercise, you may want to consult the following references in *Human Physiology,* seventh edition, by Stuart I. Fox:

- *The Eyes and Vision.* Chapter 10, pp. 260–268.
- *Retina.* Chapter 10, pp. 268–274.

Those using different physiology textbooks may want to consult the corresponding information in those books.

aphragm that regulates the entry of light into the eye through its aperture (*pupil*). The semigelatinous *vitreous humor* (or *vitreous body*) occupies the posterior chamber and lends structural support to the eye.

A. REFRACTION: TEST FOR VISUAL ACUITY AND ASTIGMATISM

Light rays are bent (*refracted*) when they pass from air to a medium of greater density, where their rate of transmission is slower. The light rays that diverge from an object in the visual field are refracted by the cornea, aqueous humor, lens, and vitreous humor of the eye so that the rays converge (are focused) on the retina and form an inverted image, reversed from left to right (fig. 3.17).

The refractive power of the cornea and vitreous humor is constant. The strength of the lens (i.e., its ability to refract light) can be varied by making it more or less convex. The greater the degree of convexity, the greater the strength of the lens (i.e., the greater the ability to bring parallel rays of light to a focus). A lens that brings light to a focus 0.25 m from its center is stronger (more convex) than a lens that brings light to a focus 1 m from its center. The strength of a lens is expressed in **diopters.**

$$\text{Strength (diopters)} = \frac{1}{\text{focal length (meters)}}$$

You should be familiar with the gross structure of the eye (fig. 3.16). The eye has three walls, or tunics, that form an *outer fibrous layer* (the **sclera** and **cornea**), a middle vascular layer (the **choroid**), and an inner layer (the **retina**). The **lens,** suspended by *suspensory ligaments* attached to the **ciliary body,** divides the eye into anterior and posterior *cavities*. Filled with *aqueous humor*, the anterior cavity is further divided into anterior and posterior *chambers* by the colored **iris.** The iris is a muscular di-

*See Appendix 3 for correlations to the A.D.A.M. *InterActive PHYSIOLOGY Modules*.

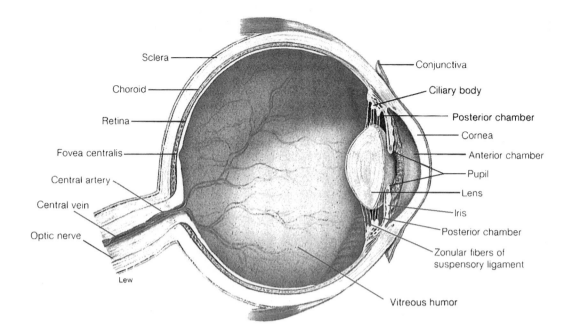

Figure 3.16 **The gross structure of the eye.**

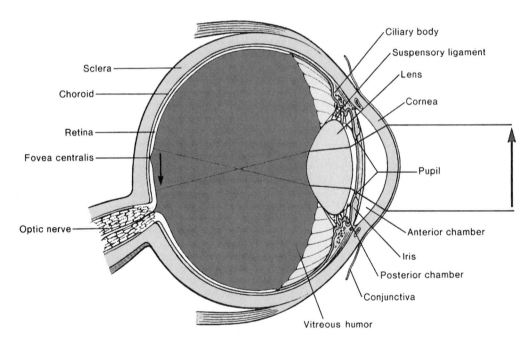

Figure 3.17 **The refraction of light waves within the eyeball.** This focuses the image on the retina, but causes the image of an object to be inverted.

The lens that brings parallel waves of light to a focus 0.25 m from its center has a *focal length* of 0.25 m and a strength of 4 diopters, whereas the lens that brings light to a focus 2 m from its center has a focal length of 2 m and a strength of 0.5 diopters. The refractive power of the normal eye when an object is 20 feet or more away is 67 diopters.

When the light rays that diverge from two adjacent points in the visual field are each brought to a perfect focus on the retina, two points will clearly be perceived. If, however, the light rays converge on a point in front of or behind the retina, the two points in the visual field will be perceived as one fuzzy or blurred point. To correct this defect in **visual acuity** (sharpness of vision), the individual must either adjust the distance between the eye and the object or wear corrective lenses that change the degree of refraction.

When a distant object (20 feet or more) is brought to a focus in front of the retina, the individual is said to have **myopia** (*nearsightedness*). Myopia is usually due to an elongated eyeball (excessive distance from lens to retina) and is corrected by a concave lens. In the opposite condition, **hyperopia** (hypermetropia, or *farsightedness*), the image is brought to a focus behind the retina. This condition is usually due to an eyeball that is too short. In this case, an increase in refractive power is needed, and a convex lens is used. Normal visual acuity is called **emmetropia.**

Visual acuity is frequently tested by means of the **Snellen eye chart.** A person with normal visual acuity can read the line marked 20/20 from a distance of 20 feet. An individual with 20/40 visual acuity must stand 20 feet away from a line that a normal person can read at 40 feet. An individual with 20/15 visual acuity can read a line at a distance of 20 feet that the average, normal young adult could not read at a distance greater than 15 feet. The person with 20/40 vision has myopia, but the person with 20/15 vision does not necessarily have hyperopia. The farsighted person has a decreased ability to see near objects but cannot see distant objects any better than a person with normal vision.

An **astigmatism** is a visual defect produced by an abnormal curvature of the cornea or lens, or by an irregularity in their surface. Because of this abnormality, the refraction of light rays in the horizontal plane is different from the refraction in the vertical plane. At a given distance from an astigmatism chart, therefore, lines in the visual field oriented in one plane will be clear, while lines oriented in the other plane will be blurred. Astigmatism is corrected by means of a cylindrical lens.

The strength of corrective lenses prescribed is given in diopters, preceded by either a plus sign (convex lens for hyperopia; e.g., +14 diopters) or a minus sign (concave for myopia; e.g., –5 diopters). The correction for astigmatism indicates both the strength of the cylindrical lens (e.g., +2) and the axis of the defect (90° for vertical plane, 180° for horizontal plane). A correction for both myopia and astigmatism may be indicated, for example, as –3 + 2 axis 180°.

PROCEDURE

1. Stand 20 feet (6 m) from the *Snellen eye chart.* Covering one eye, attempt to read the line with the smallest letters you can see (with glasses off, if applicable). Walk up to the chart and determine the visual acuity of that eye.
2. Repeat this procedure using the other eye (with glasses off, if applicable).
3. Repeat this procedure for each eye with glasses on (if applicable).
4. Stand about 20 feet away from an *astigmatism chart* and cover one eye (glasses off). This chart consists of a number of dark lines radiating from a central point, like spokes on a wheel. If astigmatism is present, some of the spokes will appear sharp and dark, whereas others will appear blurred and lighter because they are coming to a focus either in front of or behind the retina. Still covering the same eye, slowly walk up to the chart while observing the spokes.
5. Repeat this procedure using the other eye.
6. Repeat the test for astigmatism for both eyes with glasses on (if applicable).
7. To verify that astigmatism has been corrected with glasses, hold the glasses in front of your face while standing 10 feet from the chart and rotate the glasses 90°. The shape of the wheel should change when the glasses are rotated.

B. ACCOMMODATION

If the refractive power (strength) of a lens is constant, the distance between the lens and the point of focus (focal length) will increase as an object moves closer to the lens. For example, if the image of an object that is 20 feet away is in focus on the retina (or on the photosensitive film of a camera), the image of an object 10 feet away will be focused *behind* the retina (or the camera film) and will appear blurred. A camera can be adjusted to focus on an object 10 feet away by moving the lens outward until the focal length of the image equals the distance between the lens and the film. The object 10 feet away will now be in focus, but the object 20 feet away will be blurred because its image will now come to a focus *in front of* the film.

The human eye differs from the camera in that the distance between the retina and the lens of the eye cannot be changed to bring objects into focus. Since the human lens is elastic, however, its degree of convexity (and therefore its refractive power) can be altered by changing the tension placed on it by the suspensory ligament; this, in turn, is regulated by the degree of contraction of the ciliary muscle. When the ciliary muscle is relaxed, the suspensory ligament pulls on the lens, thereby decreasing its convexity and power; distant objects (more than 20 feet away) are thus brought to a focus on the retina. Near objects are brought to a focus on the retina by contraction of the ciliary muscle. The contraction reduces the tension on the suspensory ligament, allowing the lens to assume a more convex shape. This ability of the eye to focus the images of objects that are at different distances from the lens is called **accommodation.**

The convexity of the normal lens can be adjusted to give it a range of power from 67 diopters (for distant vision; least convex) to 79 diopters (for near vision; most convex). The elasticity of the lens and the degree of convexity it can assume for near vision decreases with age, a condition called **presbyopia,** or *old eyes*. Lens elasticity can be tested by measuring the **near point of vision** (the closest an object can be brought to the eyes while still maintaining visual acuity). The near point of vision changes dramatically with age, averaging about 8 cm at age 10 and 100 cm at age 70. Presbyopia is corrected with *bifocals*, which contain two lenses of different refractive strengths.

In addition to tests of refraction, measurements of intraocular (eyeball) pressure are frequently performed with a device known as a *tonometer*. About 6 ml of aqueous humor is formed per day by the ciliary body. This fluid is drained by the *canal of Schlemm* (or *venous sinus*). If the drainage of aqueous humor is blocked, the intraocular pressure may rise; this is a condition known as **glaucoma,** which may cause damage to the optic nerve and blindness. Glaucoma may also damage the cornea, resulting in replacement of the normally transparent tissue with opaque scar tissue. When this happens the cornea can be surgically removed and replaced with either a contact lens or a grafted cornea. Because the cornea is avascular, corneal grafts can be performed with less risk of the transplanted tissue being rejected by the immune system.

PROCEDURE

1. Place a square of wire screen about 10 inches in front of your eyes, and observe a distant object through the screen.
2. After closing your eyes momentarily, open them and note whether the screen or the distant object is in focus.
3. Repeat this procedure, this time focusing the eyes on the screen *before* opening them.
4. To measure the near point of vision, place one end of a meter stick under one eye and extend it outward. Holding a pin at arm's length, gradually bring the pin toward the eye.
5. Record the distance at which the pin first appears blurred or doubled.
   Near point of vision __15__ cm *Left eye*

6. Repeat this procedure, determining the near point of vision for the other eye.
   Near point of vision __16__ cm *Right*

Note: *If the average near point of vision at age 10 is 8 cm, and at age 70 is 100 cm, what is your expected near point of vision?*

C. EXTRINSIC MUSCLES OF THE EYE AND NYSTAGMUS

The six extrinsic muscles of the eye are shown in figure 3.18. The cranial nerve innervations and the actions of these muscles are summarized in table 3.2. These muscles allow the eyes to follow a moving object by maintaining the image on the same location of the retina of each eye, the *fovea centralis*, which provides maximum visual acuity. These muscles also allow the visual field of each eye to maintain the correct amount of central overlap. (The medial regions of each visual field overlap, while the more lateral regions are different for each eye; this **retinal disparity** helps in three-dimensional vision and depth perception.) When an object is brought closer, the correct amount of overlap and retinal disparity is maintained by the medial movement, or *convergence*, of the eyes.

The actions of antagonistic ocular muscles normally maintain the eyes in a midline position. If the tone of one muscle is weak as a result of muscle or nerve damage, the eyes will drift slowly in one direction followed by a rapid movement back to the correct position. This phenomenon is known as **nystagmus.**

In a typical examination of the ocular muscles, the subject is asked to follow an object (such as a pencil) with his or her eyes as it is moved up and down, right and left. Continued oscillations of the eye (slow phase in one direction, fast phase in the opposite direction) indicate the presence of nystagmus. Inability to move the eye outward indicates damage either to the abducens (sixth cranial) nerve or the lateral rectus muscle (table 3.2). Inability to move the eye downward when it is moved inward indicates damage to the trochlear (fourth cranial) nerve or the superior oblique muscle. All other defects in eye movement may be due either to damage of the oculomotor (third cranial) nerve or to damage of the specific muscles involved. *No movement seen but felt it for ruler?*

PROCEDURE

1. Observe retinal disparity by holding a pencil in front of your face with one eye closed and then quickly changing eyes and noting the apparent position of the pencil.
2. Observe convergence by asking a subject to focus on the tip of a pencil as it is slowly brought from a distance of 2 feet in front of the face to the bridge of the nose. Notice the change in the diameter of the subject's pupil during this procedure.

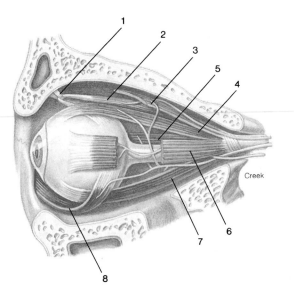

1. Trochlea
2. Superior oblique
3. Levator palpebrae superioris (cut)
4. Superior rectus
5. Medial rectus
6. Lateral rectus
7. Inferior rectus
8. Inferior oblique

Creek

Figure 3.18 **A lateral view of ocular muscles of the left eyeball.**

Table 3.2 The Ocular Muscles

Muscle	Cranial Nerve Innervation	Movement of Eyeball
Lateral rectus	Abducens	Lateral
Medial rectus	Oculomotor	Medial
Superior rectus	Oculomotor	Superior and medial
Inferior rectus	Oculomotor	Inferior and medial
Inferior oblique	Oculomotor	Superior and lateral
Superior oblique	Trochlear	Inferior and lateral

3. Hold a pencil about 2 feet away from the bridge of the subject's nose. Then move the pencil to the left, to the right, and up and down, leaving the pencil in each position at least 10 seconds. Observe the movement of the eyes and note the presence or absence of nystagmus. *as pencil gets close lens gets smaller*

Movement

D. PUPILLARY REFLEX *then opposite*

The correct amount of light is admitted into the eye through an adjustable aperture, the pupil, surrounded by the iris. The iris consists of two groups of smooth muscles with opposing actions. Operating like sphincters, the *circular muscles* constrict the pupil in bright light, whereas the *radial muscles* work to dilate the pupil in dim light. These responses are mediated by the autonomic nervous system. Sympathetic nerves stimulate the radial muscles to dilate the pupil, and parasympathetic nerves stimulate the circular muscles to constrict the pupil.

PROCEDURE

1. Stay with the subject in a darkened room for at least 1 minute, allowing his or her eyes to adjust to the dim light. This adjustment is known as *dark adaptation*.

2. Shine a narrow beam of light (from a pen flashlight or an ophthalmoscope, for example) from the right side into the subject's right eye. Observe the pupillary reflex in the right eye and also in the left eye the pupillary reflex in the other (left) eye is called the *consensual reaction*.

3. Repeat this procedure (first dark-adapting the eyes again) from the left side with the left eye. *constricts w light without light dilates.*

E. EXAMINATION OF THE EYE WITH AN OPHTHALMOSCOPE

An **ophthalmoscope** is a device used to observe the posterior inner part of the eye (the *fundus*). A mirror positioned at the top of the instrument deflects light at a right angle into the eye, enabling an observer to see the interior of the eye through a small slit in the mirror. Different depths of focus are attained by changing the lenses that are positioned in the slit. The lenses are carried on a wheel in regular order according to their focal lengths. The strength of each lens is given in diopters preceded by a plus (+) for a convex lens or a minus (−) for a concave lens, with 0 indicating no lens.

Fovea centralis

Macula lutea

Optic disc
(blind spot)

Arteriole

Veins

Figure 3.19 **A view of the interior of the eye using an ophthalmoscope.**

In this exercise, an opthalmoscope will be used to observe the arteries and veins of the fundus and two regions of the retina: the **optic disc** (the region where the optic nerve exits the eye, otherwise known as the *blind spot*) and the **macula lutea** (fig. 3.19). The macula lutea is a yellowish region containing a central pit, the **fovea centralis,** where is found the highest concentration of the photoreceptors responsible for visual acuity (the cones). When the eyes are looking directly at an object, the image is focused on the fovea.

 Clinical examination of the fundus (**ophthalmoscopy**) can aid the diagnosis of a number of ocular and systemic (body) diseases. The features noted in these examinations include the condition of the blood vessels; the color and shape of the disc; the presence of particles, exudates, or hemorrhage; the presence of edema and inflammation of the optic nerve (*papilledema*); and myopia and hyperopia.

PROCEDURE

1. Have the subject sit in a darkened room and look at a distant object (blinking is normal).
2. Position your chair so that you are close to and facing the subject. Hold the ophthalmoscope with your right hand and use your right eye when observing the subject's right eye. (The situation is reversed when viewing the left eye.)
3. With your forefinger on the lens adjustment wheel and your eye as close as possible to the small hole in the ophthalmoscope (glasses off if applicable), bring the instrument as close as possible to the subject's eye. (Steady your hand by resting it on the subject's cheek.)
4. Examine the subject's eye from the front to the back. Looking slightly from the side of the eye (not directly in front), examine the iris and lens using a +20 to +15 lens. The lens selected will vary among examiners who normally wear glasses. If both your eyes and the subject's eyes are normal, you will be able to see clearly without a lens. (On the 0 setting, the refractive strength of the subject's eye will be sufficient to focus the light on the eye.)
5. Rotate the wheel counterclockwise to examine the fundus.
   (a) If a positive (convex) lens is necessary to focus on the fundus, and your eyes are normal, the subject has hyperopia (hypermetropia).
   (b) If a negative (concave) lens is necessary to focus on the fundus, and your eyes are normal, the subject has myopia.
6. Observe the arteries and veins of the fundus and follow them to their point of convergence. This will enable you to see the optic disc (fig. 3.19).
7. Finally, at the end of the examination, observe the macula lutea by asking the subject to look directly into the light of the ophthalmoscope.

F. THE BLIND SPOT

The retina contains two types of *photoreceptors*, **rods** and **cones.** Rods and cones synapse with other cells (*bipolar neurons*), which in turn synapse with *ganglion cells* whose axons form the optic nerve transmitting sensory information out of the eye to the brain. In the fovea, only one cone will synapse with one bipolar cell, whereas several rods may converge on a given bipolar cell (fig. 3.20). Thus, the rods are more sensitive to low levels of illumination, whereas the cones require more light but provide greater visual acuity. The rods, therefore, are responsible

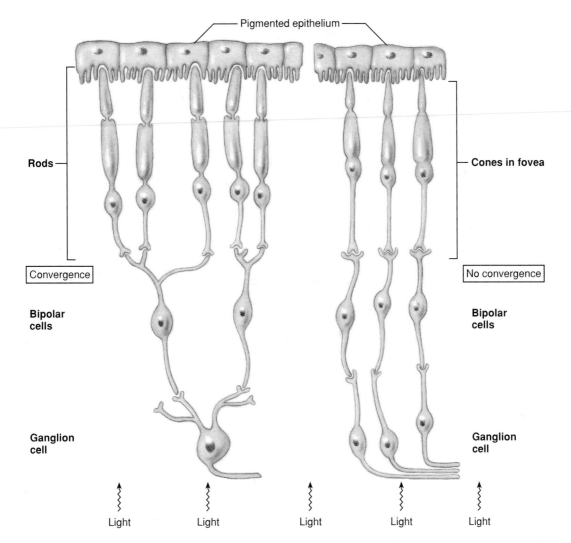

Pigmented epithelium

Rods

Cones in fovea

Convergence

No convergence

Bipolar cells

Bipolar cells

Ganglion cell

Ganglion cell

Light   Light   Light   Light   Light

Figure 3.20   The organization of cells in the retina. Note that the light must pass through layers of nerve fibers, ganglion cells, and bipolar cells before reaching the photoreceptors (rods and cones). The cones synapse in a 1:1 ratio with bipolar cells, allowing greater visual acuity, whereas a number of rods can synapse with a single bipolar cell, allowing greater visual sensitivity.

for night (*scotopic*) vision, when sensitivity is most important. The cones are responsible for day (*photopic*) vision, when visual acuity is most important. The cones also provide color vision—colors are seen during the day, whereas night vision is in black and white.

The axons of all ganglion cells in the retina gather together to become the optic nerve that exits the eye at the optic disc. This is also called the **blind spot** because there are no rods or cones in the optic disc, so an object whose image is focused here will not be seen.

PROCEDURE

1. Hold the drawing of the circle and the cross (fig. 3.21) about 20 inches from your face with the left eye covered or closed. Focus on the circle; this is most easily done if the circle is positioned in line with the right eye.

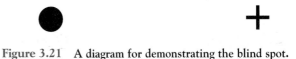

Figure 3.21   A diagram for demonstrating the blind spot.

2. Keeping the right eye focused on the circle, slowly bring the drawing closer to your face until the cross disappears. Continue to move the drawing slowly toward your face until the cross reappears.
3. Repeat this procedure with the right eye closed or covered and the left eye focused on the cross. Observe the disappearance of the circle as the drawing is brought closer to your face.

Right Eye   33 cm
Left Eye   33 cm

H. Dominant Eye
Right side dom
could not see   after covering Right eye

123

G. The Afterimage

The light that strikes the receptors of the eye stimulates a photochemical reaction in which the pigment **rhodopsin** (within the rods) dissociates (breaks apart) to form the pigment *retinene* and the protein *opsin*. This chemical dissociation produces electrical changes in the photoreceptors, which trigger a train of action potentials in the axons of the optic nerve. These events cannot be repeated in a given rod receptor until the rhodopsin is regenerated. This requires a series of chemical reactions in which one isomer of retinene is converted to another through the intermediate compound *vitamin A_1*. In other words, after the rhodopsin visual pigment in the rod has been "bleached" or dissociated by light from an object, a certain period of time is required before that receptor can again be stimulated.

When an eye that has adapted to a bright light, such as a light bulb, is closed or quickly turned towards a wall, the bright image of the light bulb will still be seen. This is called a **positive afterimage** and is caused by the continued "firing" of the photoreceptors. After a short period, the dark image of the light bulb, called the **negative afterimage,** will appear against a lighter background due to the "bleaching" of the visual pigment of the affected receptors.

According to the **Young-Helmholtz** theory of color vision, there are three systems of cones that respond respectively to *red, green,* and *blue* (or violet) light, and all other colors are seen by the brain's interpretation of mixtures of impulses from these three systems. Color discrimination will of course be impaired if one system of cones is defective (color blindness), or if one system of cones has been "bleached" by the continued viewing of an object. In the latter case, the positive afterimage of the object will appear in the complementary color.

 It is important to remember that the eye is a receptor, transducing light into electrical nerve impulses. We actually see with our brain. Impulses from the retina pass, via the *lateral geniculate bodies,* to the *visual cortex* of the occipital lobe, where the patterns of impulses are integrated to produce an image. The importance of the visual cortex in vision is illustrated by **strabismus,** a condition in which weak extrinsic eye muscles prevent the two eyes from converging on an object and fusing the images. To avoid confusion, the cortical cells eventually stop responding to information from one eye, making that eye functionally blind. Visual information is integrated with input from the other senses in the cortex of the *inferior temporal lobe.* If this area is damaged (the **Klüver-Bucy syndrome**), visual recognition is impaired. Although the image is seen, it lacks meaning and emotional content.

PROCEDURE

1. After staring at a light bulb, suddenly shift your gaze to a blank wall. Observe the appearance of the *negative afterimage.*
2. For 1 minute, stare at a dot made in the center of a small red square that has been pasted on a larger sheet of black paper.
3. Suddenly shift your gaze to a sheet of white paper and note the color of the *positive afterimage.*
    Positive afterimage (red) is _____ .
4. Repeat this procedure using blue squares and yellow squares.
    Positive afterimage (blue) is _____ .
    Positive afterimage (yellow) is _____ .

REVIEW ACTIVITIES FOR EXERCISE 3.5

Test Your Knowledge of Terms and Facts

1. The photoreceptors responsible for color vision are the _____; of these, there are _____ different kinds.
2. The region of the retina in which there are no photoreceptors is called the _____; this is also known as the _____.
3. Retinene (retinaldehyde) is derived from which vitamin? _____
4. When light enters the retina, it first passes through the _____ cell layer, then the _____ cell layer, before reaching the photoreceptors.
5. The axons of _____ cells gather together to produce the optic nerve.
6. Define the following terms:
   (a) *visual acuity* _____
   (b) *accommodation* _____

Match the following terms with the appropriate description:
_____ 7. myopia   (a) abnormal curvature of the cornea or lens
_____ 8. hyperopia   (b) eye too long
_____ 9. presbyopia   (c) abnormally high intraocular pressure
_____10. astigmatism   (d) eye too short
_____11. glaucoma   (e) loss of lens elasticity

Test Your Understanding of Concepts

12. Describe the muscle layers of the iris and the innervation to these muscles. Use this information to explain how pupils constrict in bright light and dilate in dim light.

13. Describe how the curvature of the lens is regulated, and use this information to explain how images are kept in focus on the retina as a distant object is brought closer to the eyes.

14. Explain the reason for blurred vision in a person with myopia, and describe how this person's vision is improved by the lenses in a pair of glasses.

Test Your Ability to Analyze and Apply Your Knowledge

15. Carrots contain large amounts of *carotene*, a precursor of vitamin A. Can eating carrots help improve blurred vision from myopia or hyperopia? Explain. Also, explain how eating carrots may influence scotopic (night) vision.

16. "You see with your brain, not with your eyes." Defend this statement with reference to the Young-Helmholtz theory of color vision and to the optic disc.

Ears: Cochlea and Hearing

⚗ MATERIALS

1. Tuning forks
2. Rubber mallets

Sound is conducted by the middle ear to the inner ear, where events within the cochlea result in the production of nerve impulses. Clinical tests of middle ear (conductive) and inner ear (sensory) function aid in the diagnosis of hearing disorders.

OBJECTIVES

1. Describe the structure of the middle ear and explain how the ossicles function.
2. Describe the structure of the inner ear and explain how the cochlea functions.
3. Demonstrate Rinne's test and Weber's test and explain the significance of each.
4. Explain how the source of a sound is localized.

📖 Textbook Correlations*

Before performing this exercise, you may want to consult the following references in *Human Physiology*, seventh edition, by Stuart I. Fox:

- *The Ears and Hearing.* Chapter 10, pp. 253–259.

Those using different physiology textbooks may want to consult the corresponding information in those books.

*See Appendix 3 for correlations to the *A.D.A.M. InterActive PHYSIOLOGY Modules*.

You should be familiar with the gross structure of the ear (fig. 3.22). Sound waves are conducted through the **outer ear** (the pinna and the external auditory meatus) to the tympanic membrane (eardrum), causing it to vibrate. The vibration of the tympanic membrane causes the three ossicles of the **middle ear**—the *malleus* (hammer), *incus* (anvil), and *stapes* (stirrup)—to vibrate, thus pushing the footplate of the stapes against a flexible membrane, the *oval window*. Vibration of the oval window produces compression waves in the fluid-filled cochlea of the **inner ear.**

The compression waves of cochlear fluid flow over a thin, flexible membrane within the cochlea called the *basilar membrane*, causing it to vibrate. Within the **organ of Corti,** the basilar membrane is coated with sensory *hair cells*, which are displaced upward by this vibration into a stiff overhanging structure called the *tectorial membrane* (fig. 3.23). The distortion of the hair cells produced by this action stimulates a train of action potentials that travels along the cochlear branch of the vestibulocochlear (eighth cranial) nerve to the brain. Here, the action potentials are interpreted as the sound of a specific **pitch,** which is determined by the location of the stimulated hair cells on the basilar membrane, and of a specific **loudness,** which is coded by the frequency of action potentials.

A. CONDUCTION OF SOUND WAVES THROUGH BONE: RINNE'S AND WEBER'S TESTS

Although hearing is normally produced by the vibration of the oval window in response to sound waves conducted through the movements of the middle-ear ossicles, the *endolymph* fluid of the cochlea can also be made to vibrate in response to sound waves conducted through the skull bones directly, thereby bypassing the middle ear. This makes it possible to differentiate between deafness resulting from middle-ear damage (**conduction deafness,** such as from damage to the ossicles in *otitis media* or immobilization of the stapes in *otosclerosis*) and deafness resulting from damage to the cochlea or vestibulocochlear nerve (**sensory deafness,** such as from infections, streptomycin toxicity, or prolonged exposure to loud sounds).

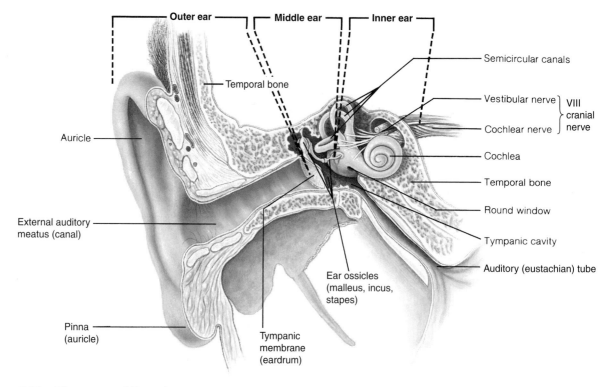

Figure 3.22 The outer, middle, and inner ear.

 Conduction deafness may be caused by infections of the middle ear **(otitis media)**, infections of the tympanic membrane **(tympanitis)**, or an excessive accumulation of ear wax (*cerumen*). People with conduction deafness often wear hearing aids over the mastoid process of the temporal bone. These devices amplify sounds and transmit them by bone conduction to the cochlea. Hearing aids do not help in cases of complete sensory (nerve) deafness. The first cochlear implant device received FDA approval in 1984. Designed to bypass damaged or destroyed hair cells, cochlear implants stimulate the auditory nerve directly. Although this procedure has met with limited success, recent technological breakthroughs may offer future remedy for deafness.

PROCEDURE

Rinne's Test

1. Strike a tuning fork with a rubber mallet to produce vibrations.
2. Perform Rinne's test by placing the *handle* of the vibrating tuning fork against the mastoid process of the temporal bone (the bony prominence behind the ear), with the tuning fork pointed down and behind the ear. When the sound has almost died away, move the tuning fork (by the handle) near the external auditory meatus. If there is no damage to the middle ear, the sound will reappear.
3. Simulate conduction deafness by repeating Rinne's test with a plug of cotton in the ear. Notice that in conductive deafness, conduction by bone (via the mastoid process) is more effective than conduction by air.

Weber's Test

1. Perform Weber's test by placing the handle of the vibrating tuning fork on the midsagittal line of the head, and listen. In conduction deafness, the sound will seem louder in the affected ear (room noise is excluded but bone conduction continues), whereas in sensory deafness, the cochlea is defective and the sound will be louder in the normal ear.
2. Repeat Weber's test with one ear plugged with your finger. The sound will appear louder in the plugged ear because external room noise is excluded.

B. BINAURAL LOCALIZATION OF SOUND

Just as binocular vision provides valuable clues for viewing scenes in three dimensions, binaural hearing helps to localize sounds (people have stereoscopic vision and stereophonic hearing). The ability to localize the source of a sound depends partly on the difference

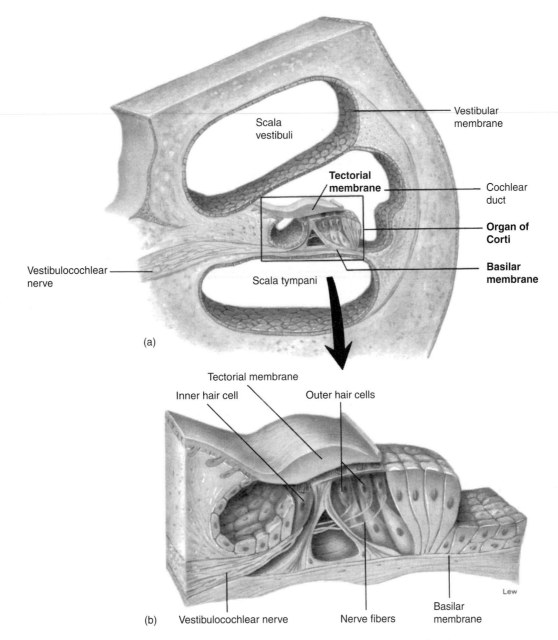

Scala vestibuli

Vestibular membrane

Tectorial membrane

Cochlear duct

Organ of Corti

Vestibulocochlear nerve

Scala tympani

Basilar membrane

(a)

Tectorial membrane

Inner hair cell

Outer hair cells

(b)   Vestibulocochlear nerve   Nerve fibers   Basilar membrane

Lew

Figure 3.23   The organ of Corti. (a) The organ of Corti within the cochlea and (b) in greater detail.

in loudness of the sound that reaches the two ears and partly on the difference in the time of arrival of the sound at the two ears. The difference in loudness is more important for high-pitched sounds, where the sound waves are blocked by the head. The difference in the time of arrival is more important for low-pitched sounds, whose wavelengths are large enough to bend around the head.

PROCEDURE

1. With both eyes closed, the subject is asked to locate the source of a sound (e.g., a vibrating tuning fork).
2. The vibrating tuning fork is placed at various positions (front, back, and sides about a foot from the subject's head), and the subject is asked to describe the location of the tuning fork.
3. Repeat the above procedures with one of the subject's ears plugged.

Laboratory Report 3.6

REVIEW ACTIVITIES FOR EXERCISE 3.6

Test Your Knowledge of Terms and Facts

1. List the three ossicles of the middle ear in sequence, from the outermost to the innermost.
   Outermost _____
   Middle _____
   Innermost _____
2. The scientific name for the eardrum is the _____.
3. The middle chamber of the cochlea is the _____.
4. Sensory hair cells in the cochlea are located on a membrane called the _____.
5. The innermost middle ear ossicle presses against a flexible membrane called the _____.
6. The sensory structure of the inner ear responsible for transducing vibrations into nerve impulses is known as the
   _____.

Test Your Understanding of Concepts

7. In sequence, describe the events that occur between the arrival of sound waves at the tympanic membrane and the production of nerve impulses.

8. Explain how different pitches of sound affect the basilar membrane, and how this relates to the information sent to the brain.

9. Explain the result that might be obtained by performing the Rinne's and Weber's test on a patient with otosclerosis. How might these results compare with those obtained from a patient with sensory deafness? Explain.

10. Hair cells are mechanoreceptors that are employed in the inner ear of mammals and also in the lateral line organs on the body of fish. How might the bending of hair cell processes lead to depolarization? Propose a general description of the function of hair cells in the animal kingdom.

Ears: Vestibular Apparatus— Balance and Equilibrium

MATERIALS

1. Swivel chair

The vestibular apparatus provides a sense of balance and equilibrium. As a result of inertia acting on the structures within the vestibular apparatus, changes in the position of the head result in the production of afferent nerve impulses that are conducted to the brain along the eighth cranial nerve. This information results in eye movements and other motor activities that help to orient the body in space.

OBJECTIVES

1. Describe the structure of the semicircular canals and explain how movements of the head result in the production of nerve impulses.
2. Describe vestibular nystagmus and explain how it is produced.

Textbook Correlations*

Before performing this exercise, you may want to consult the following references in *Human Physiology,* seventh edition, by Stuart I. Fox:

- *Vestibular Apparatus and Equilibrium.* Chapter 10, pp. 249–253.

Those using different physiology textbooks may want to consult the corresponding information in those books.

---

*See Appendix 3 for correlations to the A.D.A.M. InterActive PHYSIOLOGY Modules.

The **vestibular apparatus,** located in the inner ear above the cochlea, consists of three *semicircular canals* (oriented in three planes), the *utricle,* and the *saccule* (fig. 3.24). The utricle and saccule are together called *otolith organs,* and provide a sense of linear acceleration. These structures, like the cochlea, are filled with *endolymph* and contain sensory cells activated by bending. These sensory hair cells of the semicircular canals support numerous hairlike extensions, which are embedded in a gelatinous "sail" (the *cupula*) that projects into the endolymph (fig. 3.25). Movement of the endolymph fluid, induced by acceleration or deceleration, bends the extensions of the hair cells. This sends a train of impulses to the brain along the *vestibulocochlear* (*eighth* cranial) *nerve.* The sensory hair cells of the utricle and the saccule serve to orient the head with respect to the gravitational pull of the earth.

Afferent impulses from the vestibular apparatus help make us aware of our position in space and affect a variety of efferent somatic motor nerves (such as those regulating the voluntary extrinsic eye muscles). Under intense vestibular activity, autonomic motor nerves may also become stimulated, producing involuntary responses that can include vomiting, perspiration, and hypotension. This exercise will test the effect of vestibular activity on the extrinsic muscles of the eye by producing involuntary eye oscillations known as **vestibular nystagmus.**

The semicircular canals can be stimulated by rotating a subject in a chair. Stimulation of the lateral canal can occur when the head is flexed 30° forward, almost touching the chest. When the subject is first rotated to the right, the cupula will be bent to the left because of the inertial lag of the endolymph. This will cause nystagmus in which the eyes drift slowly to the left followed by a quick movement to the right (midline position). Nystagmus will continue until the inertia of the endolymph has been overcome and the cupula returns to its initial unbent position. At this time, the endolymph and the cupula are moving in the same direction and at the same speed. When the rotation of the subject is abruptly stopped, the greater inertia of the endolymph will cause bending of the cupula in the previous direction of spin (to the right), producing nystagmus with a slow drift phase to the right and a rapid phase to the left. This activation of

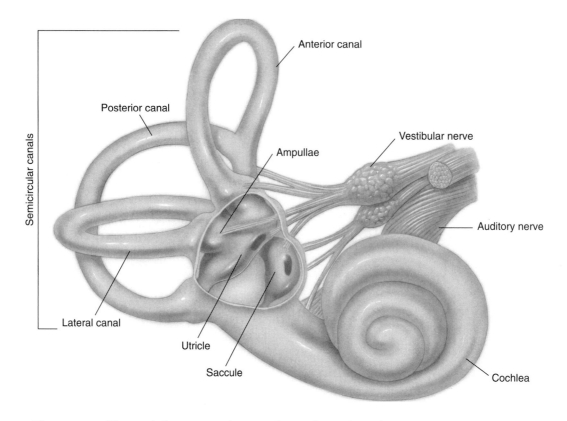

Figure 3.24   The inner ear. The vestibular apparatus (semicircular canals, utricle, and saccule), required for a sense of equilibrium, and the cochlea, required for hearing, make up the inner ear.

the vestibular apparatus is often accompanied by **vertigo** (an illusion of movement, or spinning) and a tendency to fall to the right.

Vertigo may be accompanied by dizziness, but the two sensations are not the same. A person may be dizzy without experiencing the more severe effects of vertigo. In the procedure described below, activation of the vestibular apparatus by rotation of the subject produces oscillatory eye movements (nystagmus). The eyes in turn can activate the vestibular apparatus and produce vertigo, as occurs in motion sickness (seasickness or car sickness, for example). Vertigo can also accompany diseases unrelated to the special senses, such as cardiovascular disease. Many of the unpleasant symptoms associated with vertigo, such as nausea and vomiting, are the result of activation of the autonomic motor system. Drugs taken for motion sickness (e.g., Dramamine or scopolamine) act by suppressing these autonomic responses.

PROCEDURE

1. Have the subject sit in a swivel chair with the eyes open and the head flexed 30° forward (chin almost touching chest). Rotate the chair quickly to the right for 20 seconds (about 10 revolutions). After noting the initial nystagmus, have the subject close his or her eyes.

    Note: *Only subjects who are not subject to motion sickness should be used. The exercise should be stopped immediately if the subject feels sick.*

2. Abruptly stop the chair and have the subject open his or her eyes as wide as possible. Note the direction of nystagmus (left to right or right to left, circle one).

3. Repeat this procedure (using different subjects), alternating with the head resting on the right shoulder and the left shoulder (this stimulates the vertical canals), and note the direction of the postrotational nystagmus.

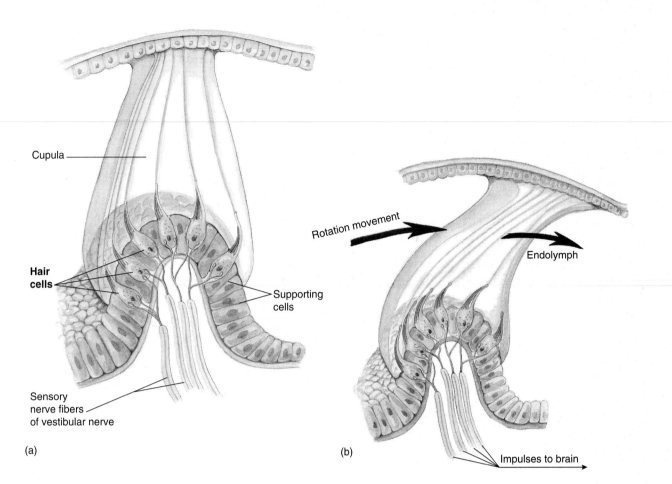

Cupula

Hair cells

Supporting cells

Sensory nerve fibers of vestibular nerve

(a)

Rotation movement

Endolymph

Impulses to brain

(b)

Figure 3.25   Cupula and hair cells in the ampulla of each semicircular canal. (a) The sensory structures in an unactivated state, when the body is not accelerating. (b) Rotation causes the cupula to bend, stimulating the hair cells and sensory neurons.

REVIEW ACTIVITIES FOR EXERCISE 3.7

Test Your Knowledge of Terms and Facts

1. The organ of equilibrium is called the _____.
2. The structures sensitive to angular acceleration in three planes are the _____.
3. The structures sensitive to linear acceleration are the _____ and
   _____. Together, they are called the _____ organs.
4. The fluid within the organs of equilibrium is known as _____.
5. The sense of equilibrium is transmitted by the _____ (its number) cranial nerve,
   also known as the _____ (its name) nerve.
6. An illusion of movement or spinning is called _____.

Test Your Understanding of Concepts

7. Describe the meaning of the term *inertia* and explain how inertia affects the hair cell processes of the inner ear during acceleration and deceleration.

8. Including all of the relevant anatomical structures (and ending with the activation of the cranial nerve), explain how we sense spinning in a chair. What happens when the chair is suddenly stopped?

9. Which structures of the inner ear would be activated by a somersault? By a cartwheel? By fast acceleration in a race car?

Test Your Ability to Analyze and Apply Your Knowledge

10. Does nystagmus occur after a person in a rotating chair has achieved constant velocity? Explain?

11. Propose an explanation of the causes of vertigo and nausea in a seasick person.

Taste Perception

MATERIALS

1. Cotton-tipped applicator sticks
2. Solutions of 5% sucrose, 1% acetic acid, 5% NaCl, and 0.5% quinine sulfate

There are four modalities of taste perception: sweet, sour, bitter, and salty. Taste buds sensitive to each of these modalities have a characteristic distribution on the tongue. Therefore, each of these taste modalities is perceived most acutely in a particular tongue region.

OBJECTIVES

1. Describe the structure of a taste bud and the location of taste buds on the tongue.
2. List the *four* primary taste modalities, and describe their distribution on the tongue.

Textbook Correlations*

Before performing this exercise, you may want to consult the following references in *Human Physiology*, seventh edition, by Stuart I. Fox:

- *Taste and Smell.* Chapter 10, pp. 246–249.

Those using different physiology textbooks may want to consult the corresponding information in those books.

*See Appendix 3 for correlations to the *A.D.A.M. InterActive PHYSIOLOGY Modules*.

The **taste buds** consist of specialized epithelial cells arranged in the form of barrel-shaped receptors (fig. 3.26c), associated with sensory (afferent) nerves. Long microvilli extend through a pore at the external surface of the taste bud and are bathed in saliva. Although not considered neurons, taste cells can be depolarized to release chemical neurotransmitters that, in turn, stimulate associated sensory neurons leading to the brain. In adults, these receptors are located primarily on the surface of the tongue, with a lesser number on the soft palate and epiglottis. In children the sense of taste is more diffuse, with additional receptors located on the inside of the cheeks.

Taste perception is due to different combinations of only four basic taste stimuli—*sweet, sour, bitter,* and *salty*—together with nuances provided by the sense of smell (olfaction). Although a given taste bud can be stimulated by all four modalities, different taste buds will respond more readily (i.e., have a lower threshold) to one or two stimuli than to the others. A sweet stimulus, therefore, may be most easily perceived at the tip of the tongue, bitter at the back, sour at the sides, and salty over most of the tongue. This mapping of taste receptors, however, has been questioned by more recent research.

The *sour* taste of solutions is due to the presence of acid or hydrogen ions (H^+), and the *salty* taste is due to the presence of Na^+ ions (but modified by the anion—sodium chloride tastes saltier than other sodium salts, such as sodium acetate). The chemical basis for *bitter* and *sweet* taste is largely unknown, since these can be produced by a variety of seemingly unrelated compounds. (Fructose tastes the sweetest, followed by sucrose, and then by glucose. The artificial, nonsugar sweeteners, however, taste sweeter than any of these.)

The afferent pathway from the taste buds to the brain involves primarily two cranial nerves (fig. 3.27). Taste buds on the anterior two-thirds of the tongue are served by the *facial* (seventh cranial) *nerve;* whereas those on the posterior third of the tongue have a sensory pathway along the *glossopharyngeal* (ninth cranial) *nerve.* The vagus (tenth cranial) nerve also has limited sensory innervation in the epiglottis area. Originating from stimulated taste buds, impulses in these nerves are conducted through the thalamus to the postcentral gyrus of the cerebral cortex (ending in the region represented by the

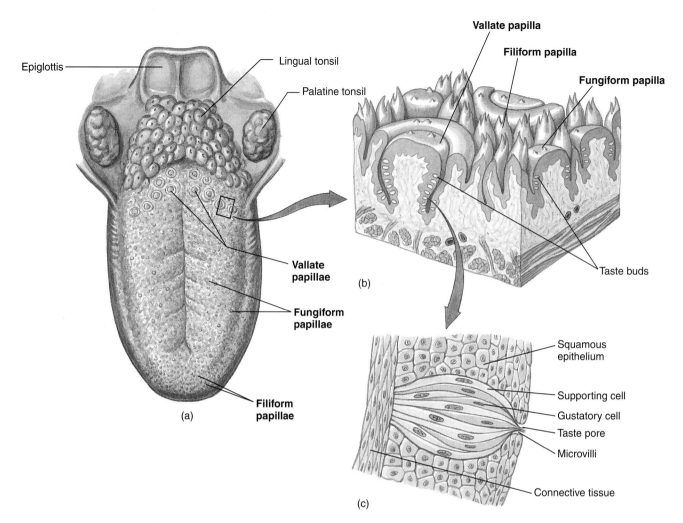

Epiglottis

Lingual tonsil

Palatine tonsil

Vallate papillae

Fungiform papillae

Filiform papillae

(a)

Vallate papilla

Filiform papilla

Fungiform papilla

Taste buds

(b)

Squamous epithelium

Supporting cell

Gustatory cell

Taste pore

Microvilli

Connective tissue

(c)

Figure 3.26   Papillae of the tongue and taste buds. (a,b) The structure of the tongue, with its papillae. (c) The structure of a taste bud. Taste buds are found throughout the tongue within the vallate and fungiform papillae.

tongue), where they are interpreted. The afferent pathway from taste buds to brain is ipsilateral (traveling along the same side of the head), as opposed to most sensory tracts, which cross over.

PROCEDURE

1.   Dry the tongue with a paper towel and, using an applicator stick, apply a dab of 5% sucrose solution to the tip, sides, and back of the tongue.
2.   Repeat this procedure using 1% acetic acid, 5% NaCl, and 0.5% quinine sulfate, being sure to rinse the mouth and dry the tongue between applications.

Note: *Apply quinine sulfate last; the effect is dominant and often lingering.*

3.   Using the sketch provided in the laboratory report, record the location where you tasted each solution. Use the symbols *sw* for sweet, *sl* for salty, *sr* for sour, and *b* for bitter.

The sensations of *gustation* (taste) and *olfaction* (smell) are often grouped together in a single category—the *chemical senses*. The molecular basis of taste and smell is complex. Apparently, both sweet and bitter tastes (as well as perception of particular odorant molecules) are mediated by receptors that are coupled to membrane G-proteins. Dissociation of the G-protein subunit activates second-messenger systems that lead to depolarization of the receptor cell. Together, these chemoreceptors function to provide the proper nuances of "taste," which are extremely well developed in some people, such as wine tasters. Humans can distinguish up to 10,000 different odors, with a sensitivity capable of detecting a billionth of an ounce of perfume in the air. Since the sense of smell is so important in tasting, a stuffy nose from a cold or allergy can greatly affect the taste of foods.

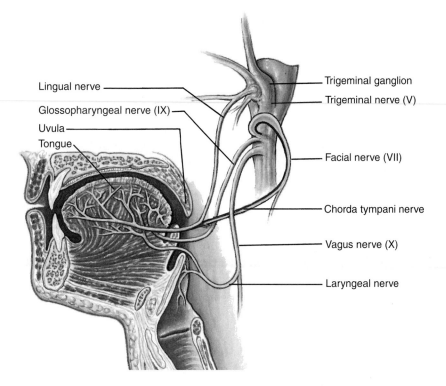

Figure 3.27   The innervation of the tongue.

DATA FROM EXERCISE 3.8

Map the areas of the tongue that seem to be most sensitive to sweet, salty, bitter, and sour.

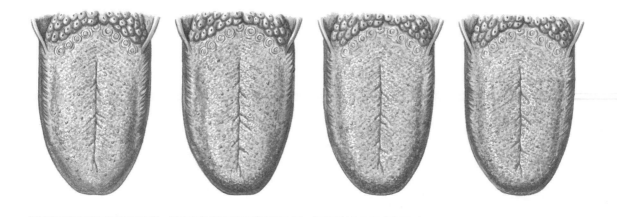

_____ _____ _____ _____

REVIEW ACTIVITIES FOR EXERCISE 3.8

Test Your Knowledge of Terms and Facts

1. What aspect of a solution causes it to taste sour? _____
2. What aspect of a solution causes it to taste salty? _____
3. Describe the areas of the tongue where the following tastes are best perceived:
   (a) sweet _____
   (b) salty _____
   (c) sour _____
   (d) bitter _____

Test Your Understanding of Concepts

4. Describe the common mechanism of action of molecules that taste sweet or bitter.

5. What does the term *chemical senses* mean? Explain how the sense of olfaction and gustation interact.

6. Using the traditional taste map or the map you obtained in this exercise, identify the taste modalities that would be most affected by the destruction of the glossopharyngeal nerve. Which would be most affected by the destruction of the facial nerve? Explain.

7. Did your map of taste perception support the traditional mapping of taste on the tongue? Propose scientific tests to resolve this issue.

8. Humans are said to be able to distinguish 10,000 different odors. Therefore, do you think we have 10,000 different receptors, each specialized for a different odor? Why or why not? Propose another mechanism that might account for the ability to distinguish so many different odors.

The Endocrine System

Glands are clusters of secretory cells derived from glandular epithelial membranes that invaginate into the underlying connective tissues (fig. 4.1). When the invagination persists, a duct is formed leading from the secretory cells to the epithelial membrane surface and to the outside of the body. Ducts may thus lead directly to an external body surface or indirectly to the luminal lining of tubes within the digestive, respiratory, urinary, or reproductive tract, which ultimately lead to the outside. These glands are called **exocrine glands,** and include the sebaceous (oil) glands, sweat glands, mammary glands, salivary glands, and the pancreas. In contrast, if the invagination disappears, the chemical product of the gland, a **hormone,** is secreted internally into the blood capillaries. Ductless glands producing hormones are called **endocrine glands.**

Hormones secreted by endocrine glands regulate the activities of other organs (table 4.1). This regulation complements that of the nervous system and serves to direct the metabolism of the hormone's target organs and cells along paths that benefit the body as a whole. By regulating the activity of enzymes within their target cells, hormones primarily regulate total body metabolism and the function of the reproductive system.

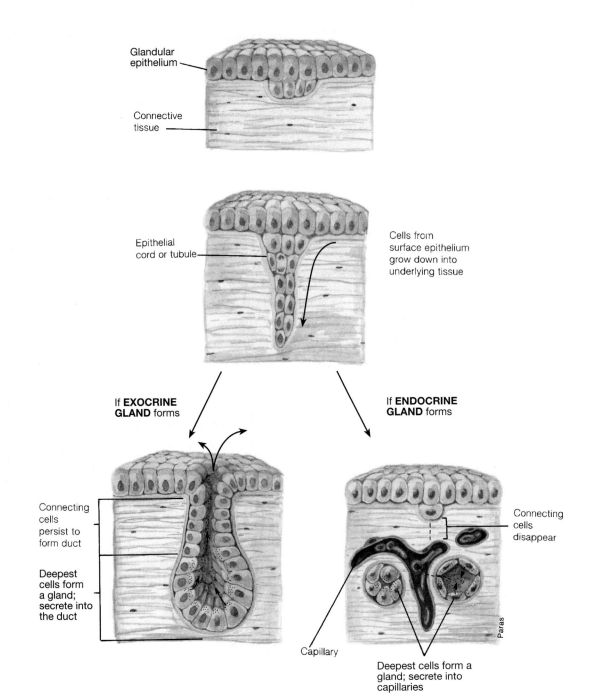

Glandular epithelium

Connective tissue

Epithelial cord or tubule

Cells from surface epithelium grow down into underlying tissue

If **EXOCRINE GLAND** forms

If **ENDOCRINE GLAND** forms

Connecting cells persist to form duct

Deepest cells form a gland; secrete into the duct

Connecting cells disappear

Capillary

Deepest cells form a gland; secrete into capillaries

Figure 4.1 The formation of exocrine and endocrine glands.

Table 4.1 A Partial Listing of the Endocrine Glands

Endocrine Gland	Major Hormones	Primary Target Organs	Primary Effects
Adipose tissue	Leptin	Hypothalamus	Suppresses appetite
Adrenal cortex	Glucocorticoids Aldosterone	Liver and muscles Kidneys	Glucocorticoids influence glucose metabolism; aldosterone promotes Na^+ retention, K^+ excretion
Adrenal medulla	Epinephrine	Heart bronchioles and blood vessels	Causes adrenergic stimulation
Heart	Atrial natriuretic hormone	Kidneys	Promotes excretion of Na^+ in the urine
Hypothalamus	Releasing and inhibiting hormones	Anterior pituitary	Regulates secretion of anterior pituitary hormones
Small intestine	Secretin and cholecystokinin	Stomach, liver, and pancreas	Inhibits gastric motility and stimulates bile and pancreatic juice secretion
Islets of Langerhans (pancreas)	Insulin Glucagon	Many organs Liver and adipose tissue	Insulin promotes cellular uptake of glucose and formation of glycogen and fat; glucagon stimulates hydrolysis of glycogen and fat
Kidneys	Erythropoietin	Bone marrow	Stimulates red blood cell production
Liver	Somatomedins	Cartilage	Stimulates cell division and growth
Ovaries	Estradiol 1.7 β and progesterone	Female reproductive tract and mammary glands	Maintains structure of reproductive tract and promotes secondary sex characteristics
Parathyroid glands	Parathyroid hormone	Bone, small intestine, and kidneys	Increases Ca^{2+} concentration in blood
Pineal gland	Melatonin	Hypothalamus and anterior pituitary	Affects secretion of gonadotrophic hormones
Pituitary anterior	Trophic hormones	Endocrine glands and other organs	Stimulates growth and development of target organs; stimulates secretion of other hormones
Pituitary posterior	Antidiuretic hormone Oxytocin	Kidneys and blood vessels Uterus and mammary glands	Antidiuretic hormone promotes water retention and vasoconstriction; oxytocin stimulates contraction of uterus and mammary secretory units
Skin	1.25-Dihydroxy vitamin D_1	Small intestine	Stimulates absorption of Ca^{2+}
Stomach	Gastrin	Stomach	Stimulates acid secretion
Testes	Testosterone	Prostate seminal vesicles, and other organs	Stimulates secondary sexual development
Thymus	Thymopoietin	Lymph nodes	Stimulates white blood cell production
Thyroid gland	Thyroxine (T_4) and (T_3) calcitonin	Most organs	Thyroxine and triodothyronine promote growth and development and stimulate basal rate of cell respiration (basal metabolic rate or BMB); calcitonin may participate in the regulation of blood Ca^{2+} levels

Histology of the Endocrine Glands

MATERIALS

1. Microscopes
2. Prepared slides

Endocrine glands vary greatly in structure but have some features in common because they all secrete hormones into the blood. The histological structure of the endocrine glands shows how these glands function and how they are related to surrounding tissues.

OBJECTIVES

1. Describe the structure of the ovaries and testes and the functions performed by their component parts.
2. Describe the histological structure of the pancreas and identify both its endocrine and exocrine structures.
3. Describe the histological structure of the adrenal and thyroid glands and the functions of their component parts.
4. Describe the embryological origin and structure of the anterior pituitary and posterior pituitary; and their relationships to the hypothalamus.
5. List the hormones secreted by the anterior pituitary and posterior pituitary and explain how the secretion of these hormones is regulated.

Textbook Correlations

Before performing this exercise, you may want to consult the following references in *Human Physiology*, seventh edition, by Stuart I. Fox:

- *Pituitary Gland,* Chapter 11, pp. 299–305.
- *Adrenal Glands,* Chapter 11, pp. 305–308.
- *Thyroid and Parathyroid Glands,* Chapter 11, pp. 308–312.
- *Pancreas and Other Endocrine Glands,* Chapter 11, pp. 313–316.

Those using different physiology textbooks may want to consult the corresponding information in those books.

Knowledge of the normal histology of the endocrine glands helps in understanding their normal physiology and in diagnosing various pathological states. Endocrine glands may *atrophy* (lose structure) or develop hormone-secreting tumors known as *adenomas*.

In the testes, various diseases, such as those resulting from *Klinefelter's syndrome* (XXY genotype) or from the *mumps,* are associated with atrophy of the seminiferous (sperm-producing) epithelium. In the ovaries, granulosa cell tumors may secrete excessive estrogen. In *diabetes mellitus,* the beta cells within the islets of Langerhans in the pancreas may decrease in number and have decreased numbers of insulin-containing granules per cell. *Pheochromocytoma,* a tumor of the adrenal medulla, secretes excess epinephrine. Tumors of the anterior pituitary may result in *gigantism* and *acromegaly* (from elevated levels of growth hormone), hyperpigmentation (from excessive adrenocorticotropic hormone), or persistent lactation (from elevated prolactin secretion). These examples are only a few of the disorders associated with an abnormal histology of the various endocrine glands.

Endocrine glands may be independent organs, or they may be part of an organ that also performs nonendocrine functions (table 4.1). Organs that perform both endocrine and nonendocrine functions include the adipose tissue, brain, stomach, pancreas, liver, and skin.

Hormones are carried by the blood to all organs of the body; only certain organs, however, can respond to a given hormone. These are called the **target organs** for the hormone. Hormones affect the metabolism of their target organs and in so doing help to regulate growth and development, total body metabolism, and reproduction.

(a)

Primary follicles

Secondary follicle

Vesicle

(b)

Granulosa cells

Antrum

Corona radiata

Secondary oocyte

Cumulus oophorus

Zona pellucida

Theca interna

Figure 4.2 **Ovarian follicles.** Photomicrographs of (a) a secondary follicle and (b) a graafian follicle within an ovary (450×).

A. OVARY

The ovary is an endocrine gland as well as the producer of female gametes (*ova*). The ovum (egg cell) can be thought of as an exocrine secretion because it enters a duct—the uterine tube—after leaving the ovary. The primary hormones of the ovary, *estrogen* and *progesterone,* are secreted directly into the blood of the circulatory system.

The **ovarian follicles** are brought to maturity under the influence of the *gonadotropic hormones* (FSH and LH, mentioned later) secreted by the anterior pituitary. In every cycle, one of the mature follicles, or **graafian follicle,** eventually ruptures through the surface of the ovary to release its ovum (a process called *ovulation*). The empty follicle is then converted into a new endocrine structure

called the **corpus luteum.** If fertilization does not occur, the corpus luteum regresses and the cycle is ready to begin again.

The microscopic appearance of the ovary is thus continuously changing as the cycle progresses. A single slide of the ovary will reveal many follicles at different stages of maturation, including primary, secondary, and graafian follicles (fig. 4.2).

PROCEDURE

Using the low-power objective, scan the slide of the ovary and try to locate a circular-to-elliptical structure, the follicle, that encloses a space filled with fluid and scattered cells. Identify the following parts of the follicle (fig. 4.2):

1. **Ovum.** The ovum (egg cell) is the largest cell in the follicle and, at this stage of development, is called a *secondary oocyte*.
2. **Granulosa cells.** Granulosa cells are the numerous small cells found within the follicle.
3. **Antrum.** The antrum is the central fluid-filled cavity of the follicle.
4. **Cumulus oophorus.** Cumulus oophorus means "egg-bearing hill." This is the mound of granulosa cells that supports the ovum.
5. **Corona radiata.** The corona radiata is the layer of granulosa cells that surrounds the ovum. The ovum continues to be surrounded by its corona radiata after ovulation, and this layer of cells presents the first barrier to sperm penetration during fertilization.
6. **Zona pellucida.** The zona pellucida is a clear region containing glycoproteins between the plasma membrane of the ovum and the corona radiata.

B. Testis

The testis produces both the male gametes (*sperm*) and the male sex hormone (*testosterone*). Sperm are produced within the **seminiferous tubules** and travel through these tubules to the **epididymis,** where the sperm are passed into a single tubule that becomes the **ductus (vas) defer-** ens (fig. 4.3). The ductus deferens picks up fluid from the *seminal vesicles* and the *prostate* and passes its contents, now called semen, to the *ejaculatory duct*.

The seminiferous tubules are highly convoluted and tightly packed within the testis. The small spaces, or interstices, between adjacent convolutions of the tubules are filled with connective tissue known as interstitial tissue. Within this connective tissue are the interstitial **cells of Leydig,** endocrine cells that produce the *androgens* (male sex steroid hormones). The major androgen secreted by the Leydig cells of sexually mature males is *testosterone*.

Because the seminiferous tubules are highly convoluted, the chances of seeing a longitudinal section of a tubule are remote. Most of the tubules will be cut more or less in cross section, giving a circular or oblong appearance (fig. 4.3*b*).

PROCEDURE

Using the *low-power* objective, scan the slide of the testis and locate one seminiferous tubule within a section, or lobule. Switch to a high-power objective and observe the following structures (fig. 4.3):

1. **Spermatogenic cells.** Spermatogenic cells form the *germinal epithelium* found along the outer wall of the tubules. These cells divide by **meiosis** to produce the sperm. The outermost cells are *diploid* (forty-six

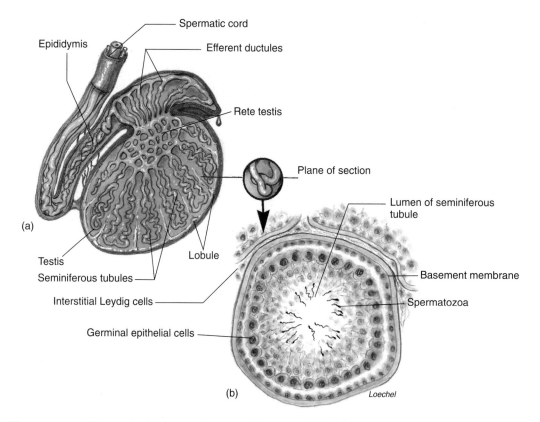

Figure 4.3 **The structure of the testis.** (a) A sagittal section of a testis and (b) a transverse section of a seminiferous tubule.

chromosomes), whereas the cells towards the lumen have completed meiotic division and are *haploid* (twenty-three chromosomes). Within the tubular epithelium, chromosomes at various stages of meiosis can be seen as darkened structures. Haploid **spermatozoa** can often be seen within the tubular lumen.

2. **Leydig cells.** These endocrine cells can be seen in the interstitial connective tissue between adjacent convolutions of the tubules.

C. PANCREATIC ISLETS (OF LANGERHANS)

The pancreas has both an exocrine and an endocrine function. The exocrine secretion (*pancreatic juice*) is produced by pancreatic cells called **acini** that are arranged in clusters around a central duct. Pancreatic juice, containing digestive enzymes and bicarbonate, drains into *interlobular ducts* located in bands of connective tissue. From here the secretion flows into the pancreatic duct and empties into the duodenum.

The endocrine secretions of the pancreas, **insulin** and **glucagon,** are produced by scattered groups of cells called the **islets of Langerhans.** These hormones do not enter the interlobular ducts, but rather leave the pancreas by way of the circulatory system.

PROCEDURE

Using the *low-power* objective, scan the slide of the pancreas and attempt to identify the following structures (fig. 4.4):

1. **Acini.** The pancreatic acini are dark-staining clusters of cells that form most of the body of the pancreas.
2. **Interlobular ducts.** The interlobular ducts may be mistaken for veins because of their large size, thin walls, and flattened, irregular shape. Unlike veins, however, their walls are composed of only a single layer of columnar epithelial cells, and no red blood cells will be seen in the lumina.
3. **Pancreatic islets (of Langerhans).** Under low power, the pancreatic islets will appear as light patches, circular in shape, against the dark background of the acini. Under high power, the *alpha cells* (which secrete glucagon) can easily be distinguished from the *beta cells* (which secrete insulin). The alpha cells are smaller and contain pink-staining granules, whereas the beta cells are larger and stain blue.

D. ADRENAL GLAND

The **adrenal gland** is actually two different glands located in the same organ. In lower organisms, these glands are separated; in higher organisms (including humans), they

Figure 4.4   The histology of the pancreas. The exocrine pancreatic acinar cells (that form secretory structures called acini) and endocrine pancreatic islet (of Langerhans) are seen.

are closely associated as the *adrenal cortex* (outer part) and the *adrenal medulla* (inner part) (fig. 4.5).

The **adrenal cortex** secretes **corticosteriod hormones.** These include hormones that regulate salt balance (*mineralocorticoids*) and hormones that regulate glucose homeostasis (*glucocorticoids*). The **adrenal medulla** secretes two hormones, **epinephrine** and **norepinephrine,** that act together with sympathetic nerve stimulation to enhance the response of the cardiovascular system to increased physical demand. The cells of the adrenal medulla are derived from the same embryonic tissue (neural crest ectoderm) as postganglionic sympathetic neurons, whereas the adrenal cortex is derived from a different embryonic tissue (mesoderm). Therefore, these two regions of the adrenal gland are different both physiologically and histologically.

PROCEDURE

Before observing the slide of the adrenal gland under the microscope, hold it up to the light and note the clear distinction between the adrenal cortex and adrenal medulla. Using the *low-power* objective, focus on the outer edge of the gland. Scan from this point inward and identify the following structures (fig. 4.5):

1. **Capsule.** The capsule is a thin, tough layer of connective tissue that surrounds the gland.
2. **Zona glomerulosa.** The zona glomerulosa is the outer layer of the adrenal cortex; its cells are tightly packed in an irregular arrangement. The z. glomerulosa secretes the mineralocorticoids (mainly *aldosterone* and *deoxycorticosterone*). The secretion of these hormones is largely under the control of angiotensin II.

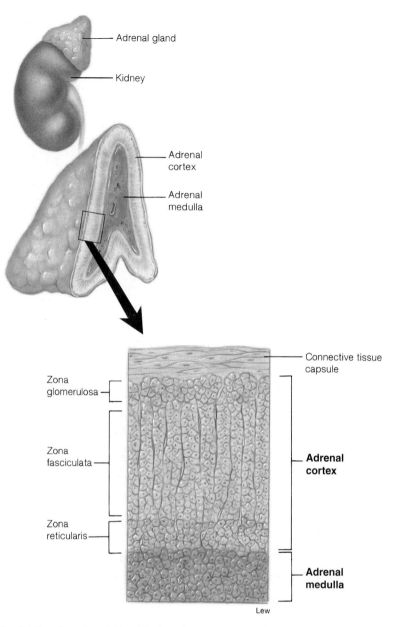

Adrenal gland

Kidney

Adrenal cortex

Adrenal medulla

Connective tissue capsule

Zona glomerulosa

Zona fasciculata

Zona reticularis

Adrenal cortex

Adrenal medulla

Lew

Figure 4.5 **The adrenal gland.** The adrenal medulla and adrenal cortex are shown in the sectioned gland, and the histology of the adrenal cortex, showing its three zones, is illustrated as it would appear at a magnification of 450×.

3. **Zona fasciculata.** The zona fasciculata is located inside the z. glomerulosa. This layer is the thickest part of the adrenal cortex, with its cells arranged in columns. They secrete the glucocorticoids when stimulated by adrenocorticotropic hormone (ACTH) secreted by the anterior pituitary. The most important glucocorticoids are *hydrocortisone* (*cortisol*) and *corticosterone*.

4. **Zona reticularis.** The z. reticularis is the innermost layer of the adrenal cortex, and is also involved in the secretion of the glucocorticoids. The epithelial cells in this layer form interconnections (anastomoses) with one another and stain a darker color than those of the z. fasciculata.

5. **Adrenal medulla.** The adrenal medulla forms the distinctive central region of the gland that stains a lighter color than the surrounding z. reticularis. It is composed of tightly packed clusters of *chromaffin cells*.

E. THYROID

Like the ovary, the functional units of the thyroid are called **follicles** (fig. 4.6). Each thyroid follicle is composed of a single layer of epithelial cells surrounding a homogenous protein-rich fluid, the *colloid*.

The hormones secreted by the thyroid follicles are *triiodothyronine* (**T**$_3$) and *tetraiodothyronine* (**T**$_4$, or thyrox-

Figure 4.6 The histology of the thyroid gland.

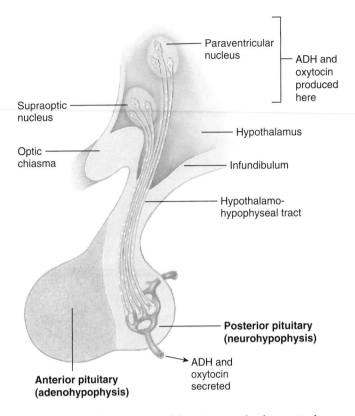

Figure 4.7 **The structure of the pituitary gland in sagittal view.** The sites of production of ADH (vasopressin) and oxytocin in nuclei of the hypothalamus, and the axon tract that transports these hormones to the posterior pituitary gland, are illustrated. The posterior pituitary stores and releases these hormones in response to appropriate stimuli.

ine). These hormones are released from the epithelial cells of the follicle into the adjacent capillaries and are important regulators of growth and metabolism.

The thyroid gland also contains **parafollicular cells** that secrete the hormone *calcitonin* (also called *thyrocalcitonin*). This hormone is believed to play a relatively minor role in the regulation of blood calcium concentrations.

PROCEDURE

Scan the slide under *low* power and observe the follicles (fig. 4.6). Note the clear space between the lighter, inner colloid and the darker, surrounding epithelial cells. This space is an artifact (produced by manipulation of the tissue when the slide was prepared) and is not present *in vivo.*

F. PITUITARY GLAND

The **pituitary,** or *hypophysis,* like the adrenal gland, is derived from two distinct embryonic origins. The **anterior pituitary,** also known as the **adenohypophysis** (*adeno* means "glandular"), originates in the embryo from a dorsal outpouching (*Rathke's pouch*) of oral epithelium.

Sometimes referred to as a master gland, the anterior pituitary secretes hormones that control other glands. These hormones include *adrenocorticotropic hormone (ACTH),* which stimulates the adrenal cortex, *thyroid-stimulating hormone (TSH),* which stimulates the thyroid, *gonadotropic hormones (FSH and LH—follicle-stimulating hormone and luteinizing hormone),* which stimulate the gonads), and *prolactin,* which stimulates the mammary glands. In addition, the anterior pituitary secretes *somatotropic hormone,* or *growth hormone (GH),* which stimulates growth in children.

In contrast, the **posterior pituitary,** also known as the **neurohypophysis,** is derived in the embryo from a ventral outpouching of the floor of the brain and secretes only two hormones: *vasopressin* (also called *antidiuretic hormone, ADH*) and *oxytocin.*

The secretion of hormones from both the anterior and the posterior pituitary is controlled by a part of the brain known as the **hypothalamus.** The posterior pituitary is derived as a downgrowth of the hypothalamus, providing the direct neural connection between them (fig. 4.7). Vasopressin (ADH) and oxytocin are manufactured in cell bodies of hypothalamic neurons and packaged into vesicles that travel down the axons of these neurons to the posterior pituitary. Here, these hormones are stored until stimulated for release by axons of the neurons in the hypothalamus. The posterior pituitary is therefore simply a storage organ.

Since the anterior pituitary is derived from oral epithelium and not from brain tissue, there is no direct neural connection between the hypothalamus and the anterior pituitary. There is, however, a special vascular connection between these two organs. A capillary bed in the hypothalamus is connected to a capillary bed in the anterior pituitary by means of venules that run between them. This vascular connection is known as the **hypothalamohypophyseal portal system** (fig. 4.8).

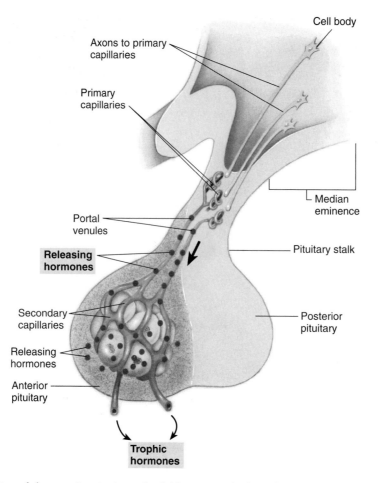

Figure 4.8 **Regulation of the anterior pituitary gland.** Neurons in the hypothalamus secrete releasing hormones (shown as dots) into the blood vessels of the hypothalamo-hypophyseal portal system. These releasing hormones stimulate the anterior pituitary to secrete its hormones into the general circulation.

Unlike the posterior pituitary, the anterior pituitary manufactures its own hormones. These hormones are released upon the arrival of specific chemical messengers, called *releasing hormones*, and secreted into the hypothalamo-hypophyseal portal system by the hypothalamus, as illustrated in figure 4.8. The anterior pituitary, then, is not actually the "master gland," since the secretion of its hormones is in turn controlled by releasing hormones secreted by the hypothalamus. Actually, both the hypothalamus and anterior pituitary are controlled by negative feedback inhibition from their target glands, so the entire concept of a "master gland" is erroneous.

PROCEDURE

Scan the slide of the pituitary gland under *low* power (fig. 4.9). Distinguish the anterior pituitary (observe capillaries and darkly stained blood cells) from the posterior pituitary (characteristic nerve tissue, lightly stained). Next, switch to the *high*-power objective and identify the following structures:

1. **Anterior pituitary**
   (a) **Sinusoids.** Sinusoids are modified capillaries that lack an endothelial wall. They can easily be identified by the presence of red blood cells.

Capillary —————

Chromophobe —————

Chromophil —————

× 160

(a)

(b)

Figure 4.9 Histology of the pituitary gland. (a) The anterior pituitary and (b) the posterior pituitary (100×).

(b) **Chromophils.** "Color-loving" chromophils are pituitary cells that readily take up stain into granules present in the cytoplasm. They are divided into two general categories on the basis of their staining properties: *acidophils* contain red-staining granules; *basophils* contain blue-staining granules. These two categories of cells produce different hormones.

(c) **Chromophobes.** The cytoplasm of chromophobes ("color-fearing" cells) does not pick up stain, and hence these cells appear quite dull next to the chromophils. It is believed that the chromophobes are not involved in hormone production.

2. **Posterior pituitary**

(a) **Nerve fibers.** Nerve fibers are the axons of neurons extending from the hypothalamus and compose most of the mass of the gland.

(b) **Pituicytes.** Pituicytes are randomly distributed among the nerve fibers and lack the bright color of the anterior pituitary cells.

REVIEW ACTIVITIES FOR EXERCISE 4.1

Test Your Knowledge of Terms and Facts

Match the gland with the hormone it secretes:

_____ 1. ovarian follicle
_____ 2. interstitial cells of Leydig
_____ 3. alpha cells of islets of Langerhans
_____ 4. beta cells of islets of Langerhans
_____ 5. zona glomerulosa of adrenal cortex
_____ 6. zona fasciculata of the adrenal cortex
_____ 7. adrenal medulla
_____ 8. posterior pituitary
_____ 9. anterior pituitary

(a) growth hormone
(b) glucagon
(c) hydrocortisone
(d) insulin
(e) testosterone
(f) aldosterone
(g) estrogen
(h) oxytocin
(i) epinephrine
(j) thyroxine

10. The fluid-filled central cavity of an ovarian follicle is called the _____ .
11. Most of the mass of the testis is composed of the _____ .
12. Another name for tetraiodothyronine is _____ ; it is secreted by the _____ gland.
13. Two glands derived from neural tissue are the _____ and the _____ .
14. Another name for vasopressin is _____ ; it is secreted by the _____ .

Test Your Understanding of Concepts

15. Distinguish between exocrine and endocrine glands.

16. Describe the structures involved in the production, transport, and secretion of oxytocin and vasopressin.

17. Where is ACTH produced? What does it do? Explain how its secretion is regulated.

18. The anterior pituitary has sometimes been called a "master gland." Why was this term used? Why is this erroneous?

Test Your Ability to Analyze and Apply Your Knowledge

19. Do you think that ligation (tying) of the vas deferens will affect the blood concentration of testosterone? Explain.

20. In *Cushing's syndrome*, the adrenal cortex secretes excessive amounts of glucocorticoids. Which hormones are these? Propose three different possible causes of excessive secretion of glucocorticoids.

Thin-Layer Chromatography of Steroid Hormones

MATERIALS

1. Thin-layer plates (silica gel, F-254 [F = fluorescent]), chromatography developing chambers, capillary tubes
2. Driers (chromatography or hair driers), ultraviolet viewing box (short wavelength), rulers or spotting template (optional)
3. Steroid solutions: 1.0 mg/mL in absolute methanol of testosterone, hydrocortisone, cortisone, corticosterone, and deoxycorticosterone; 5 mg/mL of estradiol
4. Unknown steroid solution containing any two of the steroids previously described
5. Developing solvent: 60 mL toluene plus 10 mL ethyl acetate plus 10 mL acetone, or a volume containing a comparable 6:1:1 ratio of solvents

Textbook Correlations

Before performing this exercise, you may want to consult the following references in *Human Physiology*, seventh edition, by Stuart I. Fox:

- *Chemical Classification of Hormones.* Chapter 11, pp. 287–289.
- *Mechanism of Steroid Hormone Action.* Chapter 11, pp. 292–293.
- *Functions of the Adrenal Cortex.* Chapter 11, pp. 305–307.

Gonads and Placenta. Chapter 11, p. 316.

Those using different physiology textbooks may want to consult the corresponding information in those books.

Slight differences in steroid structure are responsible for significant differences in biological effects. Differences in structure and solubility can be used to separate a mixture of steroids and to identify unknown molecules.

OBJECTIVES

1. Identify the major classes of steroid hormones and the glands that secrete them.
2. Describe the primary differences between different functional classes of steroid hormones.
3. Demonstrate the technique of thin-layer chromatography and explain how this procedure works.

A. STEROID HORMONES

The steroid hormones, secreted by the adrenal cortex and the gonads, are characterized by a common four-ring structure. The carbon atoms in this structure are numbered as follows:

Seemingly slight modifications in chemical structure result in very great differences in biological activity. On the basis of their activity and their structure, the steroid hormones can be grouped into the following functional categories: (1) androgenic hormones, (2) estrogenic hormones, (3) progestational hormones, and (4) corticosteroid hormones. These can be divided further into the subcategories of glucocorticoids and mineralocorticoids.

The **androgenic hormones** have a characteristic nineteen-carbon steroid structure and function in the development of male secondary sex characteristics. The most potent androgenic hormone is **testosterone,** secreted by the testes.

Testosterone

Although the primary source of androgens is the testes, the adrenal cortex also secretes small amounts (adrenal androgens are *dehydroepiandrosterone,* or *DHEA,* and *androstenedione*). Adrenal hyperplasia (Cushing's syndrome) and tumors of the adrenal cortex can also cause excessive levels of plasma androgen, which can have a masculinizing effect in females.

Testosterone and the other androgens are secreted by the interstitial Leydig cells of the testes. This secretion is stimulated by a gonadotropic hormone of the anterior pituitary, *interstitial cell–stimulating hormone (ICSH),* which is identical to *luteinizing hormone (LH).*

The structural difference between the androgens and **estrogens** is seemingly slight. The estrogens are eighteen-carbon steroids with three points of unsaturation (double bonds, see appendix 1) in the A ring. These two categories of steroids, however, stimulate the development of markedly different (male and female) secondary sex characteristics. The chief estrogenic hormone is **estradiol.**

Estradiol

The estrogens are normally secreted in cyclically increasing and decreasing amounts by the ovaries, reaching a peak about the time of ovulation. The cyclical secretion of estrogens is stimulated, in turn, by the cyclical secretion of follicle-stimulating hormone (FSH), a gonadotropic hormone of the anterior pituitary. Abnormally high concentrations of circulating estrogenic hormones may be due to tumors of the adrenal cortex or the ovary. Excessive levels of these hormones can have a feminizing effect in males.

In the normal female cycle, estrogens stimulate growth and development of the inner lining (endometrium) of the uterus. The final maturation of the endometrium is under the control of the hormone **progesterone,** secreted primarily in the phase of the cycle after ovulation (luteal phase) by the corpus luteum of the ovaries. The cyclical secretion of progesterone is, in turn, stimulated by the cyclical secretion of luteinizing hormone (LH) from the anterior pituitary.

During pregnancy, the placenta secretes increasing amounts of progesterone in accordance with the development of the fetus. Progesterone is a twenty-one-carbon steroid.

Progesterone

The **corticosteroid hormones** are steroid hormones of the adrenal cortex. Also composed of twenty-one carbons, corticosteroids differ from progesterone by the presence of three or more oxygen groups. These hormones are divided into two functional classes, mineralocorticoids and glucocorticoids, that are secreted by two distinct regions (zona) of the cortex.

The **mineralocorticoids,** secreted by the zona glomerulosa of the adrenal cortex, are involved in the regulation of sodium and potassium balance. Secretion of the mineralocorticoid aldosterone, for example, is stimulated by angiotensin II, which, in turn, is regulated by the secretion of renin from the kidneys. The most potent mineralocorticoids are **aldosterone** and, to a lesser degree, *deoxycorticosterone (DOC).*

Aldosterone

DOC

An abnormal secretion of the mineralocorticoids is usually associated with hypertension and may be produced by primary aldosteronism or by secondary aldosteronism due to low blood sodium, high blood potassium, hypovolemia, cardiac failure, kidney failure, or cirrhosis of the liver.

The **glucocorticoids,** secreted by the zona fasciculata and the zona reticularis of the adrenal cortex, stimulate the breakdown of muscle proteins and the conversion of amino acids into glucose (gluconeogenesis). The secretions of the z. fasciculata and the z. reticularis are stimulated by adrenocorticotropin (ACTH) from the anterior pituitary. The most potent glucocorticoid in humans is **cortisol** (or **hydrocortisone**), whereas in many other mammals it is **corticosterone.** *Cortisone* is a glucocorticoid used medically to inhibit inflammation and suppress the immune system.

An increased secretion of the glucocorticoids is associated with Cushing's syndrome (adrenal hyperplasia), pregnancy, and stress due to disease, surgery, and burns.

B. THIN-LAYER CHROMATOGRAPHY

In this exercise, an attempt will be made to identify two unknown steroids that are present in the same solution. To accomplish this task, you must (1) separate the two steroids and (2) identify these steroids by comparing their behavior with that of known steroids.

Since each steroid has a different structure, each will dissolve in a given solvent to a different degree. These differences in solubility will be used to separate and identify the steroids on a *thin-layer chromatography plate.*

The thin-layer plate consists of a thin layer of porous material (in this procedure, silica gel) that is coated on one side of a plastic, glass, or aluminum plate. The solutions of steroids are applied on different spots along a horizontal line near the bottom of the plate (a procedure called "spotting"), and the plate is placed on edge in a solvent bath *with the spots above the solvent.*

As the solvent creeps up the plate by capillary action, the steroids dissolve off their original spots (or *origins*) and are carried upward with the solvent toward the top end of the plate. Because the solubility of each steroid is different, the time required for the solvent to dissolve and transport the steroids will vary. The relative distance traveled by each steroid will also differ. If the process is halted before all the steroids have reached the top of the plate, some will have migrated farther from the origin than others.

The final pattern (**chromatogram**) obtained using the same steroids, solvent, and conditions is predictable and reproducible. In other words, the distance that a given steroid migrates in a given solvent relative to the travel of the solvent front can be used as an identifying characteristic of that steroid. Each steroid can then be identified by a characteristic numerical value (**R_f value**) determined by calculating the distance the steroid traveled (D_s) relative to the distance traveled by the solvent front (D_f), as follows:

$$R_f = \frac{\text{distance from origin to steroid spot} \left(D_s\right)}{\text{distance from origin to solvent front} \left(D_f\right)}$$

Using this method, an unknown steroid can be identified by comparing its R_f value in a given solvent with the R_f values of known steroids in the same solvent.

The chromatographic separation and identification of steroid hormones has revealed much about endocrine physiology. The placenta, for example, secretes estrogens that are more polar (more water-soluble) than estradiol, the predominant ovarian estrogen. These polar placental estrogens, *estriol* and *estetrol,* are now measured clinically during pregnancy to assess the health of the placenta.

Chromatograms of androgens recovered from their target tissues (such as the prostate) has revealed that these tissues convert testosterone into other active products. Further, these compounds appear to be more biologically active (more androgenic) than testosterone itself. Testosterone secreted by the testes, therefore, is a *prehormone* that is enzymatically converted in the target tissue into more active products, such as **dihydrotestosterone (DHT)** in many tissues. Males who have a congenital deficiency in *5 a-reductase,* the enzyme responsible for this conversion, exhibit many symptoms of androgen deficiency even though their testes secrete large amounts of testosterone.

PROCEDURE

1. Using a pencil, make a tiny notch on the left margin of the thin-layer plate, approximately 1 1/2 inches from the bottom. The origin of all the spots will lie on an imaginary line extending across the plate from this notch.

2. Using a capillary pipette, carefully spot steroid solution 1 (estradiol) about 1/2 inch in from the left-hand margin of the plate, along the imaginary line. Repeat this procedure, using the same steroid at the *same* spot, two more times. Allow the spot to dry between applications.

3. Repeat step 2 with each of the remaining steroid solutions (2, testosterone; 3, hydrocortisone; 4, cortisone; 5, corticosterone; 6, deoxycorticosterone; 7, unknown), spotting each steroid approximately 1/2 inch to the right of the previous steroid, along the imaginary line.

4. Observe the steroid spots at the origin under an ultraviolet lamp. (**Caution:** Do not look directly at the UV light.)

5. Place the thin-layer plates in a developing chamber filled with solvent (toluene/ethyl acetate/acetone, 6:1:1), and allow the chromatogram to develop for 1 hour.

6. Remove the thin-layer plate, dry it, and observe it under the UV light. Using a pencil, outline the spots observed under the UV light.

7. In the laboratory report, record the $\mathbf{R}_f$ values of the known steroids and determine the steroids present in the unknown solution.

Laboratory Report 4.2

Name _____

Date _____

Section _____

DATA FROM EXERCISE 4.2

Record your data in the table below and calculate the R_f value for each spot.

Steroid	Distance to Front	Distance to Spot	R_f Value
1. Estradiol			
2. Testosterone	same		
3. Hydrocortisone	same		
4. Cortisone	same		
5. Corticosterone	same		
6. Deoxycorticosterone	same		
7. **Unknown 1**	same		
8. **Unknown 2**	same		

The unknown solutions contained the following steroid hormones:

Unknown 1: _____

Unknown 2: _____

REVIEW ACTIVITIES FOR EXERCISE 4.2

Test Your Knowledge of Terms and Facts

1. The chief estrogenic hormone is _____ , secreted by the _____ .

2. The chief androgenic hormone is _____ , secreted by the _____ .

3. Dehydroepiandrosterone is what kind of a hormone? _____ ? Which gland secretes it? _____ .

4. Name two different endocrine glands that secrete progesterone: the _____ and the _____ .

5. The major mineralocorticoid is _____ . It is secreted by the _____ of the _____ .

6. The major glucocorticoid in humans is _____ . Another name for this hormone is _____ . It is secreted by the _____ of the _____ .

7. Name the following steroids:
   (a) eighteen-carbon sex steroid _____
   (b) nineteen-carbon sex steroid _____
   (c) twenty-one-carbon sex steroid _____

Test Your Understanding of Concepts

8. Create an outline or flow chart of the categories and subcategories of the steroid hormones. Indicate the gland that secretes each hormone.

9. Which of the steroid hormones used in this exercise was most soluble in the solvent? Which was least soluble? Explain.

Test Your Ability to Analyze and Apply Your Knowledge

10. Suppose a man took a drug that acted as a 5α-reductase inhibitor. What effects might this drug have on the prostate? Explain.

11. Suppose the 5α-reductase inhibitor drug described in question #10 also caused hair to grow in a man with male-pattern baldness. (*Note:* There is such a drug—*Propecia.*) What could you conclude about the cause of male-pattern baldness? Explain.

12. The concentration of estriol and estetrol increases in the blood of pregnant women as the pregnancy progresses. How would the migration and R_f values of these two hormones compare with the migration and R_f value of estradiol in this chromatography exercise? Explain.

Insulin Shock

MATERIALS

1. Large beaker filled with water
2. Guppy, goldfish, or another fish of comparable size
3. Insulin (insulin, zinc—100 IU), glucose

Insulin stimulates the tissue uptake of blood glucose and thus acts to lower the blood glucose concentration. Excessive insulin secretion can cause hypoglycemia, which, because of the brain's reliance on blood glucose, can affect brain function and even produce coma and death.

OBJECTIVES

1. Describe the mechanism by which insulin regulates the blood glucose concentration.
2. Demonstrate the effects of excessive insulin on a small fish and explain the clinical significance of these effects.

Textbook Correlations

Before performing this exercise, you may want to consult the following references in *Human Physiology,* seventh edition, by Stuart I. Fox:

- *Pancreatic Islets (Islets of Langerhans).* Chapter 11, pp. 313–314.
- *Energy Regulation by the Islets of Langerhans.* Chapter 19, pp. 613–616.
- *Diabetes Mellitus and Hypoglycemia.* Chapter 19, pp. 617–620.

Those using different physiology textbooks may want to consult the corresponding information in those books.

Insulin is a polypeptide hormone secreted by the *beta cells of the islets of Langerhans.* Insulin stimulates the transport of glucose from the blood into the muscles, liver, and adipose tissue, thus lowering the blood glucose concentration.

When the islets are incapable of secreting an adequate amount of insulin, a condition known as *diabetes mellitus* develops, in which the transport of glucose from the blood into the body tissues is impaired. This results in an increase in the blood sugar level (*hyperglycemia*) and the appearance of glucose in the urine (*glucosuria*).

Under these conditions, the body tissues cannot obtain sufficient glucose for cellular respiration and increasingly rely on the metabolism of fat for energy. The intermediate products of fat metabolism (ketone bodies) accumulate in the blood, resulting in *ketoacidosis.* Because one of these products is a volatile compound called *acetone* (which has a fruity odor), a person with this condition has fruity-smelling breath. The excretion of ketone bodies and glucose in the urine results is accompanied by the excretion of large amounts of water, causing dehydration. The combination of acidosis and dehydration that results from insufficient insulin may produce a diabetic coma.

People with **type I diabetes mellitus** must be given insulin injections to maintain blood glucose homeostasis. If a person with this type of diabetes is given too much insulin, however, the blood sugar level will fall below normal (*hypoglycemia*). Because the central nervous system can only use plasma glucose for energy, the lowering of blood sugar essentially starves the brain. The ensuing condition is called *insulin shock.* The symptoms of insulin shock can be illustrated by the reaction of fish to insulin in this exercise.

Hypoglycemia can, however, have other causes and can result in symptoms less severe than those accompanying insulin shock. *Reactive hypoglycemia* may occur after a meal if the beta cells secrete excessive insulin in response to carbohydrates in the digested food. This type of hypoglycemia sometimes occurs in the beginning stages of diabetes mellitus and can often be controlled by eating smaller, more frequent meals lower in carbohydrates. Hypoglycemia can also occur as a result of alcohol ingestion for reasons that are not well understood. Other causes of hypoglycemia include tumors of the beta cells (insulinomas), which secrete excessive insulin, and liver diseases in which the ability to produce glucose from glycogen and noncarbohydrate molecules is impaired.

The symptoms of hypoglycemia appear when the blood glucose concentration is about 45 mg/dL (normal serum glucose concentration ranges from 70 to 100 mg/dL). Symptoms can appear, however, at higher glucose concentrations if the cerebral circulation is impaired, such as observed in elderly people with atherosclerosis. The symptoms of glucose deficiency—faintness, weakness, nervousness, hunger, muscular trembling, and tachycardia—are symptoms similar to those seen when the brain lacks sufficient oxygen. More prolonged hypoglycemia may damage other parts of the brain, resulting in behavior resembling neuroses and psychoses. Indeed, severe brain damage may result in coma and death.

Hypoglycemia, like diabetes mellitus, may be detected by means of the oral glucose tolerance test. In this test, a patient drinks a glucose solution and blood samples are taken at fixed time periods. Hypoglycemia is suggested when the elevated blood glucose concentration returns to the normal level too quickly because of excessive insulin secretion or action.

PROCEDURE

1. Place a small fish (guppy or goldfish) in a large beaker of water to which a few hundred units of insulin have been added.
2. Observe the effects of insulin overdose. (If no effects are seen in 30 minutes, repeat this step with another fish.)
3. Remove the fish to a second beaker of water containing 5% glucose. Observe the recovery.

REVIEW ACTIVITIES FOR EXERCISE 4.3

Test Your Knowledge of Terms and Facts

1. Insulin is secreted by the _____ cells of the _____ .
2. Insulin stimulates the _____ .
3. Insulin action _____ the blood glucose concentration.
4. A deficiency of insulin causes _____ (term for blood glucose concentration); by contrast, an excess of insulin causes _____ .
5. The disease in which there is inadequate action of insulin is called _____ .
6. A person with the disease named in question 5 would, if untreated, have _____ (a term describing urine).

Test Your Understanding of Concepts

7. Explain the meaning of the term *insulin shock*, and why it is it dangerous.

8. Explain why a person with type I diabetes mellitus may have a "fruity" breath. Why is this person in danger of dehydration?

9. Explain why a person with type I diabetes mellitus can lose weight rapidly.

10. The most common type of diabetes mellitus is type II. In this condition, the person has beta cells that secrete insulin, sometime even in high amounts. Despite this, the person can still have hyperglycemia. Propose an explanation as to how this might be possible. Most people with type II diabetes mellitus are overweight. What does that suggest about large adipocytes compared to small adipocytes?

Skeletal Muscles

The basic mechanism of contraction for **striated muscles** (*skeletal* and *cardiac* muscles) can be divided into three phases: (1) electrical excitation of the muscle cell, (2) excitation-contraction coupling, and (3) sliding of the muscle filaments, or contraction.

At rest, there is an electrical *potential difference* across the *muscle fiber* (cell) membrane equal to approximately −80 mV (millivolts). The negative sign indicates that the inside of the membrane is negatively charged in comparison to the outside of the cell. When the muscle fiber is appropriately stimulated, either by a direct electric shock or by the motor nerve that innervates the muscle, the permeability of the membrane to cations changes. The diffusion of Na^+ into the cell *depolarizes* the membrane and its polarity momentarily reverses. This is immediately followed by the outward diffusion of K^+, which *repolarizes* the membrane and reestablishes the more negative resting membrane potential. This rapid depolarization and repolarization of the membrane at the stimulated point is called an **action potential.**

As action potentials are conducted along the muscle fiber membrane, they stimulate a rise in the cytoplasmic concentration of Ca^{2+}. In skeletal muscles, this Ca^{2+} comes from a system of intracellular tubules called the *sarcoplasmic reticulum.* In the resting muscle, the absence of this calcium allows two proteins, *troponin* and *tropomyosin,* which are part of the thin actin filaments within the sarcomeres (described below), to inhibit contraction. As a result of electrical stimulation, Ca^{2+} is released into the cell and becomes attached to troponin so that the troponin-tropomyosin complex no longer has an inhibitory effect. Thus, the influx of calcium ions is said to couple electrical excitation to muscle contraction.

Within the muscle fiber, there are numerous subunits (*fibrils*) that are oriented parallel to the long axis of the fiber (fig. 5.1a). Each fibril, in turn, is composed of numerous repeating subunits called **sarcomeres.** The sarcomere is the functional unit of contraction. When contraction is stimulated by Ca^{2+} the **thick** and **thin filaments** (composed of **myosin** and **actin,** respectively), within the sarcomeres slide over one another. This sliding of the filaments allows each sarcomere to shorten while its filaments remain the same length. As the sarcomeres become shorter, the fibrils and thus the entire muscle fiber shorten, resulting in muscle contraction (fig. 5.1b).

Exercise 5.1   Neural Control of Muscle Contraction

Exercise 5.2   Summation, Tetanus, and Fatigue

Exercise 5.3   Electromyogram (EMG)

Figure 5.1 The sliding filament model of muscle contraction. (a) The structure of a muscle fiber. The changes in band patterns during contraction are shown in an electron micrograph (b) and a diagram (c). As the filaments slide, the Z lines are brought closer together. The A bands remain the same length during contraction, but the I and H bands become progressively narrower and may eventually become obliterated.

170

Neural Control of Muscle Contraction

MATERIALS

1. Frogs
2. Surgical scissors, forceps, sharp probes, dissecting trays, glass probes
3. Recording equipment (either kymograph or electrical recorder, such as physiograph) and electrical stimulators
4. Straight pins (bent into a "Z" shape) and thread
5. Bone clamp (if kymograph is used) or myograph transducer (if physiograph is used)
6. Frog Ringer's solution. (Dissolve 6 g of NaCl, 0.075 g of KCl, 0.10 g of $CaCl_2$, and 0.10 g of $NaHCO_3$ in a liter of water.)

Textbook Correlations*

Before performing this exercise, you may want to consult the following references in *Human Physiology*, seventh edition, by Stuart I. Fox:

- *Skeletal Muscles.* Chapter 12, pp. 326–330.
- *Regulation of Contraction.* Chapter 12, pp. 336–339.
- *Neural Control of Skeletal Muscles.* Chapter 12, pp. 347–354.

Those using different physiology textbooks may want to consult the corresponding information in those books.

Isolated muscles from a pithed frog can be used to study the physiology of muscle contraction. Isolated frog muscles can be stimulated directly by an electric shock and indirectly through the activation of the appropriate motor nerve.

OBJECTIVES

1. Prepare a pithed frog for the study of muscle physiology.
2. Describe how muscle contraction can be stimulated by a direct electric shock.
3. Explain how motor nerves stimulate the contraction of skeletal muscles.

RECORDING PROCEDURES

The **physiograph** is a modern device for recording the mechanical aspects of muscular contraction (fig. 5.2). It is very sensitive because the mechanical movements of the muscle are first transduced (converted) into electrical current and then greatly amplified prior to recording. Mechanical events with different energies (from muscle contraction to sound waves) can also be recorded, as can nerve impulses and other primarily electrical activity, such as the *electrocardiograph* (ECG), *electromyograph* (EMG), or the *electroencephalograph* (EEG) recordings. Because a number of physiological parameters can be simultaneously recorded on different channels of the physiograph, the temporal (time) relationship between events can be studied.

The physiograph consists of four basic parts: (1) the **transducer** changes the original energy of the physiological event into electrical energy; (2) the **coupler** makes the input energy from the transducer compatible with the built-in amplifier (fig. 5.3); (3) the **amplifier** then increases the strength of the electrical current and forwards the signal to a galvanometer; (4) the **galvanometer** responds to the current generated by directing movement of a pen. The movement of the pen is proportional to the strength of the electrical current generated by the physiological event being measured. Because recording paper moves continuously at a known speed under the pen, both the *frequency* (number per unit time) and the *strength* (amplitude of the pen deflection from a baseline) of the physiological event can be continuously recorded (fig. 5.2).

---

*See Appendix 3 for correlations to the *A.D.A.M. InterActive Physiology Modules.*
See Appendix 3 for correlations to the *Virtual Physiology Laboratory* CD-ROM by McGraw-Hill and Cypris Publishing, Inc.

Figure 5.2   The Physiograph Mark III recorder.

Figure 5.3   **The Narco Mark III Physiograph with an inserted transducer coupler.** The transducer coupler is connected by means of a cable to a myograph transducer.

For Physiograph Recording

1. Insert the *transducer coupler* into the physiograph and plug the *myograph transducer* into the coupler (fig. 5.3).*
2. Raise the inkwells and lower the pens onto the paper by lowering the pen lifter (fig. 5.2). With an index finger covering the hole on the rubber bulb, squeeze to force ink into each pen.
3. Turn the physiograph on with the rocking power switch. Set the paper speed by depressing the *paper-*

*For Physiograph Mark III, Narco Bio-Systems

speed button marked 0.5 cm per second. Turn on the paper drive by depressing the *paper-advance* button and releasing it, allowing it to rise (the *up* position is on).

4. Move the *time switch* to "on." The bottom pen, labeled *time & event,* will make upward deflections every second.

Note: *At a paper speed of 0.5 cm/sec, these deflections will be separated by a distance equal to the width of one small box on physiograph recording paper. If the paper speed is increased to 1.0 cm/sec, the deflections of the time-and-event pen will be two small boxes apart.*

5. Turn the outer knob of the amplifier sensitivity control to its lowest number (this will be its greatest sensitivity) (fig. 5.3). With the *record* button off (in the up position), adjust the position of the recording pen for the appropriate channel with the *position* knob, so that the pen writes on the heavy horizontal line closest to the center of the channel being recorded.
6. Depress the *record* button (the *down* position is on), causing the pen to move away from the heavy horizontal line. Bring the pen back to the line by rotating the *balance* knob. The pen should now remain on the heavy line whether or not the record button is depressed and regardless of the setting of the sensitivity control.

A. FROG MUSCLE PREPARATION AND STIMULATION

To study the physiology of frog muscle and nerve, the frog must be killed but its tissues kept alive. This can be accomplished by destroying or **pithing** the frog's central nervous system. The frog is clinically dead (clinical death is defined

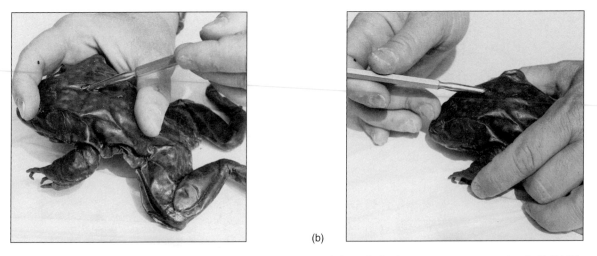

(a)   (b)

Figure 5.4   The procedure for pithing a frog. (a) A probe is first inserted through the foramen magnum into the skull. (b) The probe is then inserted through the spinal cord. (Procedure was simulated with a preserved frog.)

(a)   (b)

Figure 5.5   An alternate pithing procedure. (a) The frog is first decapitated. (b) A probe is then inserted into the spinal cord. (Procedure was simulated with a preserved frog.)

as the irreversible loss of higher brain function), but its muscles and peripheral nerves will continue to function as long as their cells remain alive. Under the proper conditions, this state can be prolonged for several hours.

There are two techniques for pithing a frog. One technique involves grasping the frog securely with one hand and flexing its head forward so that the base of the skull can be felt with the fingers of the other hand. Then perform these steps:

1. Quickly insert a sturdy metal probe into the skull through the foramen magnum (the opening in the skull where the spinal cord joins the brain stem) as in figure 5.4*a*.
2. Move the probe around in the skull, destroying the brain and preventing the frog from feeling any pain (it is now clinically dead).
3. Keeping the head flexed, partially withdraw the probe and turn it so that it points toward the hind end of the frog. Insert the probe downward into the

spinal canal (fig. 5.4*b*), destroying the spinal cord and its reflexes. The frog's legs will straighten out as the inserted probe causes reflex stimulation of the spinal nerves. When the spinal cord is destroyed, the frog will become limp.

Alternatively, the following procedure may be employed. Force one blade of a pair of sharp scissors into the frog's mouth as shown in figure 5.5*a*. Quickly decapitate the frog by cutting behind its eyes. It should be understood that the frog is dead as soon as its brain has been severed from its spinal cord. Insert a probe down into the exposed spinal cord as described above to destroy the frog's spinal reflexes (fig. 5.5*b*).

After the frog has been pithed, completely skin one of its legs to expose the underlying muscle. Discard the skin. Then run one blade of a pair of scissors under the Achilles tendon and cut it, leaving part of the tendon still attached to the gastrocnemius muscle (fig. 5.6).

(a)

(b)

(c)

Figure 5.6   Preparation of gastrocnemius. After the frog's leg is skinned (a), the Achilles tendon is cut (b) and a bent pin is inserted into it (c). A length of thread attaches this pin to the hook of the myograph transducer (not shown). In (b), the sciatic nerve is shown between two glass probes in preparation for exercise 5.1.

SETUP FOR DIRECT STIMULATION OF THE MUSCLE

1. Secure the frog to a dissecting tray by inserting sharp pins through the arms and legs.
2. Push a Z-bent pin through the Achilles tendon. Tie one end of a cotton thread to the bent pin and the other end to the hook of a myograph transducer. Position the myograph so that it is directly above the muscle and adjust the height of the myograph so that the muscle is under tension. Insert two stimulating electrodes directly into the gastrocnemius muscle (fig. 5.6).

3. Establish the **threshold stimulus** (the minimum stimulus that will evoke a particular response). To do this, set the stimulus intensity on 1.0 V and deliver a single shock. Increase the strength of the stimulus in 0.5-V increments until the muscle responds with a contraction (*twitch*) that is recorded clearly on the kymograph or physiograph. Record this voltage and enter it in your laboratory report. **Threshold:** _____ V

Note: Rinse the muscle periodically with Ringer's solution (a salt solution balanced to the extracellular fluid of the frog). Do not allow the muscle to dry out.

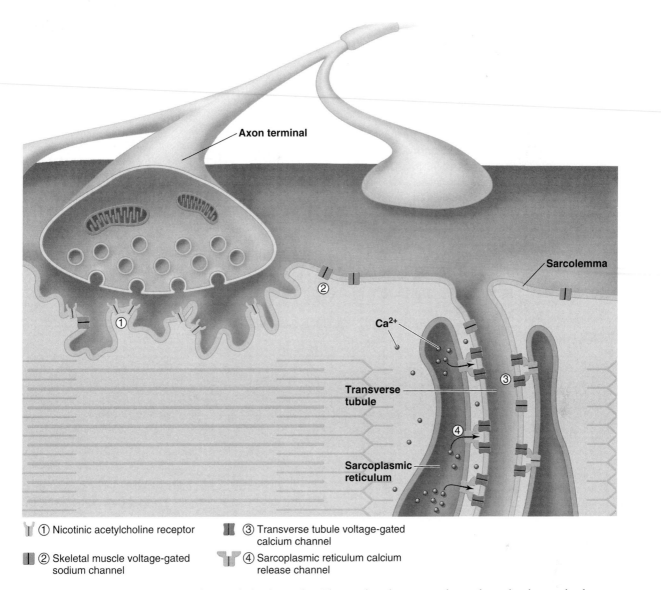

Axon terminal

Sarcolemma

Ca²⁺

②

①

③

Transverse tubule

④

Sarcoplasmic reticulum

Y ① Nicotinic acetylcholine receptor

▮▮ ② Skeletal muscle voltage-gated sodium channel

▮▮ ③ Transverse tubule voltage-gated calcium channel

Y ④ Sarcoplasmic reticulum calcium release channel

Figure 5.7 Excitation-contraction coupling in skeletal muscles. The numbered structures depict the molecules involved in *excitation-contraction coupling,* which refers to the release of Ca²⁺ from the sarcoplasmic reticulum in response to electrical excitation (action potentials), and the stimulation of muscle contraction in response to the released Ca²⁺. Voltage-gated Ca²⁺ channels in the transverse tubules (activated by action potentials) interact with Ca²⁺ release channels in the sarcoplasmic reticulum, leading to the diffusion of Ca²⁺ into the sarcoplasm. The Ca²⁺ can then bind to troponin to stimulate contraction.

B. STIMULATION OF A MOTOR NERVE

In the body (in vivo), skeletal muscles are stimulated to contract by somatic motor nerves. Action potentials in the motor nerve fibers elicit the release of a chemical neurotransmitter called **acetylcholine (ACh)** from the axon endings. This transmitter combines with a receptor protein in the muscle cell membrane and stimulates the production of new action potentials in the muscle fiber. Electrical stimulation of the muscle fibers causes Ca²⁺ to be released from the sarcoplasmic reticulum (fig. 5.7).

The release of Ca²⁺ stimulates muscle contraction following mechanisms previously described by the sliding filament model (fig. 5.1).

The electrical activity of somatic motor neurons is normally stimulated in the spinal cord by synapses with other neurons. These other neurons may be association neurons located in the brain or spinal cord, or they may be sensory, or afferent, neurons. Alternatively, action potentials in motor, or efferent, nerve fibers may be stimulated by damage to the fibers peripherally. This damage produces an *injury current* that stimulates action potentials and subsequent muscle contractions when, for example, a somatic nerve is pinched.

Some types of muscle degeneration are secondary to nerve damage or to dysfunctions at the neuromuscular junctions. Muscle degeneration follows damage to the motor nerve pathway because proper neuromuscular activity and resulting muscle tone are required for the health of the muscle. In the disease **myasthenia gravis** (*myasthenia* means abnormal muscle weakness or fatigue), antibodies secreted by the immune system block the muscle membrane receptors for acetylcholine, the neurotransmitter of somatic motor neurons. This autoimmune disease prevents the muscle from being properly stimulated by somatic motor neurons.

PROCEDURE

1. With the frog on its belly, part the posterior muscles of the thigh around the femur to reveal the white sciatic nerve (fig. 5.6a).
2. Using glass probes, gently free the nerve from its attached connective tissue and raise it between two glass probes (fig. 5.6b).
3. Place both stimulating electrodes under the sciatic nerve. Starting with the stimulator set at 0 V, gradually increase the stimulus voltage in small increments until the minimum stimulus that will produce a muscle twitch is attained. Record this threshold voltage in the space provided in your laboratory report.

   Threshold: _____ V
4. Turn off the stimulator but leave the recorder running. Using a length of cotton thread, tie a knot in the nerve. Record and observe the response of the gastrocnemius muscle.

Laboratory Report 5.1

Name _____

Date _____

Section _____

DATA FOR EXERCISE 5.1

1. Your threshold stimulus voltage when the electrodes were placed in the muscle: _____ V.
2. Your threshold stimulus voltage when the electrodes were placed on the nerve:_____ V.

REVIEW ACTIVITIES FOR EXERCISE 5.1

Test Your Knowledge of Terms and Facts

1. Arrange these structures in decreasing order of size: sarcomere, fibril, filaments, fiber:
   (a) _____
   (b) _____
   (c) _____
   (d) _____
2. The electrical events conducted along the cell (plasma) membrane that stimulate contraction are called
   _____.
3. Actin and myosin comprise the _____ and _____
   filaments, respectively.
4. What substance couples electrical excitation to muscle contraction? _____
5. The substance named in question 4 is stored in which intracellular organelle? _____
6. The substance named in question 4 binds to a regulatory protein known as _____,
   which in turn is bound to an inhibitory protein called _____.
7. The neurotransmitter chemical that stimulates contraction of skeletal muscles:_____.

Test Your Understanding of Concepts

8. Draw a sarcomere and label the parts and the bands. Then, describe and illustrate how the structure of a sarcomere changes during muscle contraction.

177

9. Trace the course of events from the release of ACh by a motor neuron to the binding of myosin cross-bridges to actin.

Test Your Ability to Analyze and Apply Your Knowledge

10. Which had the lower threshold for stimulation of muscle contraction—stimulation of the muscle directly, or stimulation of the nerve that innervates the muscle? Propose an explanation for these results.

11. Using your knowledge of the regulation of muscle contraction, predict what might happen to the beating of a heart if the blood concentration of Ca^{2+} were abnormally increased.

12. Predict the effects on muscles of a drug that blocks the action of acetylcholinesterase, an enzyme that breaks down acetylcholine. Compare that to the effects on muscles of a drug that blocks acetylcholine receptors.

Summation, Tetanus, and Fatigue

MATERIALS

1. Frogs
2. Equipment and setup used in exercise 5.1
3. Electrocardiograph plates and electrolyte gel

Twitch, summation, and tetanus can be produced by direct electrical stimulation of frog muscles *in vitro* (in glass, or the laboratory) and by stimulation of human muscles *in vivo* (within the living body). These procedures demonstrate how normal muscular movements are produced.

OBJECTIVES

1. Define the terms *twitch, summation, tetanus,* and *fatigue.*
2. Demonstrate twitch, summation, and tetanus in both frog and human muscles; and demonstrate fatigue in the frog muscle preparation.
3. Explain how a smooth, sustained contraction is normally produced.

Individual skeletal muscle fibers cannot sustain a contraction; they can only twitch. Muscle fibers likewise cannot produce a graded contraction; they can only contract maximally to any stimulus above threshold. Consequently, smooth, graded skeletal muscle contractions are produced by the **summation** of fiber twitches. Summation occurs when some fibers within a muscle are in the process of contraction before other fibers in the muscle have had time to relax completely from a previous twitch. This results in a second, stronger twitch that may partially "ride piggyback" on the first (fig. 5.8*a*). The strength of skeletal muscle contraction, therefore, depends on the number of fibers stimulated rather than on the strength of the individual muscle fiber contractions. Maintenance of a sustained muscle contraction is called **tetanus.** (The term *tetanus* should not be confused with the disease

---

Textbook Correlations*

Before performing this exercise, you may want to consult the following references in *Human Physiology,* seventh edition, by Stuart I. Fox:

- *Motor Units.* Chapter 12, pp. 327–330.
- *Contractions of Skeletal Muscles.* Chapter 12, pp. 339–341.
- *Muscle Fatigue.* Chapter 12, pp. 345–346.

Those using different physiology textbooks may want to consult the corresponding information is those books.

---

tetanus, which is accompanied by a painful state of muscle contracture, or *tetany*).

Tetanus can be demonstrated in the laboratory by setting the stimulator to deliver shocks automatically to the muscle at an ever-increasing frequency until the twitches fuse into a smooth contraction (fig. 5.8*b*). This is similar to what occurs in the body when different motor neurons in the spinal cord are activated to stimulate a muscle at slightly different times.

If the stimulator is left on so that the muscle remains in tetanus, a gradual decrease in contraction strength will be observed. This is due to **muscle fatigue.** Fatigue during a sustained maximal contraction, as when lifting a very heavy weight, appears to be due to a accumulation of extracellular K^+. This depolarizes the membrane potentials and interferes with the ability of the muscle fiber to produce action potentials. Fatigue (and muscle pain) during more moderate exercise is related to the production of lactic acid during anaerobic respiration, which decreases muscle fiber pH. The fall in pH is

---

*See Appendix 3 for correlations to the *A.D.A.M. InterActive Physiology Modules.*
See Appendix 3 for correlations to the *Virtual Physiology Laboratory* CD-ROM by McGraw-Hill and Cypris Publishing, Inc.

See Appendix 3 for correlations to the *Intelitool Physiology Laboratory Exercises* and the *Biopac Student Lab Exercises.*

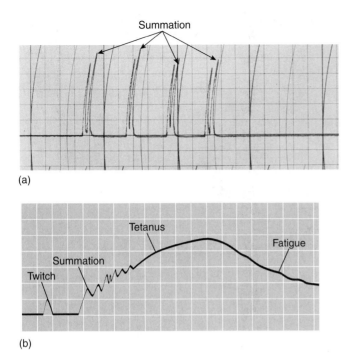

Summation

(a)

Tetanus

Fatigue

Summation

Twitch

(b)

Figure 5.8 **Recordings of muscle contractions.** (a) A recording of the summation of two muscle twitches on a physiograph recorder. Notice that contraction to the second stimulus is greater than contraction to the first stimulus (the intensity of the first and second stimuli is the same). (b) Diagram illustrating twitch, summation, tetanus, and fatigue.

believed to interfere with the storage or release of Ca^{2+} from the sarcoplasmic reticulum, and thus to interfere with excitation-contraction coupling.

A. Twitch, Summation, Tetanus, and Fatigue in the Gastrocnemius Muscle of the Frog

Summation, tetanus, and fatigue can be demonstrated with the frog gastrocnemius muscle preparation used in exercise 5.1.

PROCEDURE

1. Set the stimulus voltage above threshold and press down on the switch that delivers a single pulse to the muscle two or three times in rapid succession. If this is done with enough speed, successive twitches can be made to "ride piggyback" on preceding twitches, demonstrating *summation* (fig. 5.8a).
2. Set the stimulus switch to deliver shocks to the muscle automatically at a frequency of about one per second. Gradually increase the frequency of

stimulation until the twitches fuse into a smooth, sustained contraction, demonstrating *tetanus* (fig. 5.8b).
3. Maintain stimulation until the strength of contraction gradually diminishes as a result of muscle *fatigue* (fig. 5.8b).
4. Enter your recordings in the laboratory report.

B. Twitch, Summation, and Tetanus in Human Muscle

The properties of frog muscle contractions observed *in vitro* reflect the behavior of human muscle *in vivo* in many ways. A single pulse of electrical stimulation produces a single short contraction (twitch), and many pulses of stimulation delivered in rapid succession produce a summation of twitches resulting in a smooth, graded muscular contraction, and eventually in tetanus.

Sustained muscular spasm (*tetany*) may be produced by hypocalcemia and by alkalosis. (The most common cause of tetany is alkalosis produced by hyperventilation.) Cramps may be due to a variety of conditions, including salt depletion. General muscle weakness may be caused by alterations in plasma potassium levels (due, for example, to excessive diarrhea or vomiting).

Muscular dystrophy refers to any of a variety of diseases in which there is a progressive lack of support (dystrophy) and weakness of skeletal muscles (although heart muscle may also be involved) that does not seem to be caused by inflammation or neural disease. In severe forms of these diseases, large numbers of filaments and sarcomeres are replaced with fibrous connective tissue and fatty deposits.

PROCEDURE

1. Rub a small amount of electrolyte gel on the skin near the wrist and attach an ECG electrode plate to this area with an elastic band. Rub electrolyte gel on a second ECG electrode plate and place it on the anterior, medial area of the forearm, just below (distal to) the elbow. Do not attach this electrode to the arm, as this will be the *exploring electrode* (fig. 5.9).
2. Connect the electrode plates to a stimulator, making sure that the stimulator is *off* at this time.
3. Set the stimulus intensity at 15 V and deliver a single pulse of stimulation. If no twitch is observed or felt in the fingers, move the exploring electrode

Figure 5.9 Placement of electrodes for eliciting finger twitches in response to electrical stimulation.

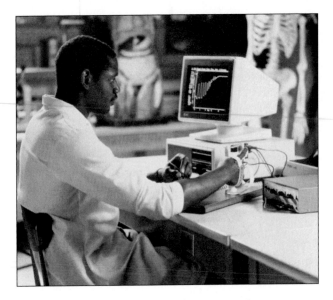

Figure 5.10 Student using the Physiogrip to demonstrate muscle twitches, summation, and tetanus.

around the medial area of the forearm until an effect is seen or felt. (See fig. 5.9 for the approximate position of the electrode.)

Caution: The stimulus intensity may have to be increased for some people, but do not exceed 30 V! An effect can generally be obtained at a lower voltage by moving the exploring electrode to a slightly different position or by adding electrolyte gel. A tingling sensation means that the stimulus intensity is adequate, although the position may have to be changed.

4. Once a muscle twitch has been observed, set the stimulator so that it automatically delivers one pulse of stimulation per second. Adjust the exploratory plate so that only one finger twitches.
5. Keeping the stimulus intensity constant, gradually increase the frequency of stimulation until a maximum contraction is reached. Gradually decrease the stimulus frequency until the individual twitches are reproduced.

ALTERNATIVE PROCEDURE FOR INTELITOOL PHYSIOGRIP

1. As in the previous procedure, determine the correct points for placement of the electrodes on the forearm so that the flexor digitorum superficialis muscle is stimulated. This will result in flexion of the finger that grips the trigger of the Physiogrip.
2. Hold the Physiogrip (fig. 5.10) with light pressure while another student delivers electrical shocks using the stimulator. Start with about 15 V at a duration of 1 millisecond and gradually increase the voltage until threshold is observed.
3. Continue to increase the voltage in small increments, demonstrating the graded increase in contraction strength in response to stronger electrical stimuli.
4. With the voltage constant, gradually increase the frequency of stimulation to demonstrate tetany.

ALTERNATIVE PROCEDURE FOR *BIOPAC* SYSTEM

1. The hand dynamometer (fig 5.11) used in the *Biopac* lesson 2 (EMG II) can be adapted for this exercise, once the stimulating electrodes are placed on the forearm to stimulate the digitorum superficialis muscle.

2. As per the previous instructions, deliver the electrical shocks using the stimulator. Start with about 14 V at a duration of 1 millisecond and gradually increase the voltage until a threshold is observed.

3. Continue to increase the voltage in small increments, demonstrating the graded increase in contraction strength in response to stronger electrical stimuli.

4. With the voltage constant, gradually increase the frequency of stimulation to demonstrate summation and tetany.

Dynagrip crossbar

Hand close to bracket but not touching

Figure 5.11 The *Biopac* hand dynamometer.

Laboratory Report 5.2

Name _____

Date _____

Section _____

DATA FROM EXERCISE 5.2

A. **Twitch, Summation, Tetanus, and Fatigue in the Gastrocnemius Muscle of the Frog**

1. In the space below, tape your recording or draw a facsimile.

2. Label twitch, summation, tetanus, and fatigue in your recording above; and compare your results with that shown in figure 5.8.

B. **Twitch, Summation, and Tetanus in Human Muscle**

1. Describe the results of your procedure in the space below. Alternatively, if you used the Intelitool or Biopac systems and have your data in a computer file, print a hard copy and paste it below. Indicate twitch, summation, tetanus and fatigue (if obtained).

REVIEW ACTIVITIES FOR EXERCISE 5.2

Test Your Knowledge of Terms and Facts

1. Define the following terms:
   (a) twitch _____
   (b) summation _____
   (c) tetanus _____
2. Sustained muscular spasm is called _____. The two most common
   causes of this are _____ and _____.

Test Your Understanding of Concepts

3. Describe how the summation of twitches is produced. Using this information, explain how variations in the strength of muscle contractions are produced.

4. Explain how tetanus was produced *in vitro* (in this laboratory exercise). Contrast this with the way that sustained contractions are normally produced *in vivo* (in the body).

5. Suppose you hold a 10-pound weight steadily in your right hand with your elbow slightly bent, so that you maintain a contraction of your biceps brachii muscle. After a period of time, you experience pain and it becomes increasingly difficult to keep the weight at the same level. Explain why this is so.

6. Continuing with the situation described in question 5, suppose your hand starts to shake. Describe the changes in the state of your skeletal muscle fibers as your hand goes from being held steady to shaking. Relate these changes to the recordings obtained in this exercise.

Electromyogram (EMG)

MATERIALS

1. Physiograph or another electrical recorder and high-gain coupler
2. EMG plates and disposable adhesive paper washers for EMG plates
3. Electrolyte (ECG) gel or paste and alcohol swabs
4. Alternative: *Biopac* system equipment for EMG I and II (lessons 1 and 2).

The electrical activity produced by muscles can be recorded using surface electrodes. This recording may demonstrate the action of antagonistic muscles and may be useful in biofeedback training of muscles.

OBJECTIVES

1. Demonstrate the antagonism between the action of the biceps and triceps muscles using the electromyogram (EMG).
2. Distinguish between isotonic and isometric muscle contractions, with examples of each.
3. Describe the EMG of the biceps and triceps during flexion and extension of the arm.
4. Explain the importance of antagonist inhibition in skeletal movements.
5. Explain how the EMG can be used in biofeedback techniques.

Skeletal muscle contraction occurs in response to the electrical stimulation of the muscle fibers. In the previous exercises, following stimulation the mechanical response of the muscles was observed and recorded on a myograph transducer as tension exerted. Although the muscles were stimulated by electric shocks, the electrical activity of the individual muscle cells was not recorded. Notice that electric shocks delivered to the muscle fibers induce the formation of membrane action potentials, and that it is these action potentials (acting via the release of

Textbook Correlations*

Before performing this exercise, you may want to consult the following references in *Human Physiology*, seventh edition, by Stuart I. Fox:

- *Motor Units.* Chapter 12, pp. 327–330.
- *Contractions of Skeletal Muscles.* Chapter 12, pp. 339–341.
- *Alpha and Gamma Motoneurons.* Chapter 12, pp. 348–349.

Those using different physiology textbooks may want to consult the corresponding information in those books.

Ca^{2+} from the sarcoplasmic reticulum) that lead to contraction of the muscles.

In the body (*in vivo*), muscle fibers are stimulated to contract by motor neurons. The axon of a motor neuron branches to innervate a number of muscle fibers, all of which contract when the axon is stimulated. The axon and the muscle fiber it stimulates is known as a **motor unit** (fig. 5.12). When more strength is required for a muscle contraction, more motor units are enlisted, or *recruited*, into the contraction.

The contraction that results in muscles shortening is called **isotonic** ("equal tension") because the force of contractions remains relatively constant throughout the movement. Isotonic contractions are easily observed since the muscle becomes shorter and the corresponding limb or object attached is moved, such as movements when walking, lifting a chair, or doing push-ups (fig 5.13*b*). In **isometric** ("equal measure") contractions, the length of a muscle remains constant and no movement is seen because the force in the muscle being contracted is opposed by an equal opposing force, such as gravity. For example,

---

*See Appendix 3 for correlations to the A.D.A.M. InterActive Physiology Modules.

See Appendix 3 for correlations to the *Virtual Physiology Laboratory CD-ROM* by McGraw-Hill and Cypris Publishing, Inc.

 See Appendix 3 for correlations to the *Intelitool Physiology Laboratory Exercises.*

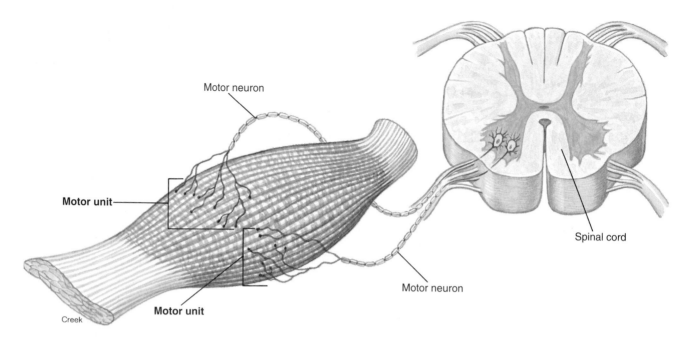

Figure 5.12 Motor units. A motor unit consists of a motor neuron and the muscle fibers it innervates. This diagram illustrates the innervation of muscle fibers by different motor units. (Actually, many more muscle fibers would be included in a single motor unit than are shown here.)

(a)

(b)

Figure 5.13 Isometric (a) and isotonic (b) muscle contractions.

an isometric contraction occurs when a person supports an object in a fixed position, such as in postural muscles when standing or sitting motionless (fig. 5.13a). An isometric contraction can be converted to an isotonic contraction when an increased force generated within the muscle overcomes the opposing resistance and results in muscle movement. This occurs, for example, when a straining body successfully clears the floor during the first push-up exercise.

A. ELECTROMYOGRAM RECORDING

When somatic motor nerves stimulate skeletal muscles to contract, the action potentials produced by the muscles transmit potential differences to the overlying skin that can be recorded by a pair of surface electrodes on the skin. The recording obtained is called an **electromyogram (EMG).** When the pair of electrodes are placed on the anterior surface of the upper arm, they

Figure 5.14 Electrode plates and adhesive washers needed for electromyograph (EMG) procedure.

(a)

(b)

Figure 5.15 Placement of the EMG electrodes. The positions are shown for recording from (a) the biceps and (b) the triceps muscles.

record potentials generated by the *biceps brachii* muscle; and when they are placed over the posterior surface of the upper arm, they record the electrical activity of the *triceps brachii* muscle. Contraction of the biceps flexes the arm, whereas contraction of the triceps extends the arm. These two groups of muscles, therefore, are **antagonistic.** Notice that during arm flexion the activity of the triceps muscle is inhibited, and during arm extension the activity of the biceps muscle is inhibited. In this way, flexion stretches the extensor muscles, whereas extension stretches the flexor muscles.

PROCEDURE: RECRUITMENT OF MOTOR UNITS

1. Using cotton or a paper towel soaked in alcohol, cleanse the skin over the biceps and triceps muscles.
2. Apply the self-sticking paper washers to the raised plastic area surrounding the electrode plates (fig. 5.14). Squeeze *electrolyte gel* onto the metal electrode plates. Use a paper towel to smooth the gel so that it completely fills the well between the electrode and the surrounding plastic.
3. Remove the paper coverings over the adhesive area of the washers and apply the electrodes to the skin over the biceps muscle. Apply one electrode to the skin over the proximal portion of the biceps and the other over the distal portion, aligned with the first (fig. 5.15). Apply the ground electrode over the triceps muscle.
4. Plug the electrodes into the *high-gain coupler* module of the physiograph. Set this module to a *gain of*

×100, a *time constant* of 0.03, and a *sensitivity* between 20 and 100.

5. Set the *chart speed* at 0.5 cm/sec. With the arm relaxed and hanging down, establish a baseline, or control, in the recording. Then flex the arm (bringing the hand upward), and observe the recording of an isotonic contraction. Extend the arm back to its previous position, and then flex it again so that the difference between flexion and extension can be seen in the recording.
6. Flex the arm again, this time lifting a chair or other weight. Observe the effect of this activity on the height (amplitude) of the recording. Now, "make a muscle" of increasing strength up to a maximum, demonstrating the recruitment of motor units into the isometric contraction.

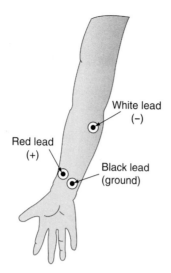

Figure 5.16 *Biopac* electrode positions for the EMG.

The activity of antagonistic muscle groups is controlled in the central nervous system, so that when one group of muscles (the *agonist*) is stimulated to contract, the *antagonist* muscle group is inhibited, and will be stretched. This inhibition of antagonistic muscle groups occurs largely through the action of descending motor tracts that originate in the brain. When a person has spinal cord damage that blocks these descending inhibitory influences, the antagonistic muscles may contract when they are stretched by the movement of a limb. Without inhibitory influences, stretch reflexes in a person with spinal cord damage may cause antagonistic muscles to alternately stretch and contract, producing a **flapping tremor,** or **clonus.**

ALTERNATIVE PROCEDURE USING THE *BIOPAC* SYSTEM

1. Stick three electrodes on the forearm in the positions indicated in figure 5.16, and attach the correctly colored electrodes as shown in that figure.
2. Proceed with the set-up instructions as per lesson 1 (EMG 1) for the *Biopac* student lab.
3. Demonstrate recruitment of motor units by clenching your fist tighter and tighter, up to your maximum strength.
4. Repeat this procedure after moving the electrodes to your other forearm.

PROCEDURE: EMG DURING ARM FLEXION AND EXTENSION

1. With the electrodes placed over the biceps brachii muscle as previously described, flex your arm as if you were trying to lift the table.
2. Change the position of the electrodes so that the two recording electrodes are lined up over the triceps muscle (one proximal and one distal) and the ground electrode is over the biceps muscle. Flex and extend the arm as before and observe the recording.
3. Place the hand on a table with the elbow bent and slowly extend the arm, as if doing a pushup. Observe the effect of this action on the EMG. Note the change in the recording as an isometric contraction becomes isotonic. Enter your recordings in the laboratory report.

B. BIOFEEDBACK AND THE ELECTROMYOGRAPH

Our behavior changes as a result of the pleasant or unpleasant consequences of our actions. That is, positive and negative reinforcements modify behavior; this represents a type of learning that experimental psychologists call *operant conditioning.* The pairing of a particular behavior, such as cigarette smoking, with unpleasant sensations has been used successfully to shape human behavior through *aversion conditioning.*

Biofeedback techniques similarly effect learning, usually through feedback provided by electronic monitors of specific physiological states. The electromyogram, for example, provides a visual display of muscle stimulation that can be used as a psychological reward to reinforce effort spent attempting to contract specific muscle groups. In this exercise, the EMG will be used to demonstrate biofeedback techniques involved in learning how to increase the strength of contraction of the triceps muscle.

Biofeedback techniques serve a variety of clinical functions. The EMG is sometimes used to train people with neuromuscular disorders to regain use of affected limbs. Physiological monitoring of the heart rate and blood pressure has enabled patients with high blood pressure to lower their pulse; and the production of alpha rhythms on *electroencaphalogram (EEG)* recordings has been used to teach people with stress techniques for relaxation.

PROCEDURE

1. Prepare the EMG electrodes as described in the previous procedure. Cleanse the skin with alcohol and place the two recording electrodes over the triceps muscle and the ground electrode over the biceps muscle.

2. Set the *high-gain coupler* to a *gain* of ×100, a *time constant* of 0.03, and a *sensitivity* between 20 and 100. Set the *chart speed* of the recorder to 0.5 cm/sec.

3. Extend the arm and observe the highest amplitude of the recording. Attempt a forced extension and observe the amplitude (height) of the recording. Attempt to increase the amplitude of the EMG by various procedures. (*Hint:* Try to extend the arm with the back of the hand against a table.)

4. Alternatively, the *Biopac* system can be used with the same set-up previously described. Demonstrate biofeedback by attempting to reproduce a particular muscle tension while watching the recording on the computer screen.

Name _____

Date _____

Section _____

DATA FROM EXERCISE 5.3

A. Electromyogram Recording

1. Recruitment of Motor Units
In the space below, tape your recordings or draw facsimiles. Label the parts of your recording.

2. Electromyogram During Arm Flexion and Extension
In the space below, tape your recordings or draw facsimiles.
(a) Label flexion and extension for both the biceps and triceps muscles in your recording.
(b) Compare your reading of the *isotonic* contraction to that of the *isometric* contraction.

B. Biofeedback and the Electromyograph

In the space below, tape your recordings or draw a facsimiles. Label the region of your recording that demonstrates biofeedback.

REVIEW ACTIVITIES FOR EXERCISE 5.3

Test Your Knowledge of Terms and Facts

1. During arm flexion, the biceps brachii muscle is the _____ (agonist/antagonist) and the triceps brachii is the _____ (agonist/antagonist).
2. Define an *isotonic contraction*. _____
3. Define an *isometric contraction*. _____
4. Define the term *motor unit*. _____
5. The process of enlisting more and larger motor units to produce a stronger contraction is called _____.
6. The recording of the electrical currents produced by contraction of skeletal muscle is called a(n): _____.

Test Your Understanding of Concepts

7. Draw two motor units, one smaller and one larger. With reference to your illustration, explain how these motor units could produce three different strengths of contraction.

8. Distinguish between isometric and isotonic contraction, and give examples of each.

9. Describe the activity of the biceps and triceps brachii during arm flexion and extension. Also, describe their activity when you "make a muscle."

Test Your Ability to Analyze and Apply Your Knowledge

10. Suppose you want to lift an object very slowly and carefully. Using the concepts of motor unit size and biofeedback, explain how you could do this.

11. Suppose a person had a stroke and could no longer write with their dominant hand. How could you use the equipment and techniques in this lab exercise to help train this person to write with the nondominant hand?

Blood: Gas Transport, Immunity, and Clotting Functions

The various components of the blood serve different physiological functions. The **plasma,** or fluid portion of the blood (fig. 6.1), provides the major means for distributing chemicals between organs. For example, it transports food molecules absorbed through the small intestine, hormones secreted by the endocrine glands, and antibodies produced by certain white blood cells. The plasma also helps to eliminate metabolic wastes by carrying these unwanted molecules to the liver (for excretion in the bile), to the kidneys (for excretion in the urine), and, in the case of CO_2 gas, to the lungs (for excretion in the exhaled air).

In addition to plasma, the blood also contains two major types of cells: **red blood cells** (*erythrocytes*) and **white blood cells** (*leukocytes*) (fig. 6.1*a*). Red blood cells contribute to the respiratory function of the blood by providing transport for oxygen and, to a lesser degree, for carbon dioxide. Blood is "typed" based upon the presence or absence of specific molecules displayed by red blood cells (the ABO system and the Rh factor are examples). White blood cells and their products help to provide immunity from infection by recognizing and attacking foreign molecules and cells. The blood also contains **platelets,** which are membrane-bound fragments derived from a bone marrow cell called a *megakaryocyte*. Platelets, together with proteins in the plasma, help to maintain the integrity of blood vessels by forming *blood clots.*

Figure 6.1 Composition of blood. (a) Centrifugation of whole blood causes the red blood cells to become packed at the bottom of the tube; this layer is covered by a thin, buffy coat of leukocytes and platelets, and all of these formed elements of blood are separated from the plasma. (b) The structure of hemoglobin within red blood cells, and (c) the structure of each heme group. There are four heme groups per hemoglobin molecule.

Red Blood Cell Count, Hemoglobin, and Oxygen Transport

MATERIALS

1. Hemocytometer
2. Unopettes (Becton-Dickinson) for manual red blood cell count and hemoglobin measurements
3. Heparinized capillary tubes, clay capillary tube sealant (Seal-ease), microcapillary centrifuge, hematocrit reader
4. Microscope
5. Sterile lancets and 70% alcohol
6. Colorimeter and cuvettes
7. Container for disposal of blood-containing items

Textbook Correlations

Before performing this exercise, you may want to consult the following references in *Human Physiology*, seventh edition, by Stuart I. Fox:

- *The Formed Elements of Blood.* Chapter 13, pp. 368–370.
- *Hemoglobin and Oxygen Transport.* Chapter 16, pp. 506–512.

Those using different physiology textbooks may want to consult the corresponding information in those books.

Almost all of the oxygen transported by the blood is carried within the red blood cells attached to hemoglobin. Measurements of the oxygen-carrying capacity of blood include the red blood cell count, hemoglobin concentration, and hematocrit. Anemia results when one or more of these measurements is abnormally low.

OBJECTIVES

1. Describe the composition of blood.
2. Describe the composition of hemoglobin and explain how hemoglobin participates in oxygen transport.
3. Demonstrate the procedures for taking the red blood cell count and hemoglobin and hematocrit measurements, and list the normal values for these measurements.
4. Explain how measurements of the oxygen-carrying capacity of blood can be used to diagnose anemia and polycythemia.

Each ventilation cycle delivers a fresh supply of oxygen to the alveoli of the lungs. The amount of oxygen that leaves the lungs *dissolved in plasma* is equal to 0.3 mL of O_2 per 100 mL of blood. The amount of oxygen leaving the lungs in whole blood, however, is equal to 20 mL of O_2 per 100 mL of blood. Most of the oxygen (19.7 mL O_2 100 mL blood), therefore, must be carried within the cellular elements of the blood. This oxygen is carried by **hemoglobin** molecules within the red blood cells (fig. 6.1*b*). In this way the oxygen is transported to the body cells and used for aerobic respiration and the production of cellular energy.

Each hemoglobin molecule consists of two pairs of polypeptide chains (one pair called the *alpha chains* and one pair called the *beta chains*) and four disc-shaped organic groups called *heme groups*. Each heme group contains one central ferrous ion (Fe^{2+}) capable of bonding with one molecule of oxygen (fig. 6.1*b*). Thus, one molecule of hemoglobin can combine with four molecules of oxygen.

The hemoglobin within the red blood cells load up with oxygen in the capillaries of the lungs and unload oxygen in the tissue capillaries. In both cases, oxygen moves according to its diffusion gradient. Since red blood cells always respire anaerobically (so they do not consume the oxygen they carry), a maximum diffusion gradient for oxygen is maintained between the red blood cells and the tissues.

The **oxygen-carrying capacity** of the blood is dependent on the total number of red blood cells and, consequently, on the total amount of hemoglobin. The total number of red blood cells is dependent on a balance between the rates of red blood cell production and destruction. The rate of red blood cell production by the bone marrow is regulated by the hormone **erythropoietin,** secreted by the kidneys. Erythropoietin is secreted when blood oxygen levels fall, such as when traveling in high-altitude environments. The rate of renal erythropoietin secretion is, therefore, regulated by the oxygen requirements of the body.

Older red blood cells (those that are approximately 120 days old) are routinely destroyed by the action of phagocytic cells fixed to the sides of blood channels (sinusoids) by a meshwork (reticulum) of fibers. Located in the spleen, liver, and bone marrow, these fixed phagocytes compose the **reticuloendothelial system.** These reticuloendothelial cells digest the hemoglobin within the old red blood cells into the component parts of protein, iron, and the *heme* pigment. The protein is hydrolyzed and returned to the general amino acid pool of the body, the iron is recycled to the bone marrow, and the heme is changed into a new pigment called **bilirubin.**

Bilirubin is released into the blood by the reticuloendothelial cells, then picked up by the liver and secreted into the bile as bile pigment. An abnormal increase in the amount of bilirubin in the blood, due to an increased rate of red blood cell destruction, liver dysfunction, or bile duct obstruction, results in the condition known as *jaundice* (yellowing of the skin and sclera of the eyes).

A. RED BLOOD CELL COUNT

The object of this exercise is to determine the number of red blood cells in a cubic millimeter of blood. Because this number is very large, it is practical to dilute a sample of blood with an isotonic solution, count the number of red blood cells in a fraction of this diluted blood, and then multiply by a correction factor. This procedure is accurate only when (1) the blood diluted is a representative fraction of all the blood in the body, (2) the dilution volumes are accurate, and (3) the sample counted is representative of the total volume of diluted blood.

PROCEDURE

Obtaining and Diluting Blood Samples

1. The Unopette reservoir (fig. 6.2) contains a premeasured amount of diluting solution (Hayem's solution). Use the pointed end of the shielded capillary tube (fig. 6.2) to puncture the plastic top of this reservoir. Turn the shielded capillary tube and attach it to the reservoir opening until needed.
2. Swing your hand around until your fingers become engorged with blood (*hyperemia*). Cleanse the tip of your index or third finger with 70% alcohol and prick it with a sterile lancet.

⚠ **Caution:** *Because of the danger of exposure to the AIDS virus and other harmful agents when handling blood, each student should perform this and other blood*

Figure 6.2   The Unopette system (Becton-Dickinson). This consists of a reservoir containing the premeasured amount of diluent (left) and a plastic capillary tube within a shield (right) for puncturing the reservoir top and delivering a measured amount of whole blood to the reservoir.

exercises with his or her own blood only. All objects that have been in contact with blood must be discarded in a container indicated by the instructor.

3. Discard the first drop of blood and point your finger downward to collect the next large drop of blood. Remove and discard the shield over the capillary pipette of the Unopette, and simply touch the tip of the pipette to the drop of blood. Allow the pipette to fill by capillary action (fig. 6.3*a*).
4. Squeeze the previously punctured reservoir with the fingers of one hand and, while squeezing, insert the pipette of blood into the punctured top of the reservoir. Releasing the pressure on the reservoir will draw the blood into the premeasured Hayem's solution within the reservoir (fig. 6.3*b*).
5. Gently mix the blood with the Hayem's solution for approximately 1 minute.

PROCEDURE

Filling the Hemocytometer and Determining Red Blood Cell Count

1. Place a coverslip on the hemocytometer so that it covers one of the silvered areas.
2. Remove the capillary pipette from the reservoir, turn it around, and reinsert it backwards into the reservoir so that the capillary is pointing out of the reservoir (like the needle of a syringe—see fig. 6.4). Discard the first 3 drops of blood from the Unopette (properly onto a piece of cotton, and disposed into a designated container). Place the next drop of

(a)

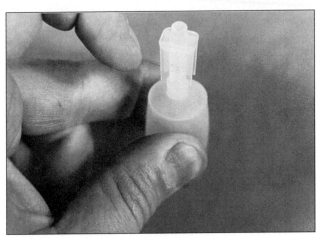

(b)

Figure 6.3   The Unopette method for measuring a red blood cell count or hemoglobin concentration. (a) Fill the plastic capillary pipette with fingertip blood. Then (b), squeeze the reservoir to draw blood out of the pipette into diluent within the reservoir.

diluted blood in the "V" region of the hemocytometer, at the edge of the coverslip. The diluted blood will be drawn underneath the coverslip by capillary action (fig. 6.4).

3. Locate the grid on the hemocytometer using the 10× objective. Focus first, then change to 45× and count the total number of red blood cells in the squares numbered 1 through 5 *only* (fig. 6.5).

Note: *If a red blood cell lies on the upper or the left-hand lines of the square, include it in your count. Do not count those that lie on the lower or the right-hand lines.*

4. The central grid of twenty-five squares is 1 square mm (mm^2) in area and 0.10 mm deep. The dilution factor is 1:200. To convert the number of red blood cells that you counted in 5 squares to the number of red blood cells per cubic millimeter, multiply your count by 10,000 (the product of 5 × 10 × 200).

Figure 6.4   Procedure for filling a hemocytometer. A Unopette reservoir is used to fill the hemocytometer with diluted blood. The squeezing of the reservoir places a drop of diluted blood at the edge of the coverslip, whereupon the drop of blood moves under the hemocytometer grid by capillary action.

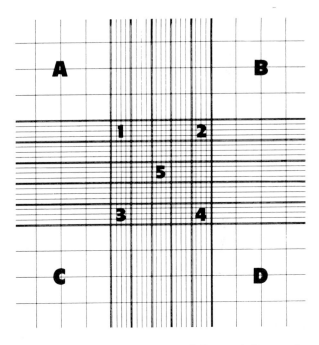

Figure 6.5   The hemocytometer grid. Squares 1–5 are used for red blood cell counts; squares A–D are used for white blood cell counts.

5. Record your total count of red blood cells in 5 indicated squares.
    Red blood cells/5 squares: _____
    Calculate the number of red blood cells in a cubic millimeter of your blood; and enter this number in the laboratory report.

Normal red blood cell counts: For an adult **male** is 4.5–6.0 million per cubic mm (mm^3); and for an adult **female** is 4.0–5.5 million/mm^3.

Figure 6.6 The method for filling a capillary tube with fingertip blood.

B. Hematocrit

When whole blood is centrifuged, the red blood cells become packed at the bottom of the tube, leaving the plasma at the top. The ratio of the volume of packed red blood cells to the total blood volume is called the **hematocrit**.

PROCEDURE

1. Prick your finger with a sterile lancet to obtain a drop of blood as described in the previous procedure. Again, discard the first drop onto an alcohol swab, and dispose of this properly in a designated container.

2. Obtain a heparinized capillary tube (**heparin** is an *anticoagulant*). Notice that one end of the tube is marked with a red band. Touch the end of the capillary tube opposite the marked end to the drop of blood, allowing blood to enter the tube by capillary action and gravity (fig. 6.6). The tube does not have to be completely full (half full or more is adequate), and air bubbles are not important (they will disappear during centrifugation).

3. Seal the red-banded (fire-polished) end of the capillary tube by gently pushing it upright into clay capillary sealant. Carefully, rotate and remove the tube.

4. Place the sealed capillary tube in a numbered slot of the microcapillary centrifuge, with the plugged end of the capillary tube facing outward against the rubber gasket. Screw the top plate onto the centrifuge head and centrifuge for 3 minutes. At the end of the centrifugation, determine the hematocrit with the hematocrit reader provided, and enter this value in your laboratory report.

> Normal hematocrit values: For an adult **male**, is 47 ± 7%; and for an adult **female** is 42 ± 5% (of the total blood volume).

C. Hemoglobin Concentration

Hemoglobin absorbs light in the visible spectrum and hence is a *pigment* (a colored compound). It should therefore be possible to measure the concentration of hemoglobin in a hemolyzed sample of blood by measuring the intensity of its color. This procedure, however, is complicated by the fact that red blood cells contain different types of hemoglobin, and each type absorbs light in a slightly different region of the visible spectrum (i.e., has a slightly different color).

When the oxygen concentration of the blood is high, such as in the capillaries of the lungs, normal **deoxyhemoglobin** combines with oxygen to form the compound **oxyhemoglobin**. When the concentration of oxygen in the blood is low, such as in the capillaries of the tissues, the oxyhemoglobin dissociates (comes apart) to form reduced hemoglobin and oxygen.

$$\text{deoxyhemoglobin} + \text{oxygen} \underset{\text{tissues}}{\overset{\text{lungs}}{\longleftrightarrow}} \text{oxyhemoglobin}$$

Arterial blood is bright red due to the predominance of the oxyhemoglobin pigment, whereas venous blood has the darker hue characteristic of deoxyhemoglobin. It should be emphasized, however, that venous blood, although darker in color, still contains a large amount of oxyhemoglobin; this functions as an oxygen reserve.

A less common though clinically important form of hemoglobin is **carboxyhemoglobin**, which is a complex of hemoglobin and *carbon monoxide*. This complex, unlike oxyhemoglobin, does not readily dissociate; thus, the hemoglobin that is bonded to carbon monoxide cannot participate in oxygen transport. The carboxyhemoglobin complex has a bright, cranberry red color.

A small percentage of hemoglobin contains iron that is oxidized to the *ferric* state (Fe^{3+}) instead of being in the normal *ferrous* state (Fe^{2+}). Hemoglobin in this oxidized state is called **methemoglobin** and is incapable of bonding with either oxygen or carbon monoxide. An increase in the amount of methemoglobin is associated with some genetic diseases, or it may result from the action of certain drugs, such as nitroglycerin.

The following exercise describes how the concentration of hemoglobin in a solution of hemolyzed blood can be assessed by measuring the intensity of its color with a spectrophotometer. To do this accurately, all the hemoglobin must be converted into one form (methemoglobin) and then combined with cyanide to make it more stable. The unknown hemoglobin concentration can then be determined by comparing its absorbance with the absorbance of a standard hemoglobin solution of known concentration.

PROCEDURE

Measurement of Blood Hemoglobin Concentration

Note: *Alternative procedures for estimating hemoglobin concentration in blood include the Tallquist paper blot method and the use of the hemoglobinometer.*

1. A different Unopette will be used for this procedure. This Unopette has a solution of cyanmethemoglobin reagent (yellow color); and has a capillary pipette that delivers twice as much blood as the one used for the red blood cell count procedure. The correct capillary pipette can be identified by the yellow number 20 (for 20 μl) on its side. As in the previous procedure, puncture the top of the reservoir with the shielded capillary pipette.

2. Wipe the tip of your third finger with 70% alcohol and puncture it with a sterile lancet. Discard the first drop of blood onto an alcohol swab, and properly dispose in a designated container. Fill the plastic capillary pipette in the Unopette with 0.02 ml of blood. Fill the capillary tube by simply touching the tip of the pipette to the drop of blood and allowing the pipette to completely fill by capillary action (see fig. 6.3*a*).

3. Squeeze the reservoir first, then insert the pipette and release the reservoir, aspirating the blood into the reservoir (see fig. 6.3*b*). Squeeze and release the reservoir a few more times to completely aspirate blood from the pipette. Mix and allow it to stand at room temperature for 10 minutes.

4. Some standard hemoglobin solutions come full strength and must be diluted with cyanmethemoglobin reagent to be at the same dilution as the unknown. Use the same procedure as in steps 2 and 3 to make this dilution (using a new Unopette).

   Some purchased hemoglobin standards arrive already diluted. Using these may require calculating the hemoglobin concentration of blood that would be equivalent to that of the diluted standard solution.

5. Set the colorimeter at a wavelength of 540 nm and standardize the instrument using plain cyanmethemoglobin solution as the blank. Record the absorbance values of the unknown and standard.
   Absorbance of **unknown:** _____
   Absorbance of **standard:** _____

6. Calculate the hemoglobin concentration of the unknown using the *formula*

$$\text{Concentration}_{\text{unknown}} = \frac{\text{Concentration}_{\text{standard}} \times A_{\text{unknown}}}{A_{\text{standard}}}$$

   Enter your hemoglobin concentration in the laboratory report.

Normal hemoglobin concentrations for an adult **male** is 13–16 g/dL, and for an adult **female** is **12–15** g/dL.

D. Calculation of Mean Corpuscular Volume (MCV) and Mean Corpuscular Hemoglobin Concentration (MCHC)

An abnormally low hemoglobin, hematocrit, or red blood cell count may indicate a condition known as anemia. Anemia may be caused by iron deficiency, vitamin B_{12} and folic acid deficiencies, bone marrow disease, hemolytic disease (e.g., sickle-cell anemia), loss of blood through hemorrhage, or infections. Diagnosis of a specific type of anemia is aided by relating the measurements of hemoglobin, hematocrit, and red blood cell count to derive the **mean corpuscular volume (MCV)** and the **mean corpuscular hemoglobin concentration (MCHC).**

Anemia is subdivided into a number of categories on the basis of the **MCV** and **MCHC.** *Macrocytic anemia* (MCV greater than 94, MCHC within normal range) may be caused by folic acid deficiency or by vitamin B_{12} deficiency associated with the disease *pernicious anemia.* In this condition, a polypeptide "intrinsic factor" that is necessary for vitamin B_{12} absorption is not secreted as it normally is by the stomach. *Normocytic normochromic anemia* (normal MCV and MCHC) may be due to acute blood loss, hemolysis, aplastic anemia (damage to the bone marrow), or a variety of chronic diseases. *Microcytic hypochromic anemia* (abnormally low MCV and low MCHC), the most common type, is caused by inadequate amounts of iron.

PROCEDURE

1. Calculate the **mean corpuscular volume** (MCV) according to the following formula:

$$\text{MCV} = \frac{\text{hematocrit} \times 10}{\text{RBC count (millions per mm}^3 \text{ blood)}}$$

Example:
Hematocrit = 46
RBC count = 5.5 million

$$\text{MCV} = \frac{46 \times 10}{5.5} = 84$$

Calculate your mean corpuscular volume (MCV) and enter it in the laboratory report.

The normal adult **male** and **female** mean corpuscular volume (MCV) ranges from **82–92** cubic microns.

2. Calculate your **mean corpuscular hemoglobin concentration** (MCHC) according to the following formula:

$$\text{MCHC} = \frac{\text{Hemoglobin}\left(\text{g/dl}\right) \times 100}{\text{Hematocrit}}$$

Example
Hematocrit = 46
Hemoglobin = 16 g/dl

$$\text{MCHC} = \frac{16 \times 100}{46} = 35$$

Calculate your mean corpuscular hemoglobin concentration (MCHC) and enter it in the laboratory report.

The average normal adult **male** and **female** mean corpuscular hemoglobin concentration (MCHC) is **32–36** (in percent).

DATA FROM EXERCISE 6.1

A. Red Blood Cell Count

Write your red blood cell count per cubic millimeter (mm^3) of blood in the space below.

_____ per mm^3 of blood

B. Hematocrit

Write your hematocrit (in percent) in the space below.

_____

C. Hemoglobin Concentration

Enter your hemoglobin concentration in g per 100 mL (or deciliter, dL) of blood in the space below.

_____ g/dL of blood

D. MCV and MCHC

Calculate your mean corpuscular volume (MCV) and mean corpuscular hemoglobin concentration (MCHC) and enter these values in the spaces below.

MCV: _____ cubic microns (μ^3)

MCHC: _____ percent (%)

Compare your values to the normal values and write your conclusions in the space below.

REVIEW ACTIVITIES FOR EXERCISE 6.1

Test Your Knowledge of Terms and Facts

1. One hemoglobin molecule contains _____ heme groups; each heme group normally combines with one molecule of _____ .
2. The hormone _____ stimulates the bone marrow to produce red blood cells; this hormone is secreted by the _____ .
3. Old red blood cells are destroyed by the _____ system, which includes the following three organs:
   (a) _____
   (b) _____
   (c) _____
4. Heme derived from hemoglobin, minus the iron, is converted into a different pigment, known as _____; an accumulation of this pigment can cause a yellowing known as _____.
5. Define the term *hematocrit*. _____
6. The molecule formed by the binding of oxygen to deoxyhemoglobin: _____ .
7. A hemoglobin molecule containing oxidized iron (Fe^{3+}) is called _____ .
8. A molecule formed from the combination of hemoglobin and carbon monoxide is _____ .
9. A general term for an abnormally low red blood cell count or hemoglobin concentration: _____ .
10. The most common cause of the condition described in question 9 is _____ .

Test Your Understanding of Concepts

11. Describe some of the causes of anemia. Why is anemia dangerous?

12. Newborn babies, particularly premature ones, often have a rapid rate of red blood cell destruction and have jaundice. What is the relationship between these two conditions?

13. Could a person have a low hematocrit yet have a normal red blood cell count? Explain what might cause this condition.

Test Your Ability to Analyze and Apply Your Knowledge

14. Results of blood tests performed in this exercise would be different for anemia and for carbon monoxide poisoning, yet in one respect these two conditions are similar. Explain why this statement is true.

15. People who live at high altitudes often have a high red blood cell count, a condition called *polycythemia*. Explain the cause of the polycythemia, and its possible benefit. Do you think it could have any adverse effects? Explain.

White Blood Cell Count, Differential, and Immunity

MATERIALS

1. Microscopes, hemocytometer slides
2. Thoma diluting pipettes, lancets, alcohol swabs
3. For total white blood cell count: Methylene blue in 1% acetic acid; and for differential count: Wright's stain (or Harleco Diff-Quik or VWR Statstain)
4. Heparinized capillary tubes and glass slides

Textbook Correlations

Before performing this exercise, you may want to consult the following references in *Human Physiology*, seventh edition, by Stuart I. Fox:

- *The Formed Elements of Blood.* Chapter 13, pp. 368–370.
- *Defense Mechanisms.* Chapter 15, pp. 448–454.
- *Functions of B Lymphocytes.* Chapter 15, pp. 455–458.
- *Functions of T Lymphocytes.* Chapter 15, pp. 459–464.

Those using different physiology textbooks may want to consult the corresponding information in those books.

White blood cells—lymphocytes, monocytes, neutrophils, eosinophils, and basophils—are agents of the immune system. Lymphocytes provide immunity against specific antigens, whereas the other leukocytes are phagocytic. The total white blood cell count and the relative proportion of each type of white blood cell (differential count) change in a characteristic way in different disease states.

OBJECTIVES

1. Distinguish the different types of leukocytes by the appearance of their nuclei and their cytoplasm.
2. Describe the origin and function of B and T lymphocytes.
3. List the phagocytic white blood cells and explain their functions during local inflammation.
4. Perform a total and a differential white blood cell count and explain the importance of this information in the diagnosis of diseases.

The **white blood cells (leukocytes)** are divided into two general categories on the basis of their histological appearance: granular (or polymorphonuclear) and agranular. Leukocytes in the granular category have granules in the cytoplasm and lobed or segmented nuclei, whereas those in the agranular category lack visible cytoplasmic granules and have unlobed nuclei (see **plate 1**).

The *granular leukocytes* are distinguished by their affinity for specific stains. The cytoplasmic granules of **eosinophils** stain bright red (the color of eosin stain), and the granules of **basophils** stain dark blue (the color of basic stain). The granules of **neutrophils** have a low affinity for stain; therefore, the cytoplasm of these cells appears relatively clear.

The *agranular leukocytes* include **lymphocytes** and **monocytes.** Lymphocytes are the smaller of these two cell types and are easily identified by their round nuclei and scant cytoplasm. Larger monocytes have kidney-bean-shaped nuclei, often with brainlike convolutions, and their cytoplasm has a ground glass appearance. Monocytes may also be identified occasionally by the appearance of short, blunt, cytoplasmic extensions (pseudopods).

Leukocytes can leave the vascular system and enter the connective tissues of the body by squeezing through capillaries (a process known as *diapedesis* or *extravasation*). During an inflammation response, the release of *histamine* from tissue mast cells and basophils increases the permeability of the capillaries, and consequently promotes the process of diapedesis. This sequence of events also produces the local edema, redness, and pain associated with inflammation.

Neutrophils and, to a lesser degree, eosinophils, destroy the invading pathogens by phagocytosis. The battle is then joined by monocytes, which also enter the connective tissues and are transformed into voracious phagocytic cells known as *tissue macrophages.* The engorged white blood cells form pus in the inflamed area.

If these nonspecific immunological defenses are not sufficient to destroy the pathogens, lymphocytes may be recruited, and their specific actions used to reinforce the nonspecific immune responses. Lymphocytes are first produced in the embryonic bone marrow, which then seeds the other lymphopoietic sites: the *thymus, lymph nodes,* and *spleen.* The thymus, in turn, sends cells to other locations and apparently regulates the general rate of lymphocyte production at all these sites through the release of a hormone. Therefore, all lymphocytes may be categorized in terms of their ancestry as either bone marrow-derived **B cells** or thymus-derived **T cells.**

Antigens are molecules that activate the immune system. A specific *receptor protein* displayed on the outer membrane of each lymphocyte is capable of recognizing and binding to a specific antigen. By means of this bonding, the antigen selects the lymphocyte that is capable of attacking it. Bonding of the antigen to its membrane receptor protein stimulates that lymphocyte to divide numerous times, until a large population of genetically identical cells (a *clone*) is produced. This **clonal selection theory** accounts for the fact that the immune response to a second and subsequent exposures to an antigen is greater than the immune response to the initial exposure to the antigen.

When stimulated by antigens (generally bacterial), B lymphocytes (or B cells) develop into **plasma cells** that secrete large numbers of **antibody** molecules into the plasma, thus providing **humoral immunity.** These antibodies may destroy bacteria in one of two ways: (1) the antibodies coat the bacterial cell, making it more easily attacked by the phagocytic neutrophils and tissue macrophages; (2) the attachment of antibody to antigen on the bacterial surface activates a system of plasma proteins—*complement*—that lyses the bacterial cell. These two systems work together, since a chemical released from complement attracts the phagocytic white blood cells and increases capillary permeability. Inflammation can later be suppressed by eosinophils, which engulf free antigen-antibody complexes, thus preventing the complement reaction.

T lymphocytes (or T cells) do not secrete antibodies. Instead, they must move into close proximity with their victim cells in order to destroy them. Consequently, T lymphocytes are said to provide **cell-mediated immunity,** often involving the secretion of chemicals, *lymphokines,* released by some T cells. This cell-mediated immunity against cells infected with viruses, cancer cells, and cells of tissue transplants is again directed against specific antigens on the victim cell surface. Therefore, both T and B lymphocytes are specific in their immune attack, and indeed cooperate with each other in the immune defense against disease.

A. TOTAL WHITE BLOOD CELL COUNT

In this procedure, a small amount of blood is diluted with a solution that disintegrates the red blood cells (hemolysis) and lightly stains the white blood cells (WBC or leukocytes). The stained white blood cells are counted in the four large corner squares of a hemocytometer (see exercise 6.1, red blood cell count).

Because the dilution factor is 20 and each of the four squares counted has a volume of 0.1 cubic millimeters (mm^3), the number of white blood cells per cubic millimeter of blood can be calculated as follows:

$$WBC \text{ per } mm^3 = \frac{\text{white cells} \times 20}{4 \times 0.1 \ mm^3}$$

or,

$$WBC \text{ per } mm^3 = \text{white cells} \times 50$$

PROCEDURE

⚠️ *Caution: Because of the danger of exposure to the AIDS virus and other harmful agents when handling blood, each student should perform this and other blood exercises with his or her own blood only. All objects that have been in contact with blood must be discarded in a container indicated by the instructor.*

1. As described in exercise 6.1A, obtain a drop of blood after discarding the first drop, and fill the diluting pipette (the one with the white bead) to the *0.5 mark.* Avoid air bubbles; if too much blood is drawn into the pipette, remove it by touching the tip of the pipette to a filter paper.
2. Draw the diluting fluid to the *11 mark* on the pipette.
3. Shake or roll the pipette for 3 minutes.
4. Discard the first 4 drops, and fill the hemocytometer as described in exercise 6.1.
5. Allow the cells to settle for 1 minute; then using the low-power objective, count the number of white blood cells in the four large corner squares (labeled A, B, C, and D in fig. 6.5). If a cell is lying on the upper or left-hand line, include it in your count, but do not include cells that are touching the lower or right-hand line.
6. Calculate the number of white blood cells per cubic millimeter of blood and enter this value in the laboratory report.

The normal white blood cell count is 5,000–10,000 cells per cubic millimeter (mm^3) of blood.

B. DIFFERENTIAL WHITE BLOOD CELL COUNT

Clinically, it is also important to determine the relative quantity (percentage) of each leukocyte type within a population of white blood cells. This percentage is obtained by microscopic identification (differentiation) of each leukocyte type within a total count of 100 white blood cells (see plate 1).

PROCEDURE

Making a Blood Smear

1. Fill a heparinized capillary tube with blood. This can serve as a reservoir of blood for making a number of slides.
2. Using the capillary tube, apply a small drop of blood on one end of a glass slide that is *absolutely clean* and free of grease (fig. 6.7a). Place this slide flat on a laboratory bench.
3. Lower a second glass slide at an angle of 30° to the first slide, so that it is lightly touching the first slide *in front of* the drop of blood (fig. 6.7b).
4. Gently pull the second slide backwards into the drop of blood, maintaining the pressure and angle that allows the blood to spread out along the edge of the second slide (fig. 6.7b).
5. Keeping the same angle and pressure, push the second slide across the first in a rapid, smooth motion. The blood should now be spread in a thin film across the first slide. Done correctly, the concentration of blood in the smear should diminish toward the distal end, producing a feathered appearance (fig. 6.7c, d).

Figure 6.7 A procedure (a–d) for making a blood smear for a differential white blood cell count.

An increase in the white blood cell count (**leukocytosis**) may be produced by an increase in any one of the leukocyte types. These include: (1) *neutrophil leukocytosis,* due to appendicitis, rheumatic fever, smallpox, diabetic acidosis, or hemorrhage; (2) *lymphocyte leukocytosis,* due to infectious mononucleosis or chronic infections (such as syphilis); (3) *eosinophil leukocytosis,* due to parasitic diseases (such as trichinosis), psoriasis, bronchial asthma, or hay fever; (4) *basophil leukocytosis,* due to hemolytic anemia, chicken pox, or smallpox; and (5) *monocyte leukocytosis,* due to malaria, Rocky Mountain spotted fever, bacterial endocarditis, or typhoid fever. In certain cases, an increase in the relative abundance of one type of leukocyte may occur in the absence of an increase in the total white blood cell count—for example, lymphocytosis due to pernicious anemia, influenza, infectious hepatitis, rubella ("German measles"), or mumps.

A decrease in the white blood cell count (**leukopenia**) is usually due to either a decrease in the number of neutrophils or a decrease in the number of eosinophils. A decrease in the number of neutrophils occurs in typhoid fever, measles, infectious hepatitis, rubella, and aplastic anemia. *Eosinopenia* is produced by an elevated secretion of the corticosteroids, which occurs under various conditions of stress, such as severe infections and shock, and in adrenal hyperfunction (Cushing's syndrome).

PROCEDURE

Staining a Slide Using Wright's Stain

1. Place the slide on a slide rack and flood the surface of the slide with Wright's stain. Rock the slide back and forth gently for 1 to 3 minutes.

Note: *The stain is dissolved in methyl alcohol, which evaporates easily. If any part of the slide should dry during this procedure, the stain will precipitate, ruining the slide.*

2. Drip buffer or distilled water on top of the Wright's stain, being careful not to wash the stain off the slide. Mixing Wright's stain with water is crucial for proper staining; this mixing can be aided by gently blowing on the surface of the stain. Proper staining is indicated by the presence of a metallic sheen on the surface of the stain. The diluted stain should be left on the slide for a full 5 minutes.

3. Wash the stain off the slide with a jet of distilled water from a water bottle and allow the slide to drain at an angle for a few minutes.
4. Using the oil-immersion objective, count the different types of white blood cells. Start at one point in the feathered-tip (less populated) area and systematically scan the slide until you have counted a total of 100 leukocytes.
5. Keep a running count of the different leukocytes in the table provided on plate 1 and indicate the total number of each. Calculate the percentage of the total count contributed by each type of leukocyte, and enter these values in your laboratory report.

PROCEDURE

Alternative: Staining a Slide Using Diff-Quik (Harleco)

1. Dip the slide in fixative solution (light blue) five times, allowing 1 second per dip.
2. Dip the slide in solution 1 (orange) five times, allowing 1 second per dip.
3. Dip the slide in solution 2 (dark blue) five times, allowing 1 second per dip.
4. Rinse the slide with distilled water and count the white blood cells using the oil-immersion objective as described in steps 4 and 5 of the previous procedure. Enter these values in your laboratory report.

Laboratory Report 6.2

DATA FROM EXERCISE 6.2

A. Total White Blood Cell Count

1. Enter your white blood cell count in the space below.

   _____WBC per mm^3

2. Compare your measured values to the normal range and write your conclusions in the space below.

B. Differential White Blood Cell Count

1. Count each type of white blood cell (leukocyte) and record the number in the table below. Next, determine your grand total of leukocytes counted (roughly 100). Now, calculate the percentage of each type of white blood cell and enter these values in the table below (see plate 1 for normal values).

Leukocytes	Cells Counted	Total WBCs Counted	Percentage
Neutrophils			
Eosinophils			
Basophils			
Lymphocytes			
Monocytes			

2. Compare your values to the normal range and write your conclusions in the space below.

REVIEW ACTIVITIES FOR EXERCISE 6.2

Test Your Knowledge of Terms and Facts

Identify the leukocyte by the following description:

_____1. polymorphonuclear with poorly staining granules
_____2. agranular with round nucleus, little cytoplasm
_____3. granules with affinity for red stain
_____4. rarest white blood cell
_____5. agranular and phagocytic

(a) eosinophil
(b) neutrophil
(c) monocyte
(d) lymphocyte
(e) basophil

6. Antibodies are produced by _____ lymphocytes; cell-mediated immunity is provided by _____ lymphocytes.
7. While blood cells leave capillaries by a process called_____ .
8. The major phagocytic white blood cells are the _____ .
9. Molecules that activate the immune system are called _____ .

Test Your Understanding of Concepts

10. Distinguish between humoral and cell-mediated immunity, identifying the cells involved, their origin, and their functions.

11. Describe the clonal selection theory and explain how it accounts for the ability to defend against subsequent exposure to a particular antigen.

12. *Active immunizations* involve the exposure of a person to a pathogen whose *virulence* (ability to cause disease) has been reduced without altering its *antigenicity* (nature of its antigens). How do you think this might be accomplished? What are the benefits and dangers of this procedure?

13. *Passive immunizations* involve injecting a person who has been exposed to a pathogen with serum containing antibodies, called *antiserum* or *antitoxin*. Antiserum is developed by injecting an animal with a pathogen. What happens in that animal? What are the benefits and shortcomings of passive immunization compared to active immunization?

Blood Types

MATERIALS

1. Sterile lancets, 70% alcohol
2. Anti-A, anti-B, and anti-Rh sera
3. Slide warmer, glass slides, and toothpicks
4. Container for the disposal of blood-containing objects

Red blood cells (RBCs) have characteristic molecules on the surface of their membranes that can be different in different people. These genetically determined membrane molecules can function as antigens—capable of bonding to specific antibodies when exposed to plasma from a person with a different blood type. The major blood group antigens are the Rh antigen and the antigens of the ABO system.

OBJECTIVES

1. Explain what is meant by the term *blood type,* and identify the major blood types.
2. Explain how agglutination occurs, and how agglutination tests can be used to determine a person's blood type.
3. Identify the different genotypes that can produce the different blood group phenotypes, and explain how different blood types can be inherited.
4. Explain how erythroblastosis fetalis is produced.
5. Explain the dangers of mismatched blood types in blood transfusions.

Textbook Correlations

Before performing this exercise, you may want to consult the following references in *Human Physiology*, seventh edition, by Stuart I. Fox:

- *Red Blood Cell Antigens and Blood Typing,* Chapter 13, pp. 372–374.

Those using different physiology textbooks may want to consult the corresponding information in those books.

When blood from one person is mixed with plasma from another person, the red blood cells will sometimes **agglutinate,** or clump together (fig. 6.8). This agglutination reaction, which is very important in determining the safety of transfusions (agglutinated cells can block small blood vessels), is due to a mismatch of genetically determined blood types.

On the surface of each red blood cell are a number of molecules that have antigenic properties, and in the plasma each antibody molecule has two combining sites for antigens. In a positive agglutination test, the red blood cells clump together because they are combined through antibody bridges.

A. THE RH FACTOR

One of the antigens on the surface of red blood cells is the **Rh factor** (named because it was first discovered in rhesus monkeys). The Rh factor is found on the red blood cell membranes of approximately 85% of the people in the United States. The presence of this antigen on the red blood cells (an **Rh positive** phenotype) is inherited as a dominant trait and is produced by both the *homozygous* (RR) genotype and the *heterozygous* (Rr) genotype. Individuals who have the *homozygous recessive* genotype (rr) do not have this antigen on their red blood cells and are said to have the **Rh negative** phenotype.

Suppose an Rh positive man who is heterozygous (Rr) mates with an Rh positive woman who is also heterozygous.

Since the mother is Rh positive, her immune system cannot be stimulated to produce antibodies by the presence of an Rh positive fetus. The development of

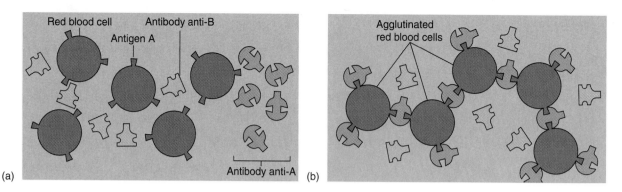

Figure 6.8 An agglutination reaction. (a) Type A red blood cells are mixed with anti-A antibodies. (b) This results in the formation of antigen-antibody bridges that cause the red blood cells to clump together (agglutinate).

immunological competence does not occur until shortly after birth, so that an Rh negative fetus in an Rh positive mother would not yet have an immune response during normal gestation (pregnancy).

However, when an Rh negative mother is carrying an Rh positive fetus, some of the Rh antigens may enter her circulation when the placenta tears at birth (red blood cells do not normally cross the placenta during pregnancy). Since these red blood cells express an antigen (the Rh factor) that is foreign to the mother, her immune system will eventually be stimulated to produce antibodies that are capable of destroying the red blood cells of subsequent Rh positive fetuses, a condition known as *hemolytic disease of the newborn,* or **erythroblastosis fetalis.** However, erythroblastosis fetalis can be prevented by the administration of exogenous Rh antibodies known as **Rho(D) immune globulin** (e.g., *RhoGAM*) to the mother within 72 hours after delivery. These antibodies destroy the fetal Rh positive red blood cells that have entered the maternal circulation before they can stimulate an immune response in the mother.

PROCEDURE

1. Place one drop of anti-Rh serum on a clean glass slide.
2. Add an equal amount of fingertip blood and mix it with the antiserum (use an applicator stick or a toothpick).
3. Place the slide on a slide warmer (45°C to 50°C) and rock it back and forth.
4. Examine the slide for agglutination. If no agglutination is observed after a 2-minute period, examine the slide under the low-power objective of the microscope. The presence of grains of agglutinated red blood cells indicates Rh positive blood.

⚠️ **Caution:** *Handle only your own blood and be sure to discard the slide, toothpicks, and lancet in the container provided by the instructor.*

5. Enter your Rh factor type (positive or negative) in the laboratory report.

Table 6.1 Incidence of Blood Types—Approximate Incidence in the U.S. (%)

Blood Types	Caucasian	Black	Asian
O	45	48	36
A	41	27	28
B	10	21	23
AB	4	4	13

B. THE ABO ANTIGEN SYSTEM

Each individual inherits two genes, one from each parent, that control the synthesis of red blood cell antigens of the ABO classification. Each gene contains the information for one of three possible phenotypes: antigen A, antigen B, or no antigen (written O). Thus, an individual may have one of six possible genotypes: **AA, AO, BB, BO, AB,** or **OO.**

An individual who has the genotype AO will produce type A antigens just like an individual who has the genotype AA; and therefore both are said to have **type A** blood. Likewise, an individual with the genotype BO and one with the genotype BB will both have **type B** blood. Since lack of antigen is a recessive trait, an individual with **type O** blood must have the genotype OO.

Unlike many other traits, the heterozygous genotype AB has a phenotype that is different from either of the homozygous genotypes (AA or BB). Since there is no dominance between A and B, individuals with the genotype AB produce red blood cells with *both* the A and B antigens (a condition known as *codominance*) and have **type AB** blood. The most common blood types are type O and type A; the rarest is type AB (table 6.1).

Also, unlike the other immune responses considered, antibodies against the A and B antigens are not induced by prior exposure to these blood types. A person with type A blood, for example, has antibodies in the plasma against type B blood even though that person may never have been exposed to this antigen. A transfusion with type B blood into the type A person would be extremely dangerous because the anti-B antibodies in the recipient's plasma would agglutinate the red blood cells in the donor's blood. The outcome would be the same if the donor were type A and the recipient type B (see **plate 2**).

Antigen on RBC Surface	Antibody in Plasma
A (type A)	Anti-B
B (type B)	Anti-A
O (type O)	Anti-A and anti-B
AB (type AB)	No antibody

⚠️ **PROCEDURE**

1. Draw a line down the center of a clean glass slide with a marking pencil and label one side **A** and the other side **B.**
2. Place a drop of anti-A serum on the side marked **A** and a drop of anti-B serum on the side marked **B.**
3. Add a drop of blood to each antiserum and mix each with a separate applicator stick.
4. Tilt the slide back and forth and examine for agglutination over a 2-minute period. *Do not heat the slide on the slide warmer.*
5. Enter your ABO blood type in the laboratory report.

Laboratory Report 6.3

Name _____

Date _____

Section _____

DATA FROM EXERCISE 6.3

1. Did your blood agglutinate with the anti-Rh serum? _____
   Are you Rh positive or negative? _____
2. Indicate below (with a yes or no) whether your blood agglutinated with the anti-A and anti-B sera.
   Anti-A: _____
   Anti-B: _____
   What is your blood type? _____

REVIEW ACTIVITIES FOR EXERCISES 6.3

Test Your Knowledge of Terms and Facts

1. Name the antigens present and absent on the surface of a red blood cell if the person is:
   (a) type A negative_____
   (b) type O positive _____
   (c) type AB negative _____
2. If a person has blood type A, the possible genotypes that the person may have are _____
   and _____ .
3. If a person who is blood type O marries a person who is blood type A, what are the possible blood types their children could have?
   _____ or _____
4. The universal blood donor is blood type _____ .
5. The rarest blood type is blood type _____ .
6. The most common Rh type is _____ .
7. The person most in danger of having a child who develops erythroblastosis fetalis is a woman who has the blood type _____ when her husband has the blood type _____ .

Test Your Understanding of Concepts

8. What are the dangers of giving a person a transfusion when the blood types don't match?

9. Explain how hemolytic disease of the newborn is produced. How may the disease be prevented?

10. Can blood types be used in paternity cases to prove or disprove possible fatherhood? Give examples to support your answer.

11. Suppose a person who has type A blood receives large amounts of whole blood from a person who has the universal donor blood type. Will that be safe? Explain.

Blood Clotting System

MATERIALS

1. Pipettes (0.10–0.20 mL), and small test tubes
2. Constant-temperature water bath set at 37°C
3. 0.02 M calcium chloride, activated thromboplastin, activated cephaloplastin (Dade), fresh plasma

Two interrelated clotting pathways—the intrinsic system and the extrinsic system—require the successive activation of specific plasma clotting factors. Defects in these factors can be detected by means of two clotting-time tests.

OBJECTIVES

1. Describe the intrinsic and extrinsic clotting systems.
2. Describe why bleeding time is prolonged in cases of hemophilia and vitamin K deficiency.
3. Demonstrate the tests for prothrombin time and for activated partial thromboplastin time (APTT). Identify the normal values for each, and explain how these tests are used to diagnose bleeding disorders.

Textbook Correlations

Before performing this exercise, you may want to consult the following references in *Human Physiology*, seventh edition, by Stuart I. Fox:

• *Blood Clotting*. Chapter 13, pp. 374–377.

Those using different physiology textbooks may want to consult the corresponding information in those books.

Damage to a blood vessel initiates a series of events that, if successful, culminate in *hemostasis* (the arrest of bleeding).

1. The first event is **vasoconstriction,** which decreases the flow of the blood in the damaged vessel.
2. The next event is the formation of a **platelet plug.** This response occurs in two steps:
   a. In the first step, platelets adhere to the exposed collagen (connective tissue protein) of the damaged vessel and then release *adenosine diphosphate (ADP)*.
   b. The second step occurs when the ADP, by making the adherent platelets sticky, causes other platelets to cling at this site, forming a platelet clump.
3. The third event is the sequential activation of **clotting factors** in the plasma, resulting in the formation of an insoluble fibrous protein, *fibrin*, around the platelet clump. This produces a blood clot.

The formation of fibrin from its precursor, *fibrinogen*, requires the presence of the enzyme *thrombin*.

$$\text{fibronogen} \xrightarrow{\text{thrombin}} \text{fibrin}$$

The insoluble fibrin is formed instantly whenever the enzyme thrombin is present, and thus the formation of thrombin must be a carefully regulated event in the body. The formation of thrombin from its precursor, *prothrombin*, requires the sequential activation of a number of other clotting factors.

When the sequence of events leading to the formation of thrombin is initiated by the release of *tissue thromboplastin* from damaged tissue cells, fibrin is rapidly formed (10–15 seconds). These events constitute the **extrinsic system** of blood clotting, since the sequence is initiated by a factor extrinsic to the blood (fig. 6.9). Alternatively, a blood clot may be initiated in the absence of tissue thromboplastin through the activation of *Hageman factor* (factor XII) by the exposure of plasma to glass, crystals, or collagen. This **intrinsic system** of blood clotting is slower (28–45 seconds) than the extrinsic system (fig. 6.9).

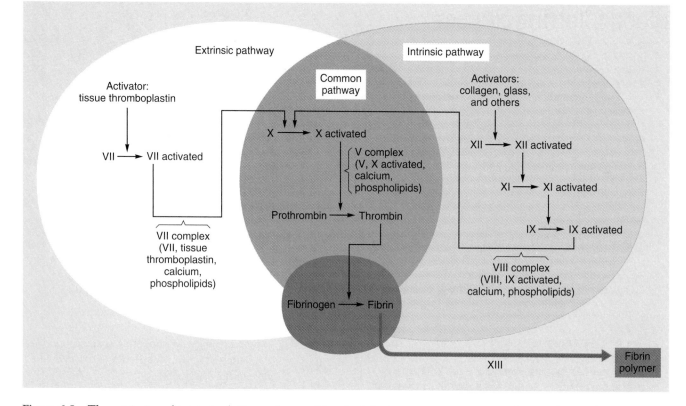

Figure 6.9 The extrinsic and intrinsic clotting pathways. These both lead to a common sequence of events that results in the formation of insoluble threads of fibrin polymers.

Many people have either an acquired or an inherited inability to form fibrin threads within the normal time interval. This inability may be due to a **vitamin K deficiency,** since vitamin K is necessary for the formation of four of the clotting factors, including prothrombin. Vitamin K deficiency may occur in the newborn who has an inadequate intake of milk, in a person with fat malabsorption due to inadequate absorption of this vitamin (bile salts facilitate absorption), as a result of antibiotic therapy destroying intestinal flora (a source of vitamin K), or as a result of oral anticoagulants such as *dicumarol.*

There are many hereditary conditions in which a clotting factor is either missing or defective. The best known of these conditions, classical **hemophilia,** is due to the genetic inability to synthesize normal factor VIII. This condition, as well as *Christmas disease* (defective factor IX), is inherited as a sex-linked recessive trait. Other genetic defects inherited as autosomal traits include those associated with factors II, VII, X, XI, and XII.

In this exercise, two tests will be performed to screen for defective clotting factors. The formation of thrombin in the plasma samples will be inhibited by an anticoagulant. The anticoagulant used, either *citric acid* or *oxalic acid,* removes calcium ion (Ca^{2+}) from the plasma. The removal of Ca^{2+} has an anticoagulant effect because calcium is a necessary cofactor in the activation of a number of the clotting factors. This inhibition can be easily reversed by adding calcium ion during the clotting tests.

The test for **prothrombin time** is used to determine deficiencies in the *extrinsic* clotting system and is prolonged when factor V, VII, or X is defective. The test for **activated partial thromboplastin time (APTT)** is used to determine deficiencies in the *intrinsic* clotting system and is sensitive to all defective factors except factor VII (tissue thromboplastin). A person with classical hemophilia (defective factor VIII), for example, would have a normal prothrombin time but an abnormal APTT. In this way, these two tests complement each other and can be used, together with other tests, to determine the exact cause of prolonged clotting time (see table 6.2).

Table 6.2 Test Results for Clotting Factors

Defective Factor	Prothrombin Time	APTT
V	Abnormal	Abnormal
VII	Abnormal	Normal
VIII	Normal	Abnormal
IX	Normal	Abnormal
X	Abnormal	Abnormal
XI	Normal	Abnormal
XII	Normal	Abnormal

A. TEST FOR PROTHROMBIN TIME

PROCEDURE

1. Pipette 0.10 mL of activated thromboplastin and 0.10 mL of 0.02 M $CaCl_2$ into a test tube. Place the tube in a 37°C water bath and allow it to warm for at least 1 minute.
2. Warm a sample of plasma in the water bath for at least 1 minute. Then use a pipette to forcibly expel 0.10 mL of plasma into the warmed tube containing the thromboplastin–$CaCl_2$ mixture. Start timing at this point.
3. Agitate this tube mixture continuously in the water bath for 10 seconds.
4. Remove the tube, quickly wipe it, and hold it in front of a bright light. Tilt the tube gently back and forth and stop timing when the first fibrin threads appear (the solution will change from a fluid to a semigel). Enter the time in your laboratory report.

The normal prothrombin time is 11 ± 1 seconds.

B. TEST FOR ACTIVATED PARTIAL THROMBOPLASTIN TIME (APTT)

PROCEDURE

1. Warm a tube of 0.02 M $CaCl_2$ by placing it in a 37°C water bath.
2. Pipette 0.10 mL of activated cephaloplastin and 0.10 mL of plasma into a test tube and allow it to incubate at 37°C for 3 minutes.
3. Using a pipette, forcibly expel 0.10 mL of warmed $CaCl_2$ into the cephaloplastin-plasma mixture. Start timing at this point.
4. Agitate this tube mixture continuously in the 37°C water bath for 30 seconds. Then remove the tube, quickly wipe it, and hold it against a bright light while rocking the tube back and forth.
5. Stop timing when the first fibrin threads appear. Enter the time in your laboratory report.

The normal APTT is less than 40 seconds.

Laboratory Report 6.4

Name _____

Date _____

Section _____

DATA FROM EXERCISE 6.4

A. Prothrombin Time

Enter your prothrombin time measurement in the space below.
_____ seconds

B. Activated Partial Thromboplastin Time (APTT)

1. Enter your APTT measurement in the space below.
_____ seconds

2. Compare your data to the normal measurements and write your conclusions in the space below.

REVIEW ACTIVITIES FOR EXERCISE 6.4

Test Your Knowledge of Terms and Facts

1. The factor that starts the extrinsic clotting pathway: _____ .
2. Which clotting pathway is faster? _____
3. The factor that converts fibrinogen into fibrin is_____ .
4. The factor name in question 3 is derived from_____ .
5. Which vitamin is needed for the formation of some of the clotting factors?_____
6. Citric acid (citrate) is an anticoagulant because it _____ .
7. The _____ test is used to detect defects in the extrinsic clotting system, whereas the _____ test is used to detect defects in the intrinsic clotting system.
8. In the formation of a platelet plug, platelets release _____, which makes other platelets sticky.
9. The general category of disorders in which a person's clotting time is abnormally long;
_____ .

10. Which four factors are common to both the intrinsic and extrinsic clotting pathway? Which clotting test(s) would be abnormally prolonged if a person had a deficiency in factor VII? Explain.

11. Which factor(s) would be defective if a person had a prolonged prothrombin time but a normal partial thromboplastin time (APTT)? Explain.

Test Your Ability to Analyze and Apply Your Knowledge

12. Heparin is a mucopolysaccharide extracted from beef lung and liver that inhibits the action of thrombin. What effect would thrombin have on the prothrombin time and APTT? Why was citric acid or oxalic acid used as an anticoagulant instead of heparin in these tests?

13. Why might a person with an abnormally slow clotting time be given vitamin K? Would treatment with vitamin K immediately improve the clotting time? Explain.

14. You might expect that almost all people with hemophilia due to factor VIII or IX deficiency would be males, whereas those with hemophilia due to factor XI or XII deficiency would be equally like to be males or females. Explain.

Neutrophils

Eosinophils

Basophils

Lymphocytes

Monocytes

Platelets
(thrombocytes)

Erythrocytes

The average differential count in the normal adult
is as follows:

Neutrophils	55%–75%
Eosinophils	2%–4%
Basophils	0.5%–1%
Lymphocytes	20%–40%
Monocytes	3%–8%

Leukocyte	Cells Counted	Total	Percent
Neutrophils			
Eosinophils			
Basophils			
Lymphocytes			
Monocytes			

Plate 1   Formed elements of blood.

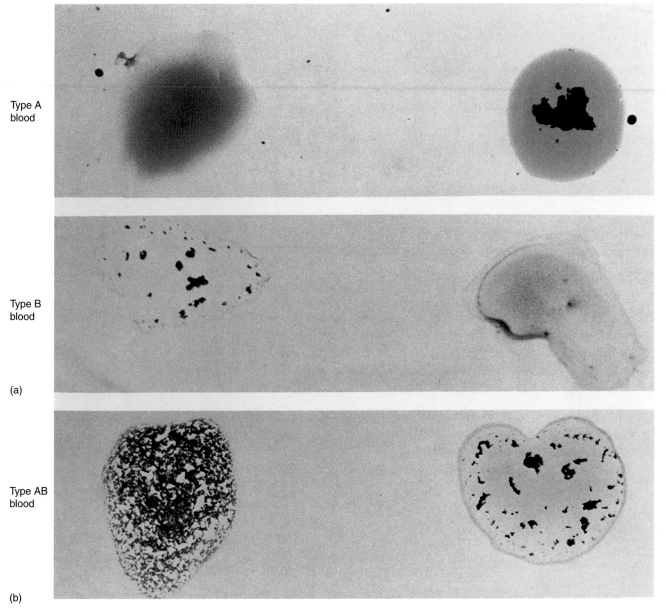

Plate 2 Results of blood typing. (a) Type A blood agglutinating (clumping) with antiserum A (above right), and type B blood agglutinating with antiserum B (below left). Type O blood (not shown) would not have agglutinated with either antiserum. (b) Type AB blood agglutinating with both antiserum B (left) and antiserum A (right).

The Cardiovascular System

The blood transports glucose, amino acids, fatty acids, and other monomers from the digestive tract, liver, and adipose tissue to all the cells of the body. The waste products of cellular metabolism are carried by the blood to the kidneys and lungs for elimination. Hormones, secreted by endocrine glands, are carried by the blood to target organs. Blood is thus the major channel of communication between the different specialized organs of the body.

The interchange of molecules between blood and tissue cells occurs across the walls of **capillaries,** which are composed of only a single layer of epithelial cells (endothelium). Blood is delivered to the capillaries in **arterioles,** microscopic vessels with walls of endothelium, smooth muscle, and connective tissue. The arterioles receive their blood from larger, more muscular **arteries.** Blood is drained from the capillaries into microscopic **venules.** The venules drain their blood into larger **veins** that are less muscular and more distensible than arteries.

Since the tissue cells must be located within 0.10 mm of a capillary for molecules to diffuse adequately, the vascular system within an organ is highly branched. The many narrow, muscular arterioles of this *vascular tree* offer great resistance to blood flow (*peripheral resistance*) through the organs. For organs to receive an adequate blood flow (*perfusion*), the arterial blood must be under sufficient pressure to overcome this resistance to blood flow. This arterial pressure is routinely measured with a device known as a *sphygmomanometer.*

The blood pressure required to overcome peripheral resistance and maintain adequate tissue perfusion is generated by a muscular pump—the *heart.* The heart has four chambers: two atria and two ventricles. The *right atrium* receives oxygen-depleted blood returning from body cells in the superior and inferior venae cavae; and the *left atrium* receives oxygen-rich blood from the pulmonary veins. The *right ventricle* pumps blood into the pulmonary arteries to the lungs where oxygen enters and carbon dioxide exits the blood. The *left ventricle* pumps blood into the large *aorta,* which by means of its many branches perfuses all the organs in the body. Since the blood pumped out of the heart by the two ventricles, the **cardiac output,** is carried by arteries to the body's organs, and since blood from the organs is returned by veins to the heart, the cardiovascular system forms a closed circle called the **circulatory system** (fig. 7.1).

The heart's ability to maintain adequate perfusion of the body's organs depends on proper electrical stimulation and muscular contraction, proper functioning of *valves* (directing blood flow within the heart), and integrity of the blood vessels. These functions can be assessed by various techniques, which will be explored in the following exercises.

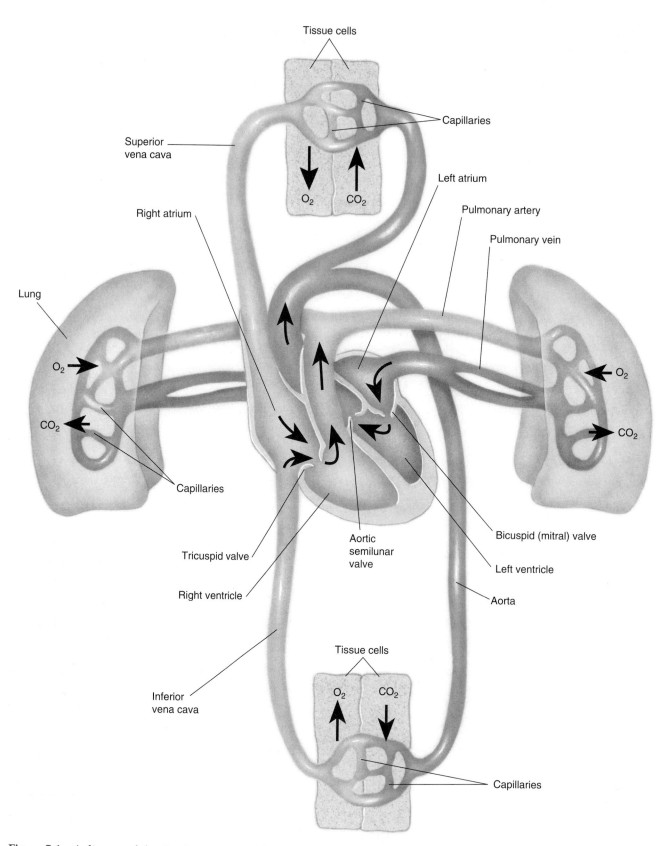

Figure 7.1   A diagram of the circulatory system. The pulmonary arteries and veins compose the pulmonary circulation, whereas other arteries and veins are part of the systemic circulation. The right ventricle pumps blood into the pulmonary circulation, while the left ventricle pumps blood into the systemic circulation.

Effects of Drugs
on the Frog Heart

MATERIALS

1. Frogs, dissecting instruments, trays
2. Copper wire or bent pin, thread
3. Recording apparatus: physiograph, transducer coupler, and myograph transducer (Narco); or kymograph, kymograph paper, and kerosene burner
4. Ringer's solution (see exercise 5.1—all drugs to be prepared using Ringer's solution as a solvent); calcium chloride (2.0 g/100 mL); digitoxin (0.2 g/ 100 mL); pilocarpine (2.5 g/100 mL); atropine (5.0 g/100 mL); potassium chloride (2.0 g/100 mL); epinephrine (0.01 g/100 mL); caffeine (0.2 g/dL); and nicotine (1.0 g/2L; or 6.16 ml liquid/dL).

> The heart of a pithed frog may continue to beat automatically after the frog's central nervous system has been destroyed. By this means, the function of the heart and the effects of various drugs on the heart can be studied.

OBJECTIVES

1. Describe the pattern of contraction in the frog heart.
2. Describe the effect of various drugs on the heart and explain their mechanisms of action.

Textbook Correlations*

Before performing this exercise, you may want to consult the following references in *Human Physiology*, seventh edition, by Stuart I. Fox:

- *Cardiac Muscle.* Chapter 12, p. 354.
- *Structure of the Heart.* Chapter 13, pp. 379–381.
- *Electrical Activity of the Heart.* Chapter 13, pp. 385–387.

Those using different physiology textbooks may want to consult the corresponding information in those books.

A **drug** is a substance that affects some aspects of physiology when given to the body. Drugs may be identical to naturally occurring substances found in the body, such as minerals, vitamins, and hormones, or they may be molecules uniquely produced by particular plants or fungi. Many drugs marketed by pharmaceutical companies are derived from natural products whose chemical structure has been slightly modified to alter the biological activity of the native compounds.

The biological effects of *endogenous* compounds (those compounds normally found in the body) vary with their concentration. A normal blood potassium concentration, for example, is necessary for good health, but too high a concentration can be fatal. Similarly, the actions exhibited by many hormones at abnormally high concentrations may not occur when the hormones are at normal concentrations. It is important, therefore, to distinguish between the **physiological effects** (normal effects) of these substances and their **pharmacological effects** (those that occur when the substances are administered as drugs). A study of the pharmacology of various substances however, can reveal much about the normal physiology of the body.

In this exercise, we will test the effects of various pharmacological agents on the heart of a pithed frog. Although the heart, like skeletal muscle, is striated, it differs from skeletal muscles in several respects. The heartbeat is automatic; it does not have to be stimulated by nerves or electrodes to contract. Action potentials begin spontaneously in the *pacemaker region*—the *sinoatrial* or *SA node* region—of the right atrium and spread through the ventricles in an automatic, rhythmic cycle. As can be seen in the exposed frog heart, this causes the atria to contract before the ventricle. (Note that, unlike mammals, frogs have only one ventricle.)

When the frog heart is connected by a thread to the recording equipment, contractions of the atria and ventricle produce two successive peaks in the recordings. The

---

*See Appendix 3 for correlations to the A.D.A.M. *InterActive PHYSIOLOGY Modules*.

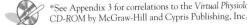 *See Appendix 3 for correlations to the *Virtual Physiology Laboratory* CD-ROM by McGraw-Hill and Cypris Publishing, Inc.

Figure 7.2 Procedure for exposing the frog heart. (a) First the skin is cut. (b) The body cavity is exposed by cutting through the muscles to the sternum, which is then split to expose the heart.

strength of contraction is related to the amplitude (height) of these peaks, and the rate of beat can be determined by the distance between the ventricular peaks if the chart speed is known. The rate of impulse conduction between the atria and ventricle is related to the distance between the atrial and ventricular peaks in the recording of each cycle. Therefore, the effects of various drugs on the strength of contraction, rate of contraction, and rate of impulse conduction from the atria to the ventricle can be determined.

PREPARATION FOR RECORDING

1. Double-pith a frog, and expose its heart (see exercise 5.1 and fig. 7.2). Skewer the apex of the heart muscle with a short length of thin copper wire or a bent pin, being careful not to let the wire enter the chamber of the ventricle. (The frog heart has only one ventricle and two atria.)
2. Bend the copper wire into a loop, and tie one end of cotton thread to this loop or to the head of a bent pin (see the enlarged insert in fig. 7.3).
3. Procedure for **kymograph** recording:
   (a) Tie the other end of the thread to a heart lever. The thread tension should be fairly taut so that contractions of the heart produce movements of the lever.
   (b) Attach kymograph paper (shiny side out) to the kymograph drum, and rotate the drum slowly over a kerosene burner until the paper is uniformly blackened. Arrange the heart lever so that it lightly drags across the smoked paper. Too much pressure of the writing stylus against the kymograph will prevent movement of the heart lever. (See fig. 7.3 for the proper setup.)

Figure 7.3 Frog heart setup. The contractions of the heart pull a lever that writes on a moving chart (kymograph). The setup using electronic recording equipment, such as a physiograph recorder, is similar to that shown here, except that the string from the heart will be connected to a myograph transducer.

4. Procedure for **physiograph** recording (see physiograph, exercise 5.1):
   (a) Tie the other end of the thread to the hook below the myograph transducer. (Make sure the myograph is plugged into the transducer coupler on the physiograph.) The heart should be positioned directly below the myograph. Adjust the height of the myograph on its stand so that the heart is pulled out of the chest cavity (fig. 7.4*b*).
   (b) Make sure the physiograph is properly balanced and set the paper speed at *0.5 cm per second*. Depress the record button, and press the gray paper advance button when ready to record.

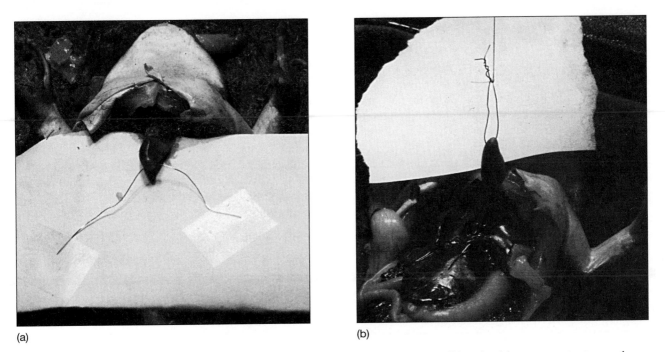

(a)   (b)

Figure 7.4 **Procedure for setting up the frog heart to record its contractions.** (a) A small length of thin copper wire is passed through the tip of the ventricle. (b) This wire is then twisted together to form a loop, which is tied by a cotton thread to the hook in the myograph transducer.

Figure 7.5 **Recordings of the frog heart contractions.** The arrow points to a recording of a smaller atrial contraction, which is followed by a recording of a larger ventricular contraction.

5. Observe the pattern of the heartbeat prior to the addition of drugs; this tracing will serve as the normal, or *control*, record (fig. 7.5). Distinguish the atrial from the ventricular beats. Measure the heart rate (in beats per minute), the strength of contraction (in millimeters deflection above baseline), and the distance (in millimeters) between atrial and ventricular peaks of the heartbeat.

Record this data in the laboratory report.

A. EFFECT OF CALCIUM IONS ON THE HEART

In addition to the role of calcium in coupling excitation to contraction, the extracellular Ca^{2+}/Mg^{2+} ratio also affects the permeability of the cell membrane. An increase in the extracellular concentration of calcium (above the normal concentration of 4.5–5.5 mEq/L) affects both the electrical properties and the contraction strength of heart muscle.

The heart is affected in a number of ways by an increase in extracellular calcium. These include (1) an increased force of contraction, (2) a decreased cardiac rate, and (3) the appearance of ectopic pacemakers in the ventricles, producing abnormal rhythms (extrasystoles and idioventricular rhythm—see exercise 7.2).

PROCEDURE

1. Obtain a control record of the normal heartbeat. Then, while the paper continues to run, use a dropper to bathe the heart in a 2.0% solution of calcium chloride ($CaCl_2$). On the moving recording paper, indicate the time at which calcium was added. Observe the effects of the added calcium solution over a period of a few minutes, then stop the recording.

2. Rinse the heart thoroughly with Ringer's solution until the heartbeat returns somewhat to normal; this new rhythm will serve as the next control.

3. In the table of your laboratory report, tape the recording or draw a facsimile of the normal heartbeat and of the changed heartbeat after the calcium solution was added.

B. Effect of Digitalis on the Heart

The effects of digitalis are believed to be due to its inhibition of the Na^+/K^+ (ATPase) pump. This inhibition results in an influx of Na^+ and an efflux of K^+ and is accompanied by an enhanced uptake of calcium ions. The effects of digitalis and of increased extracellular calcium on the heart are thus very similar.

Digitalis glycosides, such as digoxin, or digitalis, are frequently used to treat congestive heart failure, atrial flutter, and atrial fibrillation. Digitalis relieves these conditions by (1) increasing the force of contraction, (2) decreasing the cardiac rate directly by inhibiting the SA node, and (3) by slowing conduction through the bundle of His.

PROCEDURE

1. Obtain a record of the control heartbeat. Then, bathe the heart in a 2.0% solution of digitalis.
2. Rinse the heart thoroughly with Ringer's solution until the heartbeat returns somewhat to normal; this new rhythm will serve as the next control.
3. In the table of your laboratory report, tape the recording or draw a facsimile of the normal heartbeat and of the changed heartbeat after the digitalis solution was added.

C. Effect of Pilocarpine on the Heart

Pilocarpine is termed a *parasympathomimetic* drug because it mimics the effect of parasympathetic nerve stimulation. Pilocarpine acts to facilitate the release of the neurotransmitter acetylcholine from the vagus nerve, resulting in a marked decrease in the cardiac rate.

PROCEDURE

1. Obtain a record of the control heartbeat. Then, bathe the heart in a 2.5% solution of pilocarpine.

⚠ *Caution: Pilocarpine is very effective; be prepared to add atropine if necessary to counter the effects of pilocarpine (see atropine procedure, next).*

2. Rinse the heart thoroughly with Ringer's solution until the heartbeat returns somewhat to normal; this new rhythm will serve as the next control.
3. In the table of your laboratory report, tape the recording or draw a facsimile of the normal heartbeat and of the changed heartbeat after the pilocarpine solution was added.

D. Effect of Atropine on the Heart

Atropine is an alkaloid drug derived from the nightshade plant *Atropa belladonna* (the species name, *belladonna*, is often also used as the drug name). Atropine blocks the acetylcholine receptors of postganglionic parasympathetic neurons. Thus, atropine inhibits the effects of parasympathetic activity on the heart, smooth muscles, and glands. If the cardiac rate is decreased as a result of vagal stimulation (or the presence of pilocarpine), the administration of atropine will increase this rate.

The ability of **atropine** to block the effects of parasympathetic nerves is useful clinically. Atropine is used, for example, in ophthalmology to dilate the pupils (parasympathetic nerve activity causes constriction of the pupils) and in surgery to dry the mouth, pharynx, and trachea (parasympathetic nerve activity stimulates glandular secretions that wet these mucous membranes).

PROCEDURE

1. Bathe the heart in a 5.0% solution of atropine *while it is still under the influence of pilocarpine.*
2. Record the results; then stop recording and rinse the heart thoroughly with Ringer's solution.
3. In the table of your laboratory report, tape the recording or draw a facsimile of the normal heartbeat and of the changed heartbeat after the atropine solution was added.

E. Effect of Potassium Ions on the Heart

Since the resting membrane potential of all cells is dependent in large part on the maintenance of a higher concentration of potassium ions (K^+) on the inside of the cell than on the outside, an increase in the concentration of extracellular K^+ results in a *decrease in the resting membrane potential* (the potential becomes more positive). This, in turn, produces a decrease in the force of contraction and a slower conduction rate of the action potentials. In **hyperkalemia** (high blood potassium), the strength of myocardial contractions is weakened and the cardiac cells become more electrically excitable because the resting potential has risen closer to the threshold required for generating action potentials. In extreme hyperkalemia, the conduction rate may be so depressed that ectopic pacemakers appear in the ventricles and fibrillation may develop.

PROCEDURE

1. Obtain a record of the control heartbeat. Then, bathe the heart in a 2.0% solution of potassium chloride (KCl).
2. Record the results; then stop recording and rinse the heart thoroughly with Ringer's solution.
3. In the table of your laboratory report, tape the recording or draw a facsimile of the normal heartbeat and of the changed heartbeat after the potassium solution was added.

F. EFFECT OF EPINEPHRINE ON THE HEART

Epinephrine is a hormone secreted by the adrenal medulla. Together with norepinephrine, epinephrine is released in response to sympathetic nerve stimulation. Epinephrine acts to increase both the strength of contraction (contractility) of the heart and the cardiac rate. Exogenous ("from outside") epinephrine is a *sympathomimetic drug*, since it mimics the effect of sympathetic nerve stimulation.

PROCEDURE

1. Obtain a record of the control heartbeat. Then, bathe the heart in epinephrine (adrenaline).
2. Record the results; then stop recording and rinse the heart thoroughly with Ringer's solution.
3. In the table of your laboratory report, tape the recording or draw a facsimile of the normal heartbeat and of the changed heartbeat after the epinephrine solution was added.

G. EFFECT OF CAFFEINE ON THE HEART

Caffeine is a mild central nervous system (CNS) stimulant that also acts directly on the myocardium to increase both the strength of contraction and the cardiac rate. Caffeine inhibits activity of the enzyme *phosphodiesterase*, which breaks down a second messenger molecule called *cyclic AMP (cAMP)* that is present in many cells. As a result, the concentration of cAMP rises in heart cells, therefore duplicating the action of the hormone epinephrine that utilizes cAMP as a second messenger. Caffeine's usefulness as a central nervous system stimulant is limited because, in high doses, it can promote the formation of ectopic pacemakers (foci), resulting in serious arrhythmias (see exercise 7.2).

PROCEDURE

1. Obtain a record of the control heartbeat. Then, bathe the heart with a saturated solution of caffeine.
2. Record the results; then stop recording and rinse the heart thoroughly with Ringer's solution.
3. In the table of your laboratory report, tape the recording or draw a facsimile of the normal heartbeat and of the changed heartbeat after the caffeine solution was added.

H. EFFECT OF NICOTINE ON THE HEART

Nicotine promotes electrochemical transmission at the autonomic ganglia by stimulating particular nicotinic receptors for *acetylcholine* in the postganglionic neurons. When applied directly to the heart, the major effect of nicotine will be stimulation of parasympathetic ganglia located within the epicardium. Activation of postganglionic parasympathetic neurons, in turn, will cause slowing of the heart rate. When nicotine is administered into the blood in pharmacological doses (as a drug), it can stimulate sympathetic ganglia and the adrenal medulla (the sympathoadrenal system), resulting in an increase in the heart rate.

PROCEDURE

1. Obtain a record of the control heartbeat. Then, bathe the heart in a 0.2% solution of nicotine.
2. Record the results; then stop recording and rinse the heart thoroughly with Ringer's solution.
3. In the table of your laboratory report, tape the recording or draw a facsimile of the normal heartbeat and of the changed heartbeat after the nicotine solution was added.
4. Analyze your data and record your results for parts A through H in the Results table of your laboratory report.

DATA FROM EXERCISE 7.1

Condition	Effects (Tape the Recording or Draw a Facsimile)
Normal	
Calcium (Ca^{2+})	
Digitalis	
Pilocarpine	
Atropine	
Potassium (K^+)	
Epinephrine	
Caffeine	
Nicotine	

RESULTS FROM EXERCISE 7.1

Condition	Rate (beats/min)	Strength (mm above Baseline)	Distance between Atrial and Ventricular Peaks (mm)	Conclusion about Drug Effects
Normal				
Calcium (Ca^{2+})				
Digitalis				
Pilocarpine				
Atropine				
Potassium (K^+)				
Epinephrine				
Caffeine				
Nicotine				

REVIEW ACTIVITIES FOR EXERCISE 7.1

Test Your Knowledge of Terms and Facts

Match the following:

_____1. endogenous substance that makes the heart beat stronger and faster
_____2. substance that makes the beat slower and stronger
_____3. substance that facilitates the release of ACh from parasympathetic nerve endings
_____4. substance that mimics the action of epinephrine by inhibiting the action of phosphodiesterase
_____5. substance that stimulates the acetylcholine receptors of autonomic ganglia
_____6. substance that blocks the acetylcholine receptors for the target cells of postganglionic neurons

(a) digitalis
(b) nicotine
(c) caffeine
(d) epinephrine
(e) atropine
(f) pilocarpine

7. The drug used in this exercise that helps people with atrial fibrillation:_____.
8. The drug used in this exercise which is used by ophthalmologists to dilate pupils:_____.
9. The term *hyperkalemia* means_____
_____.

Test Your Understanding of Concepts

10. What are the effects of hyperkalemia on the heart? How were these effects produced?

11. What is a sympathomimetic drug? What are its effects on the heart?

12. What are the effects of digitalis on the heart? Describe the clinical uses of this drug.

Test Your Ability to Analyze and Apply Your Knowledge

13. What effect did bathing the heart with Ca^{2+} have on the strength of the contraction and the ability of the heart to relax between beats? Provide a physiological explanation of these results.

14. The drug *theophylline* is in the same chemical class as caffeine (they are methyl xanthines). Theophylline is sometimes used clinically to dilate the bronchioles in a person suffering from asthma. By analogy with caffeine, provide a physiological explanation for this clinical application.

Electrocardiogram (ECG)

MATERIALS

1. Electrocardiograph or other strip chart recorder, such as Physiograph (Narco), or Lafayette Instrument Company, Inc. recorder with EKG module (Alternatively, the *Biopac* system may be used.)
2. Electrode plates, rubber straps, electrolyte gel or paste; disposable electrodes and electrode clips

The regular pattern of electrical impulse production and conduction in the heart results in the mechanical contraction (systole) and relaxation (diastole) of the myocardium—the cardiac cycle. The recording of these electrical events, or electrocardiogram (ECG), may reveal abnormal patterns associated with abnormal cardiac rhythms.

OBJECTIVES

1. Describe the normal cyclical pattern of electrical impulse production in the heart and conduction along specialized tissues of the heart.
2. Describe the normal **electrocardiogram** (ECG, or EKG) and explain how it is produced.
3. Obtain an electrocardiogram using the limb leads, identify the waves, determine the P-R interval, and measure the cardiac rate.
4. Describe the common ECG abnormalities or arrhythmias.

Textbook Correlations*

Before performing this exercise, you may want to consult the following references in *Human Physiology*, seventh edition, by Stuart I. Fox:

- *Electrical Activity of the Heart and the Electrocardiogram.* Chapter 13, pp. 385–389.

Those using different physiology textbooks may want to consult the corresponding information in those books.

Electrical stimulation at any point in the heart musculature (myocardium) results in the almost simultaneous contraction of the individual muscle cells (myocardial cells). This allows the heart to function as an effective pump, with its chambers contracting and relaxing as integrated units. The contraction phase of the cardiac cycle is called **systole,** and the relaxation phase is called **diastole.**

Unlike skeletal muscles, cardiac muscle is able to stimulate itself electrically in the absence of neural input (termed *automaticity*). The sympathetic and para-sympathetic nerves that innervate the heart only modulate the ongoing rate of depolarization-contraction and repolarization-relaxation intrinsic to the heart. The intrinsic regulation of systole and diastole, unique to heart muscle, is termed *rhythmicity*.

Although each individual myocardial cell is potentially capable of initiating its own cycle of depolarization-contraction and repolarization-relaxation, a single group of cells usually regulates the cycle of the entire myocardium. This *pacemaker* region establishes its dominance because its cycle is more rapid than other areas, depolarizing the other myocardial cells before they can depolarize themselves (fig. 7.6). A region of the right atrium, the **sinoatrial node (SA node),** serves as the normal pacemaker of the heart. The wave of depolarization initiated by the SA node quickly spreads across the right and left atria as a result of electrical synapses (*gap junctions*) between myocardial cells. However, the depolarization wave cannot easily spread from the myocardial cells of the atria to the myocardial cells of the ventricles. For this to occur, the depolarization wave must be carried from the atria to the ventricles along the specialized conducting tissue of the heart.

---

*See Appendix 3 for correlations to the A.D.A.M. *InterActive* PHYSIOLOGY *Modules.*

See Appendix 3 for correlations to the *Virtual Physiology Laboratory* CD-ROM by McGraw-Hill and Cypris Publishing, Inc.

See Appendix 3 for correlations to the *Intelitool Physiology Laboratory Exercises.*

 See Appendix 3 for correlations with the *Biopac Student Lab Exercises.*

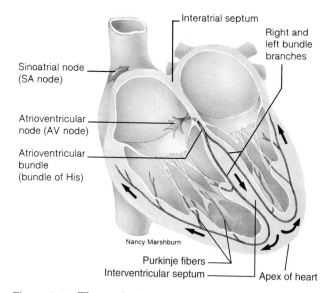

Figure 7.6 **The conduction system of the heart.**

Labels on figure:
- Interatrial septum
- Right and left bundle branches
- Sinoatrial node (SA node)
- Atrioventricular node (AV node)
- Atrioventricular bundle (bundle of His)
- Nancy Marshburn
- Purkinje fibers
- Interventricular septum
- Apex of heart

The depolarization wave that spreads over the atria stimulates a node called the **atrioventricular node (AV node),** located at the base of the interatrial septum. After a brief delay, the depolarization wave in the AV node is quickly transmitted over a bundle of specialized conducting tissue, the **atrioventricular (or AV) bundle,** otherwise known as the **bundle of His.** The bundle of His splits to form right and left "bundle branches," carrying the depolarization wave to the apex of the ventricles. The impulse is then carried to the innermost cells of the ventricles on a network of branching **Purkinje fibers** (fig. 7.6).

Because of the depolarization stimulus, the atria contract as a single unit (atrial systole), followed quickly by depolarization and contraction of the ventricles (ventricular systole). Contraction of the atria forces blood into the ventricles, and contraction of the ventricles forces blood out of the heart and into the pulmonary arteries (pulmonary circulation) and into the aorta (systemic circulation). During the period of repolarization and relaxation in diastole, the atria refill with blood, and the cycle is ready again to be initiated by the spontaneous depolarization of the SA node pacemaker.

Since the body fluids contain a high concentration of electrolytes (ions), the electrical activity generated by the heart travels throughout the body and can easily be monitored by placing a pair of electrodes on different areas of the skin (fig. 7.7). A graphic representation of these electrical activities is called an **electrocardiogram (ECG** or **EKG),** and the instrument producing this record is called an *electrocardiograph*. A normal ECG is shown in figure 7.8.

The **P wave** represents depolarization of the atria; the **QRS complex** is produced by depolarization of the ventricles; and the **T wave** represents repolarization of the ventricles at the beginning of diastole. An incompletely understood **U wave** sometimes follows the T wave.

In the following exercises we will use the *standard limb leads* I, II, and III. These leads record the difference in potential (the voltage) between two electrodes placed on the arms and legs (fig. 7.7). In clinical electrocardiography, however, *unipolar leads* are also used. These are the AVR (right arm), AVL (left arm), AVF (left leg), and the chest leads labeled V_1 to V_6.

The standard ECG running at a chart speed of 25 mm (2.5 cm) per second, shows thin vertical lines 0.04 sec apart, with the distance between every fifth, heavier line representing an interval of 0.20 sec. This produces a graphic recording in which the length of time between depolarization of the atria (P wave) and depolarization of the ventricles (QRS complex) can easily be measured. This interval is known as the **P-R interval** (although it is actually measured from the beginning of the P wave to the Q wave—fig. 7.8) and is equal to less than 0.20 sec in the normal ECG.

PROCEDURE

1. With the subject comfortably reclining (lying down), rub a silver dollar-sized amount of electrolyte gel on the medial surface, about 2 inches above the wrists and ankles. Attach electrode plates to these four spots, using the rubber straps provided (fig. 7.7).

2. Attach the four ECG leads to the appropriate plates.

3. The specific instructions for obtaining ECG tracings vary with the instrument being used. Your instructor will demonstrate the use of the recording equipment in your lab. The following instructions are valid only for a single-channel electrocardiograph.
    (a) Turn on the power switch.
    (b) Set the paper-speed selector switch to 25 mm (2.5 cm) per second.
    (c) Set the sensitivity to *1* (most sensitive).
    (d) Set the lead selector switch to the first dot to the left of the *STD* or *CAL* position.
    (e) Turn the control or knob to the *run* or *record* position.

4. Turn the position knob until the stylus is centered on the ECG paper.

5. Turn the lead selector switch to the *1* (lead I) position to measure the voltage difference between the right and left arms. As the paper is running, depress the *mark* button once—this makes a single dash at the top of the chart to indicate that the record is from lead I. Continue recording until an adequate sample of the tracing can be provided to each member of the subject's group; then stop the paper drive by turning the lead selector switch to the dot above the *1* position. Each dot is a "rest" position where the movement of the chart will stop between recording from each lead.

Figure 7.7 Electrocardiograph leads. The placement of the bipolar limb leads (RA, LA, LL) and the positions (1–6) for the unipolar chest leads (V_1–V_6) for recording electrocardiograms. (RA = right arm; LA = left arm; LL = left leg.)

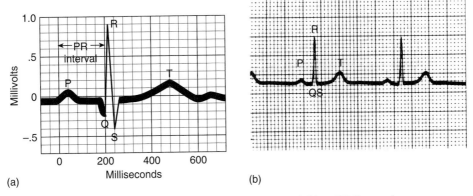

Figure 7.8 The normal electrocardiogram (ECG). (a) A labeled drawing, and (b) an ECG recording.

6. Turn the lead selector switch to the *2* (lead II) position to measure the voltage difference between the right arm and left leg. As the chart is running, depress the *mark* button twice—the two dashes produced at the top of the chart will indicate that this is the recording from lead II. Stop the chart by turning the lead selector switch to the dot above the *2* position.

7. Repeat this procedure with lead III to measure the voltage difference between the left arm and the left leg.

8. After recordings from leads I, II, and III have been obtained, turn the lead selector switch to the *STD* or *CAL* position. Run the recording out of the machine, allowing members of the group to cut sample tracings of each lead.

243

9. Remove the electrode plates from the subject's skin and thoroughly wash the electrolyte gel from both the plate and the skin.

10. Tape samples of the recordings in your laboratory report and label all the waves.

11. Determine the P-R interval of *lead II*. This can be done by counting the number of *small* boxes between the beginning of the P and the Q and multiplying this number by 0.04 sec.

Note: *If you use a multichannel recorder (such as a Physiograph) instead of an electrocardiograph, the recording paper and, consequently, the arithmetic calculation will be different. If you use a Cardiocomp, the P-R interval will be provided automatically.*

P-R interval _____ sec

The normal P-R interval is 0.12–0.20 sec.

12. Determine the cardiac rate by the following methods:

(a) Count the number of QRS complexes in a 3-sec interval (the distance between two vertical lines at the top of the ECG paper) and multiply by 20.

Note: *If you use a multichannel recorder (such as a Physiograph), the amount of chart paper corresponding to a given time interval must be calculated and will vary with the paper speed. If you use a Cardiocomp, the cardiac rate will be provided automatically.*

Beats per minute = _____

(b) Count the number of QRS complexes in a 6-sec interval, and multiply by 10.
Beats per minute = _____

(c) At a chart speed of 25 mm (2.5 cm) per second, the time interval between one light vertical line and the next is 0.04 sec. The time interval between heavy vertical lines is 0.20 sec. The cardiac rate in beats per minute can be calculated if the time interval between two R waves in two successive QRS complexes is known.

For example, suppose that the time interval from one R wave to the next is exactly 0.60 sec. Therefore,

$$\frac{1 \text{ beat}}{0.60 \text{ sec}} = \frac{x \text{ beats}}{60 \text{ sec}}$$

$$x = \frac{1 \text{ beat} \times 60 \text{ sec}}{0.60 \text{ sec}}$$

$$x = 100 \text{ beats per minute}$$

Beats per minute = _____

(d) The values obtained by *method c* can be approximated by counting the number of heavy vertical lines between one R wave and the next according to the memorized sequence: 300, 150, 100, 75, 60, 50 (at a paper speed of 25 mm per sec):

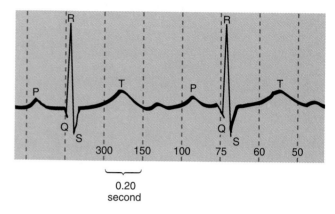

0.20 second

The cardiac rate in the sample tracing above is 75 beats per minute.
Beats per minute = _____

The normal cardiac rate is 60–100 beats per minute.

ABNORMAL ECG PATTERNS

Interpretation of the electrocardiogram can provide information about the heart rate and rhythm, as well as possible conditions of *hypertrophy*, *ischemia* (inadequate blood supply), *necrosis* (death of cells), and other conditions that may produce abnormalities of electrical conduction. According to the standards set by the National Conference on Cardiopulmonary Resuscitation and Emergency Cardiac Care, all health professionals should be able to recognize:

1. bradycardia (a ventricular rate slower than 60 beats per minute)
2. the difference between supraventricular and ventricular rhythms
3. premature ventricular contractions
4. ventricular tachycardia
5. atrioventricular block
6. atrial fibrillation and flutter
7. ventricular fibrillation

When *ectopic beats* (beats that are out of place) occur in the atria as a result of the development of an *ectopic pacemaker* (a pacemaker that develops in addition to the normal one in the SA node) or as a result of a derangement in the normal conduction pathway, a condition of atrial flutter or atrial fibrillation may be present. **Atrial flutter** is characterized by very rapid atrial waves

(a)

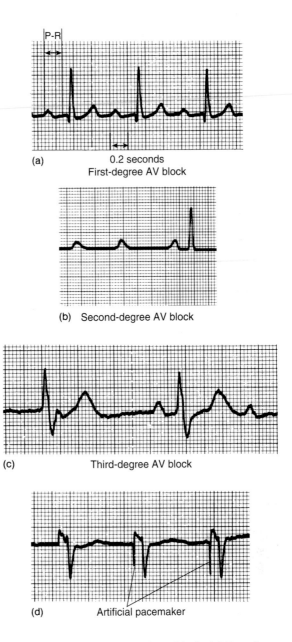

(a) 0.2 seconds
First-degree AV block

(b) Second-degree AV block

(c) Third-degree AV block

(d) Artificial pacemaker

Figure 7.9 **Abnormal atrial rhythms.** (a) Atrial flutter and (b) atrial fibrillation.

Figure 7.10 **Different stages of AV block.** (a) First-degree AV block, where the P-R interval is greater than 0.2 seconds (one large square); (b) second-degree AV block, where beats (i.e., QRS complexes) are missed (a 3:1 block is shown); and (c) third-degree, or complete, AV block, where a slower than normal heart rate is set by an ectopic focus in the ventricles. (d) An artificial pacemaker, implanted in a person with complete AV node block, produces the sharp downward spikes that precede the inverted QRS waves.

(about 300 per minute), producing a sawtoothed baseline. These atrial waves occur with such high frequency that the AV node can beat only to every second, third, or fourth wave it receives (fig. 7.9a). A person with atrial flutter may have a normal pulse rate, since the pulse is produced by contraction of the left ventricle.

In **atrial fibrillation,** the depolarization waves occur so rapidly (350–400/min) that the atria no longer function effectively, and the P waves of the ECG are replaced by a wavy baseline (fig. 7.9b). Atrial fibrillation is characteristic of atrial enlargement (hypertrophy), as might be produced by *mitral stenosis* (narrowing of the mitral valve), but may occur in all forms of heart disease and occasionally in apparently healthy individuals. *Digitalis* is a drug often used in atrial flutter and fibrillation to decrease the excitability of the AV node, thus maintaining the ventricular rate within the normal range.

A delay in the conduction of the impulse from the atria to the ventricles is known as **atrioventricular (AV) block.** When the ECG pattern is otherwise normal, a P-R interval greater than 0.20 sec represents a *first-degree AV block* (fig. 7.10a). First-degree AV block may be a result of inflammatory states, rheumatic fever, or digitalis treatment.

When the excitability of the AV node is further impaired so that two or more atrial depolarizations are required before the impulse can be transmitted to the ventricles, a *second-degree AV block* is present. This is seen on the ECG as a "dropped" beat (that is, a P wave without an associated QRS complex). In *Wenckebach's phenomenon,* the cycle after the dropped beat is normal, but the P-R interval of successive cycles lengthens until a beat (QRS complex) is again dropped (fig. 7.10b).

Third-degree, or *complete, AV block* occurs when none of the impulses from the atria reach the ventricles (fig. 7.10c). In this case, the myocardial cells of the ventricles are freed from their subservience to the SA node, causing one or more ectopic pacemakers to appear in the ventricles without corresponding P waves. The rhythm produced by these ectopic foci in the ventricles is usually very slow (20–45 beats per minute) compared to the normal pace set by the SA node (*sinus rhythm*). An artificial pacemaker may be used to compensate for this condition, as shown by the ECG tracing in figure 7.10*d*.

Premature ventricular contractions (PVCs) are produced by ectopic foci in the ventricles when the sinus rhythm is normal. This results in *extrasystoles* (extra beats or QRS complexes without preceding P waves) in addition to the normal cycle, often subjectively described by patients as "palpitations." The ectopic QRS complexes are broad and deformed and may be abnormally coupled to the preceding normal beats. This coupling is termed *bigeminy* and is often seen in digitalis toxicity (fig. 7.11*a*).

When an ectopic focus in the ventricles discharges at a rapid rate, a condition of **ventricular tachycardia** (usually 100–150 beats per minute) develops (fig. 7.11*b*). The ECG shows a widened and distorted QRS complex that often obscures the P wave (although the atria are discharging at their slower, sinus rate). This serious condition should be distinguished from the less serious **supraventricular tachycardia,** in which an ectopic focus above the ventricles results in spontaneous rapid running of the heart (150–250 beats/min) that begins and ends abruptly (paroxysmal). This condition is appropriately called *paroxysmal atrial tachycardia.*

The most serious of all arrhythmias is **ventricular fibrillation** (fig. 7.11*c*). Ventricular fibrillation may develop as ectopic foci emerge and depolarization waves circle the heart (*circus rhythm*), resulting in an impotent tremor rather than a coordinated pumping action. Under these conditions, the pumping activity of the ventricles ceases, and death occurs within minutes unless emergency measures (including electrical defibrillation) are applied successfully.

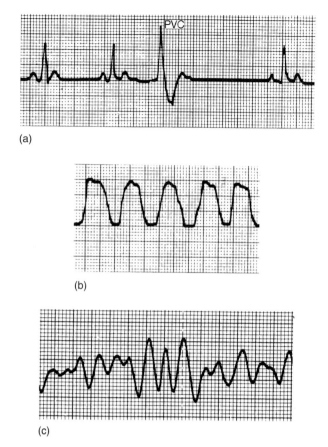

(a)

(b)

(c)

Figure 7.11   Abnormal ventricular rhythms. (a) Premature ventricular contraction (PVC), or bigeminy; (b) ventricular tachycardia; and (c) ventricular fibrillation.

Name _____

Date _____

Section _____

DATA FROM EXERCISE 7.2

Tape your recording in the spaces below.

<u>*Lead I*</u>

<u>*Lead II*</u>

<u>*Lead III*</u>

REVIEW ACTIVITIES FOR EXERCISE 7.2

Test Your Knowledge of Terms and Facts

1. The pacemaker region of the heart is the _____ .
2. The conducting tissue of the heart located in the interventricular septum is the _____ .
3. Indicate the electrical events that produce each of the following waves:
   (a) P wave _____
   (b) QRS wave _____
   (c) T wave _____
4. The electrical synapses between adjacent myocardial cells are called _____ .
5. An abnormally fast rate of beat is called _____ ; an abnormally slow rate is called _____ .
6. An abnormally long P-R interval indicates a condition called _____ .
7. Leads I, II, and III are collectively called the _____ leads.
8. Which ECG wave must occur before the ventricles can contract? _____
9. Which ECG wave must occur before the ventricles can relax? _____

Test Your Understanding of Concepts

10. What property makes the normal pacemaker region of the heart function as a pacemaker? Explain.

11. Describe the pathway of conduction from the atria to the ventricles and correlate this conduction with the ECG waves.

12. Compare supraventricular tachycardia with ventricular tachycardia in terms of its nature, ECG pattern, and seriousness.

13. What happens to the beating of the atria and ventricles during third-degree AV node block? Why does this occur?

14. On initial examination, a patient was found to have a P-R interval of 0.24 second. A year later, the patient has a resting pulse of about 40 per minute. A year after that, the patient collapsed with a pulse of 20 per minute, and required the insertion of an artificial pacemaker. Explain what caused this sequence of events.

15. Suppose a person had paroxysmal atrial tachycardia, producing "palpitations." Describe this condition, and propose ways that the condition might be treated.

Effects of Exercise
on the Electrocardiogram

MATERIALS

1. Electrocardiograph or multichannel recorder (e.g., Physiograph) with appropriate ECG module
2. ECG plates, straps, gel
3. Alternatively, the *Biopac* system may be used with the pulse transducer and electrodes for *Biopac Student Lab* lesson 7.

During exercise, there is a decrease in the activity of parasympathetic innervation to the SA node, conductive tissue, and myocardium and an increase in the activity of sympathetic innervation to the SA node, conductive tissue, and myocardium. More rapid discharge of the SA node, more rapid conduction of impulses, and a faster rate of contraction all result in an increased cardiac rate with exercise.

OBJECTIVES

1. Describe the effects of the sympathetic and parasympathetic innervation to the heart.
2. Obtain an ECG before, immediately after, and 2 minutes after exercise; and explain the differences observed.
3. Determine cardiac rate, P-R interval, and period of ventricular diastole for the ECG tracings obtained in objective 2.

Textbook Correlations*

Before performing this exercise, you may want to consult the following references in Human Physiology, seventh edition, by Stuart I. Fox:

- *Pressure Changes During the Cardiac Cycle.* Chapter 13, p. 382.
- *The Electrocardiogram.* Chapter 13, pp. 387–389.
- *Regulation of Cardiac Rate.* Chapter 14, pp. 410–411.

Those using different physiology textbooks may want to consult the corresponding information in those books.

---

tory effect on the SA node and causes even greater increases in cardiac rate.

Sympathetic fibers also innervate the conducting tissues of the heart and the ventricular muscle fibers. Through these innervations, sympathetic stimulation may increase the velocity of both impulse conduction and ventricular contraction. These effects are most evident at high cardiac rates and contribute only slightly to the increased cardiac rate during exercise. Thus, the increased cardiac rates are mainly due to a shortening of the ventricular diastole (from the peak of the T wave to the beginning of the next QRS complex) and only secondarily due to a shortening of ventricular systole (measured from the QRS peak to the peak of the T wave, fig. 7.12).

The heart is innervated by both sympathetic and parasympathetic nerve fibers. At the beginning of exercise, the activity of the parasympathetic fibers that innervate the SA node decreases. Since these fibers have an inhibitory effect on the pacemaker, a decrease in their activity results in an increase in cardiac rate. As exercise becomes more intense, the activity of sympathetic fibers that innervate the SA node increases. This has an excita-

---

*See Appendix 3 for correlations to the A.D.A.M. *InterActive PHYSIOLOGY Modules*.

 See Appendix 3 for correlations to the *Virtual Physiology Laboratory CD-ROM* by McGraw-Hill and Cypris Publishing, Inc.

See Appendix 3 for correlations to the *Intelitool Physiology Laboratory Exercises*.

See Appendix 3 for correlations with the *Biopac Student Lab Exercises*.

Figure 7.12 **Correlation of measurements during the cardiac cycle.** The relationship between the changes in intraventricular pressure and the electrocardiogram (ECG) during the cardiac cycle of systole and diastole can be seen. Notice that the QRS wave occurs at the beginning of systole, whereas the T wave occurs at the beginning of diastole.

The rate of blood flowing through the coronary circulation may be adequate to meet the aerobic requirements of the heart at rest, but may be inadequate for the increased metabolic energy demands of the heart during exercise. A portion of the cardiac muscle may receive insufficient blood flow because of a clot in a coronary artery (*coronary thrombosis*) or because of narrowing of the vessel due to *atherosclerosis*. This insufficient blood flow to the heart—**myocardial ischemia**—may, however, be relative to the aerobic demand. Supervised exercise tests such as those on a treadmill with varying levels of aerobic demand help diagnose this kind of ischemia. During these tests, changes in the *S-T segment* of the ECG (fig. 7.13) may indicate ischemia.

PROCEDURE

1. After the resting ECG has been recorded (from exercise 7.2), unplug the electrode leads from the electrocardiograph.
2. Have the subject exercise while holding the lead wires—for example, by walking up and down stairs, using a stationary bicycle, hopping, doing sit-ups or leg lifts.

Normal   Ischemia

Figure 7.13 **ECG changes during myocardial ischemia.** In myocardial ischemia, the S-T segment of the electrocardiogram may be depressed, as illustrated in this figure.

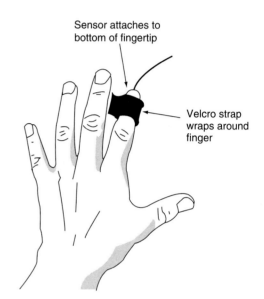

Sensor attaches to bottom of fingertip

Velcro strap wraps around finger

Figure 7.14 **Pulse transducer for *Biopac* system.**

⚠ *Caution: The intensity of exercise should be monitored so that 80% of the maximum cardiac rate is not exceeded. Students who are not in good health should not serve as subjects.*

3. Immediately after exercise, the subject should lie down and the electrode leads are plugged into the electrocardiograph. Record lead II *only*. Wait 2 minutes and record lead II again.
4. Determine the cardiac rate, period of ventricular diastole, and the duration of the QRS complex for the resting ECG and for the two postexercise ECG recordings.

   Enter these data in the table in your laboratory report.
5. Alternatively, the ECG may be obtained using the Biopac system (as set up for *Biopac Student Lab* lessons 5 and 6). If the pulse transducer is available (fig. 7.14), as used in *Biopac Student Lab* lesson 7, the pulse can be followed directly during rest, exercise, and postexercise.

DATA FROM EXERCISE 7.3

Enter your data in the table below.

ECG Tracing	Cardiac Rate (beats/min)	P-R Interval (sec) (beginning of P to Q)	Ventricular Diastole (sec) (middle of T to next Q)
Resting ECG			
Immediate postexercise ECG			
Two-minute postexercise ECG			

Which measurement changed the most as you went from resting, to exercise, to the two-minute postexercise conditions? Which changed the least? What conclusions can you draw regarding the changes in impulse conduction and the cardiac cycle as a result of exercise?

Tape your ECG recordings in the space below.

REVIEW ACTIVITIES FOR EXERCISE 7.3

Test Your Knowledge of Terms and Facts

1. The ECG wave that occurs at the beginning of ventricular systole is the _____ wave.
2. The ECG wave that occurs at the end of systole and the beginning of diastole is the _____ wave.
3. The ECG wave that is completed just before the end of ventricular diastole is the _____ wave.
4. The nerve that increases the rate of discharge of the SA node is a _____ nerve.
5. The specific nerve that, when stimulated, causes a decrease in the cardiac rate is the _____ .
6. The scientific term for insufficient blood flow to the heart muscle is _____ .

Test Your Understanding of Concepts

7. Describe the regulatory mechanisms that produce an increase in cardiac rate during exercise. Explain how these changes affect the electrocardiogram (ECG).

8. Describe the pressure changes that occur within the ventricles as each ECG wave is produced, and explain how these pressure changes are related to the ECG waves.

Test Your Ability to Analyze and Apply Your Knowledge

9. Explain how a person could have a normal electrocardiogram at rest, but show evidence of myocardial ischemia during exercise. Do you think all people should have regular treadmill (stress) electrocardiogram tests, or only certain people? Explain.

10. Provide a cause-and-effect explanation as to how the pulse rate relates to the ECG rate. How would the relationship change if pulse were measured by palpation of the carotid artery instead of by a pulse transducer in the finger?

Mean Electrical Axis
of the Ventricles

MATERIALS

1. Electrocardiograph, or multichannel recorder (e.g., Physiograph) with appropriate ECG module
2. ECG plates, straps, gel

The voltage changes in the ECG measured by two different leads can be compared and used to determine the mean electrical axis, which corresponds to the average direction of depolarization as the impulses spread into the ventricles. Significant deviations from the normal axis may be produced by specific heart disorders.

OBJECTIVES

1. Describe the electrical changes in the heart that produce the ECG waves.
2. Determine the mean electrical axis of the ventricles in a test subject, and explain the clinical significance of this measurement.

Textbook Correlations*

Before performing this exercise, you may want to consult the following references in *Human Physiology*, seventh edition, by Stuart L. Fox:

- *The Electrocardiogram.* Chapter 13, pp. 387–389.

Those using different physiology textbooks may want to consult the corresponding information in those books.

---

epolarization waves spread through the heart in a characteristic pattern. Depolarization waves begin at the SA node and spread from the pacemaker to the entire mass of both atria, producing the *P wave* in an electrocardiogram. After the AV node is excited, the interventricular septum becomes depolarized as the impulses spread through the bundle of His. Since at this point the septum is depolarized while the lateral walls of the ventricles still have their original polarity, there is a potential difference (voltage) between the septum and the ventricular walls. This produces the *R wave*. When the entire mass of the ventricles is depolarized, there is no longer a potential difference within the ventricles, and the voltage returns to zero (completing the *QRS complex*).

The direction of the depolarization waves depends on the orientation of the heart in the chest and on the particular instant of the cardiac cycle being considered. It is clinically useful, however, to determine the **mean axis** (average direction) **of depolarization** during the cardiac cycle. This can be done by observing the voltages of the QRS complex from two different perspectives using two different leads. Lead I provides a horizontal axis of observation (from left arm to right arm); lead III has an axis of about 120° (from left arm to left leg). Using the recordings from leads I and III, the normal mean electrical axis of the ventricles is found to be about 59°, as shown in figure 7.15.

---

*See Appendix 3 for correlations to the *Intelitool Physiology Laboratory Exercises*.

See Appendix 3 for correlations with the *Biopac Student Lab Exercises*.

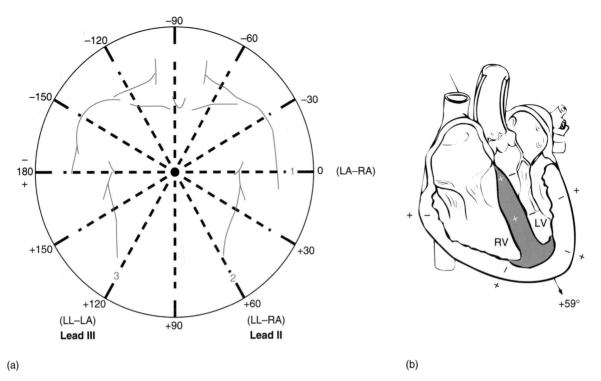

(a)

(b)

Figure 7.15 **Mean electrical axis of the heart.** (a) The convention by which the axis of depolarization is measured. The bottom half of the circle (with the heart at the center) is considered the positive role (LA = left arm, RA = right arm, LL = left leg). (b) When the interventricular septum is depolarized, the surface of the septum is electrically negative compared to the walls of the ventricles, which have not yet become depolarized. The average normal direction of depolarization, or mean electrical axis of the ventricles, is about 59° (RV = right ventricle, LV = left ventricle).

 Hypertrophy (enlargement) of one ventricle shifts the mean axis of depolarization toward the hypertrophied ventricle because it takes longer to depolarize the larger ventricle. Therefore, a left axis deviation occurs when the left ventricle is hypertrophied (as a result of hypertension or narrowing of the aortic semilunar valve). A right axis deviation occurs when the right ventricle hypertrophies. The latter condition may be secondary to narrowing of the pulmonary semilunar valve or to such congenital conditions as a septal defect or the tetralogy of Fallot.

The depolarization wave normally spreads through both the right and left ventricles at the same time. However, if there is a conduction block in one of the branches of the bundle of His—a **bundle-branch block**—depolarization will be much slower in the blocked ventricle. In left bundle-branch block, for example, depolarization will occur more slowly in the left ventricle than in the right ventricle, and the mean electrical axis will deviate to the left. In right bundle-branch block, there will be a right axis deviation. Deviations of the electrical axis also occur to varying degrees as a result of myocardial infarction.

PROCEDURE

Note: *If a Cardiocomp-7 or -12 is used, the mean electrical axis of the ventricles can be determined by examining the QRS loop of the vectorgram displayed on the computer screen. Alternatively, the following procedure can be used by examining leads I and III that were previously obtained in exercise 7.2.*

1. Analysis of lead I:
   (a) Find a QRS complex and count the number of millimeters (small boxes) that it projects above the upper edge of the baseline.
   Enter this value in the following space:
   + _____ mm
   (b) Count the number of millimeters the Q and S waves (or the R wave, if it is inverted) project below the upper edge of the baseline. Add these measurements of downward deflections.
   Enter this sum in the following space:
   _____ mm
   (c) Algebraically, add the two values from steps (a) and (b), keeping the negative sign if the sum is negative.
   Enter this sum in the laboratory report.

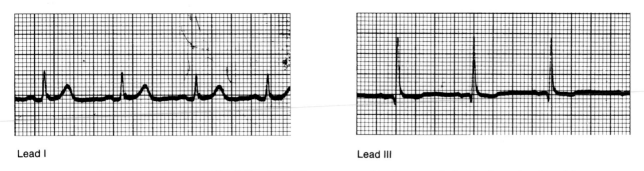

Lead I   Lead III

Figure 7.16 **Sample electrocardiograms of leads I and III.** These were used in the example for the determination of the mean electrical axis of the heart. (Note: Each small square is 1 mm on a side.)

2. Analysis of lead III:
   (a) Find a QRS complex and count the number of millimeters (small boxes) it projects above the upper edge of the baseline.
   Enter this value in the following space:
   + _____ mm
   (b) Count the number of millimeters the Q and S waves (or the R wave, if it is inverted) project below the upper edge of the baseline. Add these measurements of downward deflections.
   Enter this sum in the following space:
   _____ mm
   (c) Algebraically, add the two values from steps (a) and (b), keeping the negative sign if the sum is negative.
   Enter this sum in the laboratory report.
3. On the blank grid chart in your laboratory report, use a straight edge to make a line on the axis of *lead I* that corresponds to the sum you obtained in step 1c (see example grid, fig. 7.17).
4. Use a straight edge to make a line on the axis of *lead III* that corresponds to the sum you obtained in step 2c.
5. Use a straight edge to draw an arrow from the center of the grid chart to the intersection of the two lines drawn in steps 3 and 4. Extend this arrow to the edge of the grid chart and record the mean electrical axis of the ventricles.

Example (fig. 7.16)

Lead I of sample ECG:

Upward deflection:	+ 7 mm
Downward deflections:	− 1 mm
	6 mm

Lead III of sample ECG:

Upward deflection:	+ 14 mm
Downward deflections:	− 2 mm
	12 mm

A straight line is drawn perpendicular to the horizontal axis of lead I corresponding to position 6 on the scale. Similarly, a straight line is drawn perpendicular to the axis of lead III that corresponds to position 12 on the scale. An arrow is then drawn from the center of the circle through the intersection of the two lines previously drawn (fig. 7.17).

In this example, the mean electrical axis of the ventricles is +71°.

The normal mean electrical axis of the ventricles is between 0 and +90°.

Figure 7.17 Example of the method used to determine the mean electrical axis of the heart. This uses the data from leads I and III in the sample ECG shown in figure 7.16. In this example, the mean electrical axis is 72°.

Name _____

Date _____

Section _____

DATA FROM EXERCISE 7.4

1. Enter the value for the sum of steps 1a and 1b in the space below. Be sure to indicate whether it is a positive or negative number.

   _____

2. Enter the value for the sum of steps 2a and 2b in the space below. Be sure to indicate whether it is a positive or negative number.

   _____

3. Use the grid chart below to determine the mean electrical axis of the ventricles from your data, as described in steps 3, 4, and 5 of the procedure.

   Mean electrical axis of the ventricles: _____

Grid chart

Direction of the electrical axis of the heart as determined by the
human electrocardiogram

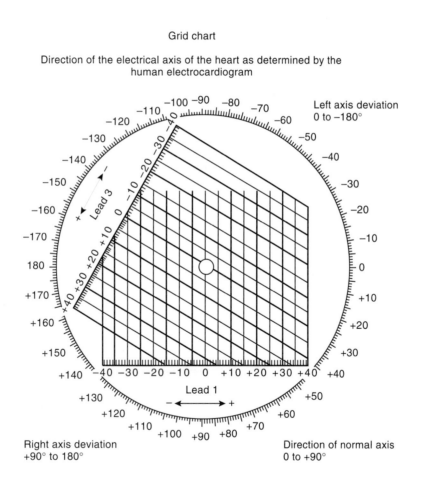

Right axis deviation
+90° to 180°

Direction of normal axis
0 to +90°

REVIEW ACTIVITIES FOR EXERCISE 7.4

Test Your Knowledge of Terms and Facts

1. Depolarization of the atria produces the _____ wave.
2. The difference in polarity between the interventricular septum and the lateral walls of the ventricles produces the _____ wave.
3. The mean electrical axis is determined using two bipolar limb leads, lead _____ and lead _____ .
4. Hypertrophy of the left ventricle would shift the mean axis of depolarization to the_____ .
5. A blockage in conduction in the right branch of the bundle of _____ would cause the mean axis of depolarization to shift to the _____ .

Test Your Understanding of Concepts

6. Explain how the normal pattern of depolarization is affected by ventricular hypertrophy. What factors may be responsible for hypertrophy of the right or left ventricle?

7. Explain how the normal pattern of depolarization is affected by bundle-branch block. What might cause bundle-branch block?

Test Your Ability to Analyze and Apply Your Knowledge

8. Explain the electrical events in the heart that produce the QRS wave. Why does the tracing go up from Q to R, and then back to baseline from R to S?

9. How do you think obesity or pregnancy might influence the results of this laboratory exercise? Explain.

Heart Sounds

MATERIALS

1. Stethoscopes
2. Physiograph, high-gain couplers, microphone for heart sounds (Narco), ultrasonic flowmeter (such as Doppler)
3. Alternatively, the *Biopac* system may be used as per student lesson 17, employing the *Biopac* electrode lead set and the amplified stethoscope.

Contraction and relaxation of the ventricles is accompanied by pressure changes that cause the one-way heart valve to close. Closing valves produce sounds that aid in the diagnosis of structural abnormalities of the heart.

OBJECTIVES

1. Describe the causes of the normal heart sounds.
2. List some of the causes of abnormal heart sounds.
3. Correlate the heart sounds with the waves of the ECG and the events of the cardiac cycle.

A. AUSCULTATION OF HEART SOUNDS WITH THE STETHOSCOPE

The cycle of mechanical contraction (**systole**) and relaxation (**diastole**) of the ventricles can be followed by listening to the heart sounds with a **stethoscope.** The contraction of the ventricles produces a rise in intraventricular pressure, resulting in the vibration of the surrounding structures as the atrioventricular valves slam shut. The valves closing produces the *first sound* of the heart, usually verbalized as **"lub."** At the end of the contraction phase, the blood in the aorta and pulmonary arteries pushes the one-way semilunar valves shut, and the resulting vibration of these structures produces the *second sound* of the heart, which is verbalized as **"dub."**

Textbook Correlations*

Before performing this exercise, you may want to consult the following references in *Human Physiology,* seventh edition, by Stuart I. Fox:

- *Cardiac Cycle and Heart Sounds.* Chapter 13, pp. 381–384.
- *The Electrocardiogram.* Chapter 13, pp. 387–389.

Those using different physiology textbooks may want to consult the corresponding information in those books.

Careful **auscultation** (listening) to the two heart sounds may reveal two further sounds. This *splitting* of the heart sounds into four components is more evident during inhalation than it is during exhalation. During deep inhalation, the first heart sound may be split into two sounds because the tricuspid and mitral valves close at different times. The second heart sound may also be split into two components because the pulmonary and aortic semilunar valves close at different times.

Auscultation of the chest aids in the diagnosis of many cardiac conditions, including **heart murmurs.** A murmur may be caused by an irregularity in a valve, a septal defect, or the persistent fetal opening (*foramen ovale*) between the right and left atria after birth, resulting in the audible regurgitation of blood in the reverse direction of normal flow. Abnormal splitting of the first and second heart sounds may be due to heart block, septal defects, aortic stenosis, hypertension, or other abnormalities.

---

*See Appendix 3 for correlations to the A.D.A.M. *InterActive PHYSIOLOGY Modules.*

See Appendix 3 for correlations to the *Virtual Physiology Laboratory* CD-ROM by McGraw-Hill and Cypris Publishing, Inc.

See Appendix 3 for correlations with the *Biopac Student Lab Exercises.*

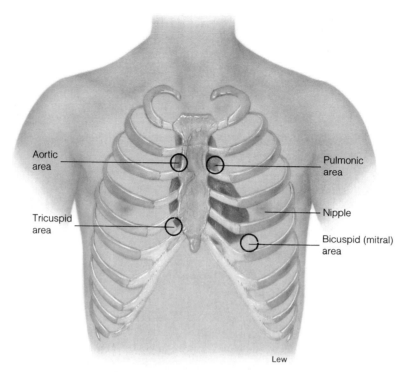

Figure 7.18 Stethoscope positions for auscultation of heart sounds.

PROCEDURE

1. Clean the earpieces of the stethoscope with an alcohol swab. To best hear the first heart sound, auscultate the apex beat of the heart by placing the diaphragm of the stethoscope in the *fifth left intercostal space* (the bicuspid area in fig. 7.18).

2. To best hear the second heart sound, place the stethoscope to the right or left of the sternum in the *second intercostal space* (the aortic or pulmonic area in fig. 7.18).

3. Compare the heart sounds in the three stethoscope positions described during quiet breathing, slow and deep inhalation, and slow exhalation.

B. CORRELATION OF THE PHONOCARDIOGRAM WITH THE ELECTROCARDIOGRAM

If the heart sounds are monitored with a device known as a **phonocardiograph** in conjunction with the monitoring of the electrical patterns of the heart with an electrocardiograph, a correlation of the two recordings will show that the first heart sound occurs at the end of the QRS complex of the ECG, and the second heart sound occurs at the end of the T wave (figs. 7.19 and 7.20). The arterial pulse is palpated (felt in either the radial artery or carotid artery) in the time interval between the two heart sounds.

Figure 7.19 The relationship between changes in intraventricular pressure and the ECG. The QRS wave (representing depolarization of the ventricles) occurs at the beginning of systole, whereas the T wave (representing repolarization of the ventricles) occurs at the beginning of diastole.

Figure 7.20 **Simultaneous recording of an electrocardiogram (ECG) and a phonocardiogram (PCG).** (a) Photograph of recordings taken with a Narco physiograph. (b) Idealized computer image of recordings taken with the *Biopac* system ECG electrodes and amplified stethoscope.

PROCEDURE

1. Insert two *high-gain couplers* into the physiograph. Plug the cable from the ECG lead selector box into the coupler for channel 1, and the cable from the phonocardiograph microphone into the coupler for channel 2.

2. Attach the ECG electrode plates to the subject in the standard limb lead positions and plug the ECG cable into the lead selector box (see exercise 7.2).

3. For the high-gain coupler in channel 1:
   (a) Turn the time constant knob to the *3.2* position.
   (b) Turn the gain knob to the *X2* position.
   (c) Turn the knob on the ECG lead selector box to the calibrate position. Turn the outer sensitivity knob on the amplifier to the *10* position. The inner knob should be turned all the way to the right until it clicks.
   (d) Lower the pen lift lever, raise the inkwells, and squeeze the rubber bulbs on the inkwells until ink flows freely. Release the paper-advance button, and position the pen for channel 1 so that it writes on the appropriate heavy horizontal line.

4. For the high-gain coupler in channel 2 adjust the time constant and gain knobs as described in step 3, and position the pen to write on the appropriate heavy horizontal line. The sensitivity knob can be adjusted for the individual subject once recording begins.

5. Place the microphone to the fifth left intercostal space (fig. 7.18) and start the recording. Continue recording during normal breathing and deep inhalation. Move the microphone to the second right (or left) intercostal space and continue recording during normal breathing and deep inhalation. Repeat the procedure on the other side of the sternum.

6. Place the ultrasonic flowmeter crystal on either the radial artery or carotid artery with a dab of electrode paste. Plug the flowmeter into another high-gain coupler (in multichannel recorders) and record the arterial pulse simultaneously with the ECG and phonocardiogram.

Tape your recordings or draw facsimiles in the laboratory report.

DATA FROM EXERCISE 7.5

In the table below, tape your phonocardiogram recordings or draw facsimiles.

Microphone Position	Normal Breathing	Deep Inhalation
Fifth left intercostal space		
Second left intercostal space		
Second right intercostal space		

REVIEW ACTIVITIES FOR EXERCISE 7.5

Test Your Knowledge of Terms and Facts

1. The scientific term for listening carefully (as with a stethoscope) is _____.
2. The first heart sound (lub) is caused by _____.
3. During which phase of the cardiac cycle (systole or diastole) does the first heart sound occur? _____
4. The second heart sound is caused by _____.
5. The second heart sound occurs at the (beginning or end) _____ of (systole or diastole) _____.
6. Abnormal heart sounds are called _____.
7. The first heart sound (lub) is correlated with which ECG wave? _____
8. The second heart sound (dub) is correlated with which ECG wave? _____

Test Your Understanding of Concepts

9. Using a complete cause-and-effect sequence, explain the correlation of the heart sounds with the ECG waves.

10. Explain how the heart sounds are normally produced, and describe some of the conditions that can cause heart murmurs.

11. What is meant by the "splitting" of the heart sounds? What can cause this splitting?

12. Can a defective valve be detected by electrocardiography, by auscultation, by palpation of the radial artery, or by all three techniques? Explain your answer.

13. Can the beating of the atria be detected by electrocardiography, by auscultation, by palpation of the radial artery, or by all three techniques? Explain your answer.

Measurements of Blood Pressure

MATERIALS

1. Sphygmomanometer
2. Stethoscope
3. Alternatively, the *Biopac* system for student lesson 16 may be employed, using the *Biopac* blood pressure cuff and *Biopac* stethoscope.

Blood exerts a hydrostatic pressure against the walls of arteries that can be measured indirectly by means of a sphygmomanometer and a stethoscope. Abnormally high or low arterial blood pressures are a health risk.

OBJECTIVES

1. Describe how the sounds of Korotkoff are produced.
2. Demonstrate the ability to take blood pressure measurements and explain why measurements of systolic and diastolic pressure correspond to the first and last sounds of Korotkoff.
3. Describe how pulse pressure and mean arterial pressure are calculated.

Textbook Correlations*

Before performing this exercise, you may want to consult the following references in *Human Physiology,* seventh edition, by Stuart I. Fox:

- *Blood Pressure.* Chapter 14, pp. 431–438.
- *Hypertension.* Chapter 14, pp. 438–439.

Those using different physiology textbooks may want to consult the corresponding information in those books.

---

*See Appendix 3 for correlations to the A.D.A.M. *InterActive Physiology Modules.* See Appendix 3 for correlations to the *Biopac Student Lab Exercises.*

The delivery of oxygen and nutrients necessary for normal metabolism depends on the adequate flow of blood through the tissues (*perfusion*). From the aorta, many narrow, muscular arterioles branch, forming an "arterial tree," which offers great resistance to blood flow through peripheral tissues. For normal perfusion of tissue capillaries, therefore, the arterial blood must be under sufficient pressure to overcome this vascular resistance.

This arterial blood pressure is directly dependent on **cardiac output** (the amount of blood pumped by the heart per minute) and **peripheral resistance** (the resistance to blood flow through the arterioles). In this way, blood pressure can be raised by constriction and lowered by dilation of the arterioles.

The arterial blood pressure is routinely measured in an indirect manner with a **sphygmomanometer,** or blood pressure cuff. This device consists of an inflatable rubber bag connected by rubber hoses to a hand pump and to a pressure gauge (manometer) graduated in millimeters of mercury (mmHg). The rubber bag is wrapped around the upper arm at the level of the heart and inflated to a pressure greater than the suspected systolic pressure, thus occluding the brachial artery. The examiner auscultates the brachial artery by placing the bell of a stethoscope in the cubital fossa and slowly opening a screw valve on the hand pump (bulb), allowing the pressure in the rubber bag to fall gradually (fig. 7.21).

At rest, the blood normally travels through the arteries in a *laminar flow*—that is, blood in the central axial stream moves faster than the blood in the peripheral layers, with little or no transverse flow (and thus little mixing) between these axial layers. Consequently, under normal conditions an artery is silent when auscultated.

When the sphygmomanometer bag is inflated to a pressure above the systolic pressure, the flow of blood in the artery is stopped and the artery is again silent. As the pressure in the bag gradually drops to levels between the systolic and diastolic pressures of the artery, the blood is pushed through the partially compressed walls of the artery, creating *turbulent flow.* Under these conditions, the layers of blood are mixed by eddies that flow at right angles to the axial stream, and the turbulence sets up vibrations in the artery that are heard as sounds in the stethoscope. These sounds are known as the **sounds of**

Figure 7.21 The use of a stethoscope and a sphygmomanometer to measure blood pressure.

Korotkoff, after Nikolai S. Korotkoff, the Russian physician who first described them.

The sounds of Korotkoff are divided into *five phases* on the basis of the loudness and quality of the sounds, as shown in figure 7.22. Start on the left of the figure with a cuff pressure at approximately 130 millimeters of mercury (mm Hg), and falling.

Phase 1. A loud, clear *tapping* (or snapping) sound is evident, which increases in intensity as the cuff is deflated. This phase begins at a cuff pressure of 120 mm Hg and ends at a pressure of 106 mm Hg.

Phase 2. A succession of *murmurs* can be heard. Sometimes the sounds seem to disappear during this time (auscultatory gap), perhaps a result of inflating or deflating the cuff too slowly. This phase begins at a cuff pressure of 106 mm Hg and ends at a pressure of 86 mm Hg.

Phase 3. A loud, *thumping* sound, similar to phase 1 but less clear, replaces the murmurs. This phase begins at a cuff pressure of 86 mm Hg and ends at a pressure of 81 mm Hg.

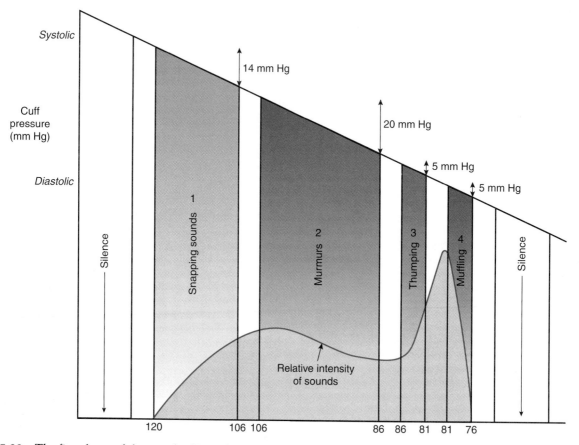

Figure 7.22 The five phases of the sounds of Korotkoff in the measurement of blood pressure.

Phase 4. A *muffled* sound abruptly replaces the thumping sounds of phase 3. This phase begins at a cuff pressure of 81 mm Hg and ends at a pressure of 76 mm Hg.

Phase 5. Silence again as all sounds disappear. This phase is absent in some people.

The cuff pressure at which the first sound is heard (the beginning of phase 1) is the **systolic pressure.** The cuff pressure at which the sound becomes muffled (the beginning of phase 4) and the pressure at which the sound disappears (the beginning of phase 5) are taken as measurements of the **diastolic pressure.** Although the phase 5 measurement is closer to the true diastolic pressure than the phase 4 measurement, the beginning of phase 4 is easier to detect and the results are more reproducible. It is often recommended that both measurements of diastolic pressure be recorded. In the example shown in figure 7.22, the pressure would be indicated as 120/81/76 mm Hg. Frequently, however, the blood pressure would simply be recorded in this example as 120/76 mm Hg, or spoken as "120 over 76". The **pulse pressure** is calculated as the difference in these two pressures; and the **mean arterial pressure** is equal to the diastolic pressure plus one-third of the pulse pressure.

Different examiners can record different values for systolic and diastolic pressures, even on the same subject. Blood pressure measurements can vary with the instruments used, the anxiety of the subject, and even body position. Although the systolic pressure (caused by the left ventricle peak ejection pressure) tends to remain fairly constant with changes in body position, the diastolic pressure is affected by gravity. The arm positioned below the heart level may cause a rise in diastolic pressure, whereas positioning the arm above the heart lowers diastolic pressure.

PROCEDURE

1. Have the subject sit with his or her right or left arm resting on a table at the level of the heart. Wrap the cuff of the sphygmomanometer around the arm about 2.5 cm above the elbow.

A normal blood pressure measurement for a given individual depends on the person's age, sex, heredity, and environment. Considering these factors, chronically elevated blood pressure measurements may indicate an unhealthy state called **hypertension,** a major risk factor in heart disease and stroke.

Hypertension may be divided into two general categories. *Primary hypertension* (95% of all cases) refers to hypertension of unknown etiology. This category is, in turn, divided into *benign hypertension* (also known as *essential hypertension*) and *malignant hypertension*. *Secondary hypertension* refers to hypertension for which the pathological process is known.

2. Palpate the brachial artery in the cubital fossa and place the bell of the stethoscope where the arterial pulse is felt. Gently close the screw valve and pump the pressure in the cuff up to about 20 mm Hg above the point where sounds disappear, or to about 20 mm Hg above the point where the radial pulse can no longer be felt.

3. Open the screw valve to allow the pressure in the cuff to fall slowly, at a rate of about 2 or 3 mm Hg per second.

4. Record the systolic pressure (beginning of phase 1) and the two measurements of diastolic pressure (beginning of phases 4 and 5). Enter these values in your laboratory report and compare your pressures with the range of normal values listed in table 7.1.

5. Calculate the subject's **pulse pressure** (systolic minus diastolic pressure). Enter this value in your laboratory report.

6. Calculate the subject's **mean arterial pressure** (equal to the diastolic pressure plus one-third of the pulse pressure). Enter this value in the laboratory report.

Table 7.1 Blood Pressure Classifications* for Adults in a Resting State

Category	Systolic (mm Hg)		Diastolic (mm Hg)	Recommended Follow-Up
Optimal	<120	and	<80	Recheck in 2 years
Normal	<130	and	<85	Recheck in 2 years
High Normal	130–139	or	85–89	Recheck in 1 year
Hypertension:				
Stage 1—mild	140–159	or	90–99	Confirm within 2 months
Stage 2—moderate	160–179	or	100–109	Evaluate within 1 month
Stage 3—severe	≥180	or	≥110	Evaluate immediately or within 1 week based on clinical situation

Note: Diagnosis of high blood pressure is based on the average of two or more readings taken at each of two or more visits after initial screening. Unusually low readings should be evaluated for clinical significance. © 1997 NIH

*From the National Institutes of Health (NIH), sixth report of the Joint Committee on Detection, Evaluation, and Treatment of High Blood Pressure (1997).

7. Repeat these measurements with the same subject in a *reclining position* (arms at sides). A few minutes later, take the measurements with the subject in a *standing position* (arms down).

8. Alternatively, the sounds of Korotkoff and their correlation with the blood pressure measurements can be visualized on the computer screen using the *Biopac* system blood pressure cuff and stethoscope (fig. 7.23). These may or may not be performed simultaneously with an ECG, as per the *Biopac* lesson 16.

Figure 7.23 *Biopac* **equipment for visualizing Korotkoff sounds and blood pressure.**

Laboratory Report 7.6

DATA FROM EXERCISE 7.6

Enter your data in the table below.

Pressure	Sitting Down	Reclining	Standing Up
Systolic pressure			
Diastolic pressure, phases 4 and 5			
Pulse pressure			
Mean arterial pressure			

REVIEW ACTIVITIES FOR EXERCISE 7.6

Test Your Knowledge of Terms and Facts

1. When blood pressure measurements are taken, the first sound of Korotkoff occurs when the cuff pressure equals the _____ pressure.
2. The last Korotkoff sound occurs when the cuff pressure equals the _____ pressure.
3. The sounds of Korotkoff are produced by _____ .
4. Suppose a person's blood pressure is 168/112.
   (a) What is the systolic pressure? _____
   (b) What is the diastolic pressure? _____
   (c) What is the pulse pressure? _____
   (d) What is the mean arterial pressure? _____
5. What condition does the person described in question 4 have? _____
6. The arterial blood pressure is directly proportional to two factors: the _____ and the _____ .
7. The scientific name of the device used to take a blood pressure reading (hint: one word) is the:
   _____ .

Test Your Understanding of Concepts

8. Explain the cause of the Korotkoff sounds, and why you normally don't hear them when the cuff is not inflated.

9. Explain why the first Korotkoff sound is heard when the cuff pressure equals the systolic pressure. Also explain why the last sound is heard when the cuff pressure equals the diastolic pressure.

Test Your Ability to Analyze and Apply Your Knowledge

10. Were the blood pressure measurements different when they were taken in the standard sitting position, reclining, and standing up? Explain why might have caused any differences observed?

11. How do you think the arterial blood pressure, pulse pressure, and mean arterial pressure would change during exercise? Explain.

12. How might a regular program of aerobic exercise influence the resting blood pressure? Explain your answer.

Cardiovascular System and Physical Fitness

MATERIALS

1. Sphygmomanometer
2. Stethoscope, watch with second hand
3. Chair, stair, or platform 18 inches high
4. Optional: pulse and blood pressure may be monitored with the *Biopac* system using the pulse transducer (see fig. 7.14 in exercise 7.3) and blood pressure cuff (see fig. 7.23 in exercise 7.6).

Physical fitness is dependent on adaptations of the cardiovascular system that include increased stroke volume, decreased resting cardiac rate, and improved cardiovascular responses to exercise. Controlled exercise can be used to assess physical fitness and also to detect heart disease.

OBJECTIVES

1. Describe the relationship between age, maximum cardiac rate, and training cardiac rate.
2. Describe the cardiovascular changes that occur when a person becomes physically fit.
3. Explain how controlled exercise protocols may be used to detect heart disease.

Textbook Correlations

Before performing this exercise, you may want to consult the following references in *Human Physiology*, seventh edition, by Stuart I. Fox:

- *Adaptations of Muscles to Exercise Training.* Chapter 12, p 346.
- *Circulatory Changes During Exercise.* Chapter 14, pp. 426–428.

Those using different physiology textbooks may want to consult the corresponding information in those books.

Although the **maximum cardiac rate** (beats per minute) can be estimated for adults in a given age group, those who are physically fit have a higher *stroke volume* (milliliters per beat) and, therefore, a greater *cardiac output* (milliliters per minute) than those who are more sedentary. Since cardiac output is equal to heart rate multiplied by stroke volume, at any given cardiac output the athlete with a higher stroke volume usually will have a lower heart rate than the nonathlete. A person in poor physical condition reaches his or her maximum cardiac rate at a lower work level than a person of comparable age who is in better shape. Maximum cardiac rates for adults can be estimated by subtracting your age from 220, and training rates are usually 60–80% of this maximum, as shown in table 7.2.

The **maximal oxygen uptake,** or **aerobic capacity,** measures the maximum rate of oxygen consumption by the body. The intensity of exercise can be rated by the percent of the aerobic capacity attained. In most healthy

Table 7.2 Maximum and Training Cardiac Rates, in Beats per Minute (bpm)

Age	Maximum Cardiac Rate	Training Cardiac Rate (% of maximum)
20	200 bpm	120 bpm (60%)
		160 bpm (80%)
25	195 bpm	117 bpm (60%)
		156 bpm (80%)
30	190 bpm	114 bpm (60%)
		152 bpm (80%)
35	185 bpm	111 bpm (60%)
		148 bpm (80%)
40	180 bpm	108 bpm (60%)
		144 bpm (80%)
45	175 bpm	105 bpm (60%)
		140 bpm (80%)
50	170 bpm	102 bpm (60%)
		136 bpm (80%)
55	165 bpm	99 bpm (60%)
		132 bpm (80%)
60	160 bpm	96 bpm (60%)
		128 bpm (80%)

people, blood levels of lactic acid (lactate) rise significantly when exercise is performed at about 50–70% of their aerobic capacity. This is called the **lactate (or anaerobic) threshold.** Endurance trained athletes have a higher aerobic capacity and may not reach their lactate threshold until they exercise at about 80% of their maximal oxygen uptake.

The primary cause of the higher aerobic capacity in endurance trained athletes is their higher maximum cardiac outputs, and thus their higher rates of oxygen delivery to the muscles. Endurance training increases the cardiac output through a stronger contraction of the ventricles (thus ejecting more blood per beat), and an increase in blood volume.

The higher stroke volume allows the same cardiac output to be achieved at a slower heart rate, as previously mentioned. The slower resting heart rate of endurance trained athletes—a condition called *athlete's bradycardia*—results from higher levels of inhibitory activity by the vagus nerve innervation to the SA node.

 Exercise testing using clinical protocols have proved extremely useful in the diagnosis of heart disease, particularly coronary artery disease and **myocardial ischemia** (inadequate blood flow to the heart). People who appear to be healthy and record normal electrocardiograms at rest may develop *angina pectoris* and abnormal ECG patterns during or after exercise. During these tests and under the watchful eye of a physician, the patient follows a standardized exercise protocol (such as the Balke, Stanford, or Naughton) while the ECG is continuously recorded and blood pressures are taken automatically at regular intervals. Most cardiologists prefer a treadmill test, during which the speed and incline are automatically varied from easy to difficult, with the patient reaching target heart rates as high as 95% of maximum for the patient's age. During the test, irregularities in the ECG or blood pressure, such as depressed or elevated S–T segments, can be observed, which may indicate myocardial ischemia.

PROCEDURE

1. Measure your reclining (lying down) pulse by placing your fingertips (not the thumb) on the radial artery in the ventrolateral region of the wrist.* Count the number of pulses in 30 seconds and multiply by two. Score points as indicated.

Reclining Pulse

Rate	Points
50–60	3
61–70	3
71–80	2
81–90	1
91–100	0
101–110	–1
Score:	

2. After the reclining pulse has been measured, stand up and measure the pulse rate *immediately* upon standing.

Standing Pulse Rate

Rate	Points
60–70	3
71–80	3
81–90	2
91–100	1
101–110	1
111–120	0
121–130	0
131–140	–1
Score:	

*The procedure for the exercise is from "A Cardiovascular Rating as a Measure of Physical Fatigue and Efficiency" by E. C. Schneider. JAMA 74(1920):1507. Copyright 1920 American Medical Association.

3. Subtract the pulse rate of step 1 from the pulse rate of step 2 to obtain the pulse rate increase on standing.

Pulse Rate Increase on Standing

Reclining Pulse	0–10 Beats	11–18 Beats	19–26 Beats	27–34 Beats	35–43 Beats
50–60	3	3	2	1	0
61–70	3	2	1	0	–1
71–80	3	2	0	–1	–2
81–90	2	1	–1	–2	–3
91–100	1	0	–2	–3	–3
101–110	0	–1	–3	–3	–3

Score: _____

4. Place your right foot on a chair or stair that is 18 inches high. Raise your body so that your left foot comes to rest by your right foot. Return your left foot to the original position, followed by the right foot. Repeat this exercise five times, allowing 3 sec for each step up. Immediately upon completion of this exercise, measure the pulse for 15 sec and multiply by four. Record this pulse rate.
Pulse: _____
Measure the pulse as described for 30, 60, 90, and 120 sec after completion of the exercise. Record the time it takes for the pulse to return to normal rate while standing (step 2). Score points as indicated.

Time for Return of Pulse to Normal, Standing Post-exercise

Seconds	Points
0–30	4
31–60	3
61–90	2
91–120	1
After 120	0

5. Subtract your normal standing pulse rate (step 2) from your pulse rate immediately after exercise (step 4).

Pulse Rate Increase Immediately Postexercise

Standing Pulse	0–10 Beats	11–20 Beats	21–30 Beats	31–40 Beats	Over 41 Beats
60–70	3	3	2	1	0
71–80	3	2	1	0	–1
81–90	3	2	1	–1	–2
91–100	2	1	0	–2	–3
101–110	1	0	–1	–3	–3
111–120	1	–1	–2	–3	–3
121–130	0	–2	–3	–3	–3
131–140	0	–3	–3	–3	–3

Score: _____

6. Calculate the change in systolic blood pressure as you go from a reclining (lying down position) to a standing position (refer to data in exercise 7.6 or take new measurements).

Change in Systolic Pressure from Reclining to Standing

Change (mm Hg)	Points
Rise of 8 or More	3
Rise of 2–7	2
No rise	1
Fall of 2–5	0
Fall of 6 or more	–1

Score: _____

7. Determine your total score for all the tests and evaluate this score on the following basis:

Excellent	18–17
Good	16–14
Fair	13–8
Poor	7 or less

Enter your score and rating in the laboratory report.

DATA FOR EXERCISE 7.7

Write your total score in the space below, and indicate whether this score is excellent, good, fair, or poor according to the rating scale in the procedure.

     Total Score: _____

     Overall Rating: _____

REVIEW ACTIVITIES FOR EXERCISE 7.7

Test Your Knowledge of Terms and Facts

1. As a person gets older, the maximum cardiac rate _____ .
2. If a person has athlete's bradycardia, the resting heart rate is _____ than the average.
3. The condition described in question 2 is caused by _____ .
4. Define the *aerobic capacity*. _____
5. Define the *lactate threshold*. _____
6. The primary cause of the higher aerobic capacity of endurance trained athletes is _____
    _____ .

Test Your Understanding of Concepts

7. What cardiovascular adaptations are associated with endurance training? How do these changes help to improve performance?

8. How does the increase in blood pressure and pulse rate after exercise, and the return of these values to baseline following exercise, compare in people who are and who are not physically fit?

9. How does a person's aerobic capacity and lactate threshold change with age? What causes these changes? How might these changes be minimized?

10. What is myocardial ischemia? How might exercise provoke this condition? How might a regular exercise program in a healthy person help reduce the chances of developing myocardial ischemia? What other measures should be taken to minimize the risk of developing myocardial ischemia?

Respiration and Metabolism

All of the chemical reactions that occur in the body are collectively termed **metabolism.** Those reactions that build larger molecules out of smaller ones are dehydration synthesis reactions and require input of energy (are *endergonic*); they are collectively termed **anabolism.** Reactions that break down larger molecules into smaller ones (by hydrolysis) release energy (are *exergonic*) and are collectively termed **catabolism.** The energy required for anabolic reactions is provided by the hydrolysis of *adenosine triphosphate (ATP).* This catabolic reaction produces *adenosine diphosphate (ADP)* and inorganic *phosphate* (PO_4^{3-}) (see reaction **a,** below).

The energy released by the hydrolysis of ATP is used to power all of the anabolic reactions in the cell. ATP is frequently referred to as the *universal energy carrier* of the cell. Since ATP is continuously hydrolyzed to release energy, energy must also be continuously supplied to resynthesize ATP by the reverse reaction, condensation of ADP with phosphate (see reaction **b,** below).

The energy required for this anabolic synthesis of ATP is provided by a sequence of catabolic reactions known as **cellular respiration.** In this complex process, monosaccharides, amino acids, and fatty acids serving as fuels are broken down to smaller molecules in a series of hydrolytic steps, with the accompanying release of energy.

Simplifying cell respiration by considering the fate of just one fuel molecule—we'll look at glucose—the final products can be indicated without showing the numerous intermediate steps. In the presence of oxygen, glucose is broken down to carbon dioxide and water, and the energy released is used to form 30 molecules of ATP. Since oxygen is required for this process, it is called *aerobic respiration* (see reaction **c**).

The continuous hydrolysis of ATP by body cells for energy requires the continuous delivery of glucose and other fuel molecules, plus oxygen for aerobic respiration. In the absence of oxygen, each molecule of glucose is broken down to lactic acid, a process called *anaerobic respiration,* with a net gain of two molecules of ATP (see reaction **d,** below).

With the production of thirty molecules of ATP per molecule of glucose, aerobic respiration is obviously more economical than anaerobic respiration. Moreover, the aerobic pathway is also favored because the accumulation of lactic acid is toxic to the body.

Oxygen is delivered to the cells of the body by *hemoglobin* molecules located within *red blood cells (erythrocytes).* Overall, the adequate delivery of oxygen to the tissue cells depends on two factors: **ventilation** and **oxygen transport.** Ventilation (breathing) results in the exchange of "old," oxygen-poor air with "new," oxygen-rich air within the alveoli of the lungs. Oxygen transport includes the uptake, transport, and delivery of oxygen by hemoglobin, as follows:

1. the diffusion of oxygen from the alveoli of the lungs into the red blood cells within pulmonary capillaries,
2. the *loading* of oxygen onto hemoglobin molecules located within these pulmonary capillary red blood cells,
3. the *unloading* of oxygen from the red blood cell hemoglobin,
4. the diffusion of oxygen from the red blood cells and tissue capillaries into the body cells.

---

a. $ATP + H_2O \longrightarrow ADP + PO_4^{3-} + energy$

b. $ADP + PO_4^{3-} + energy \longrightarrow ATP + H_2O$

c. $Glucose + O_2 + 30\ ADP + 30\ PO_4^{3-} \longrightarrow CO_2 + H_2O + 30\ ATP$

d. $Glucose + 2\ ADP + 2\ PO_4^{3-} \longrightarrow lactic\ acid + 2\ ATP$

Figure 8.1 The Collins 9-liter respirometer.

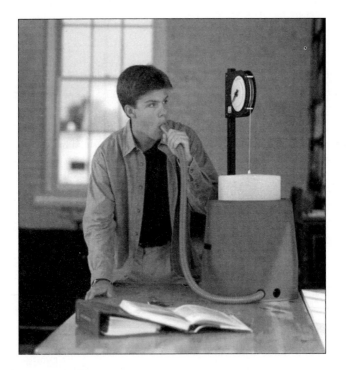

Figure 8.2 Equipment involved in using the Spirocomp program.

Measurements of Pulmonary Function

MATERIALS

1. Collins, Inc. 9-L respirometer or *Spirocomp* program and equipment
2. Disposable mouthpieces and nose clamp
3. Alternatively, the *Biopac* system may be used with the set up for *Biopac* lesson 12 (for part A of this exercise) and lesson 13 (for part B of this exercise).

Spirometry is used to measure lung volumes and capacities and to measure ventilation as a function of time. Such measurements are clinically useful in the diagnosis of restrictive and obstructive pulmonary disorders.

OBJECTIVES

1. Identify the major muscles involved in inspiration and expiration; explain the mechanics of breathing.
2. Define the different lung volumes and capacities.
3. Perform a normal spirogram, and determine measurements of the different lung volumes and capacities.
4. Describe and perform the forced expiratory volume (FEV) test; and determine FEV from the spirogram.
5. Explain how pulmonary function tests are used in the diagnosis of restrictive and obstructive pulmonary disorders.

Textbook Correlations*

Before performing this exercise, you may want to consult the following references in *Human Physiology*, seventh edition, by Stuart I. Fox:

- *Physical Aspects of Ventilation.* Chapter 16, pp. 487–490.
- *Mechanics of Breathing.* Chapter 16, pp. 490–495.

Those using different physiology textbooks may want to consult the corresponding information in those books.

---

milliliters, and the *x* axis is graduated in millimeters. Because the speed with which the chart moves is known, a graph of air volume (mL) moved into and out of the lungs in a given time interval can be obtained.

Alternatively, a computerized setup using a Phipps and Bird spirometer and a program for analyzing the data (Spirocomp) may be used (fig. 8.2). In this case, the treated data are displayed on a computer screen. Or, if the *Biopac* system is employed, a wet spirometer is avoided because the flow of air is transduced and fed into the *Biopac* MP30 unit (fig. 8.3), which then displays the data on a computer screen. By means of spirometry, many important aspects of pulmonary function can be visualized and measured, as described below and shown in figure 8.4.

The **total lung capacity (TLC)** is the total volume of gas in the lungs after a maximum (forced) inhalation.

The **vital capacity (VC)** is the maximum volume of gas that can be exhaled after a maximum inhalation.

Spirometry is a technique for measuring lung volumes and capacities. A Collins respirometer can be used for this purpose (fig. 8.1). As the subject exhales into a mouthpiece, the oxygen bell rises, causing a pen to move downward on a moving chart (kymograph). Since this is a closed system, soda lime is provided in the system to remove CO_2 from the exhaled air. As the subject inhales, the oxygen bell falls, causing the pen to move upward on the moving chart. The **y** axis of the chart is graduated in

---

*See Appendix 3 for correlations to the A.D.A.M. *InterActive PHYSIOLOGY Modules.*

See Appendix 3 for correlations to the *Virtual Physiology Laboratory* CD-ROM by McGraw-Hill and Cypris Publishing, Inc.

See Appendix 3 for correlations to the *Intelitool Physiology Laboratory Exercises.*
See Appendix 3 for correlations to the *Biopak Student Lab Exercises.*

Figure 8.3 *Biopac* system equipment for spirometry.

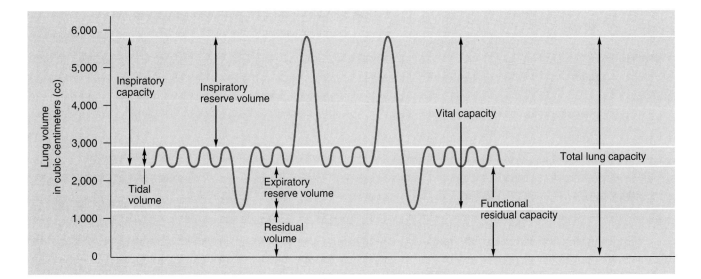

Figure 8.4 Spirogram recording of lung volumes and capacities.

The **tidal volume (TV)** is the volume of gas inspired or expired during each normal (unforced) ventilation cycle.

The **inspiratory capacity (IC)** is the maximum volume of gas that can be inhaled after a normal (unforced) exhalation.

The **inspiratory reserve volume (IRV)** is the maximum volume of gas that can be forcefully inhaled after a normal (tidal) inhalation.

The **expiratory reserve volume (ERV)** is the maximum volume of gas that can be forcefully exhaled after a normal (tidal) exhalation.

The **functional residual capacity (FRC)** is the volume of gas remaining in the lungs after a normal (unforced) exhalation.

The **residual volume (RV)** is the volume of gas remaining in the lungs after a maximum (forced) exhalation.

The movement of air into and out of the lungs (ventilation) results from a pressure difference between the pulmonary air and the atmosphere. This pressure difference is created by a change in the volume of the thoracic cavity. Since, according to **Boyle's law,** the *pressure*

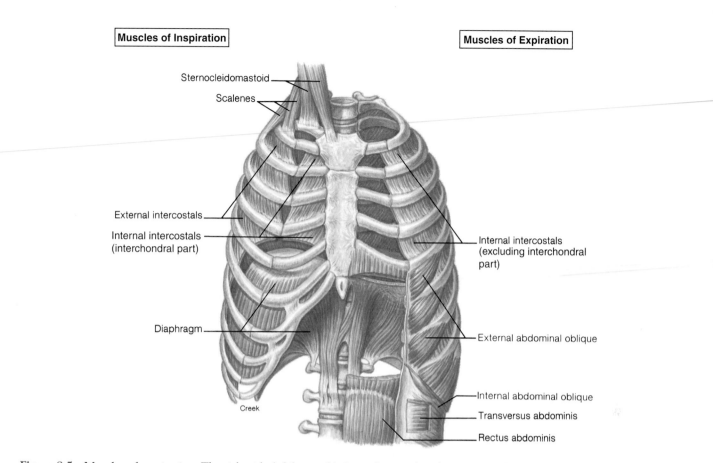

Muscles of Inspiration

Sternocleidomastoid
Scalenes

External intercostals
Internal intercostals
(interchondral part)

Diaphragm

Creek

Muscles of Expiration

Internal intercostals
(excluding interchondral
part)

External abdominal oblique

Internal abdominal oblique
Transversus abdominis
Rectus abdominis

Figure 8.5 **Muscles of respiration.** The right side (of the trunk) shows the muscles of inspiration; the left side depicts the muscles involved in forced expiration.

of a gas is inversely proportional to its volume, an increase in thoracic volume results in a decrease in intrapulmonary pressure. During inhalation, therefore, air is pushed into the lungs by the greater pressure of the atmosphere. When the thoracic volume decreases during exhalation, the intrapulmonary pressure rises above the atmospheric pressure and air is pushed out of the lungs.

In normal (unforced) ventilation, the thoracic volume is regulated by action of the *diaphragm* and the *external intercostal muscles* (fig. 8.5). At rest, the diaphragm forms the convex floor to the thoracic cavity. During inhalation, the diaphragm contracts and pulls itself into a more flattened form. This lowers the floor of the thorax and increases the thoracic volume (and pushes down on the viscera, causing the abdomen to protrude during inhalation). At the same time, the contraction of the external intercostal muscles increases the volume of the thorax by rotating the ribs upward and outward. At the end of inhalation, the diaphragm and external intercostal muscles relax, causing the thorax to resume its original volume and the air inside the lungs to be exhaled. The

amount of air inhaled or exhaled in this manner is the tidal volume.

Inhalation can become difficult if the air passages are obstructed or if the lungs lose their normal elasticity. In these cases, the affected person relies increasingly on muscles not usually used in normal (tidal) ventilation: the *scalenus, sternocleidomastoid,* and *pectoralis major muscles.* These muscles are also used in healthy people during forced inhalation to obtain the inspiratory reserve volume (IRV).

During forced exhalation, the *internal intercostal muscles* contract, depressing the rib cage, and the *abdominal muscles* contract, pushing the viscera up against the diaphragm. The push of the viscera increases the convexity of the diaphragm and decreases the thoracic volume to a lower level than that achieved in normal exhalation. The amount of air forcefully exhaled by contraction of both groups of muscles is the expiratory reserve volume (ERV).

Even after a maximum forced exhalation, there is still some air left in the lungs. This residual volume of air makes it easier to inflate the lungs during the next inhalation and oxygenates the blood between ventilation cycles.

285

A. Measurement of Simple Lung Volumes and Capacities

PROCEDURE (FOR SPIROCOMP)

1. Press the **"T"** key on the computer and the words "Breathe Normal Cycles" will appear on the computer screen. After three normal tidal volume cycles, the data will appear on the screen.
2. Press the **"E"** key on the computer and the words "Breathe Normal Cycles" will appear on the screen. At the third breathing cycle, the words "Stop After Normal Exhale" will appear.
3. After the pause in breathing, the words "Exhale Forcefully" will appear on the screen. Forcefully exhale all you can at this point.
4. Press the **"V"** key and the words "Inhale Max Then Press V Exhale Fully" will appear on the screen. After a maximal inhalation, press "V" and forcefully exhale all the air you can as fast as possible.
5. Record the data displayed on the screen and use it to complete your laboratory report.

PROCEDURE (FOR COLLINS RESPIROMETER)

1. Raise and lower the oxygen bell (fig. 8.1) several times to get fresh air into the spirometer. Notice that as the bell moves up and down, one of the pens moves a corresponding distance down and up on a shaft. By adjusting the height of the oxygen bell, position this pen so that it will begin writing in the middle of the chart paper. This pen (the *ventilometer* pen) usually has black ink; whereas the other (*respiration*) pen usually has red ink and will not be used for this exercise. (It should be covered and rocked away from the paper.)
2. With the free-breathing valve set to the *open* position, place the mouthpiece in the buccal cavity (as in breathing through a snorkel), and go through several ventilation cycles to become accustomed to the apparatus. (When the free-breathing valve is open, you will breathe room air.) If a disposable cardboard mouthpiece is used, be particularly careful to prevent air leakage from the corners of the mouth. Breathing through the nostrils can be prevented by means of a nose clamp or by pinching the nose tightly with the thumb and forefinger.
3. Turn the respirometer to the *slow* position (32 mm/min) and close the free-breathing valve so that the oxygen bell and the pen go up and down with each ventilation cycle.
4. Breathe in a normal, relaxed manner for 1–2 min. The breaths should appear relatively uniform, and the slope should go upward (see fig. 8.6).

A downward slope indicates that there is an air leak; in this event, tighten the grip of the mouth on the mouthpiece and the nose clamp on the nose, and begin again.

Note: *At this speed (32 mm/min), the distance between heavy vertical lines on the chart is traversed in 1 min.*

This procedure measures tidal volume—the amount of air inhaled or exhaled during each resting ventilation cycle.

5. With the drum still turning, perform the test for vital capacity (the maximum amount of air that can be exhaled after a maximum inhalation). At the end of a normal exhalation, inhale as much as possible, and then exhale completely to the fullest possible extent, and stop the recording. At the completion of this exercise, the chart should resemble the one shown in figure 8.6.
6. Remove the chart from the kymograph drum. Notice that the chart is marked horizontally in milliliters.

Note: *Since the temperature and pressure of the respirometer are different from those existing in the body, the volume the air occupies in the respirometer will be subject to changes in ambient (room) conditions. To standardize the volumes measured in spirometry, multiply these measured volumes by a correction factor known as the* **BTPS factor** *(body temperature, atmospheric pressure, saturated with water vapor). Since the BTPS factor is very close to* **1.1** *at normal room temperatures, we will use this figure in the calculations.*

Calculations

1. Obtain the measured tidal volume from the chart by subtracting the milliliters corresponding to the trough from the milliliters corresponding to the peak of a typical resting ventilation cycle.

Example (from fig. 8.7)

Step 1

$$
\begin{aligned}
&3{,}700 \text{ mL (inhalation)} \\
&\underline{-\,3{,}250 \text{ mL (exhalation)}} \\
&\,450 \text{ mL}
\end{aligned}
$$

Step 2

$$
\begin{aligned}
&450 \text{ mL (measured tidal volume)} \\
&\underline{\times\ \ \ 1.1 \ \ \ \text{ (BTPS factor)}} \\
&495 \text{ mL}
\end{aligned}
$$

Enter the corrected measured tidal volume (TV) in the *Measured* column of the table in your laboratory report.

2. Obtain the measured inspiratory capacity from the chart. To do this, subtract the milliliters corresponding to the last normal exhalation before performing the vital capacity maneuver from the milliliters corresponding to the maximum inhalation peak.

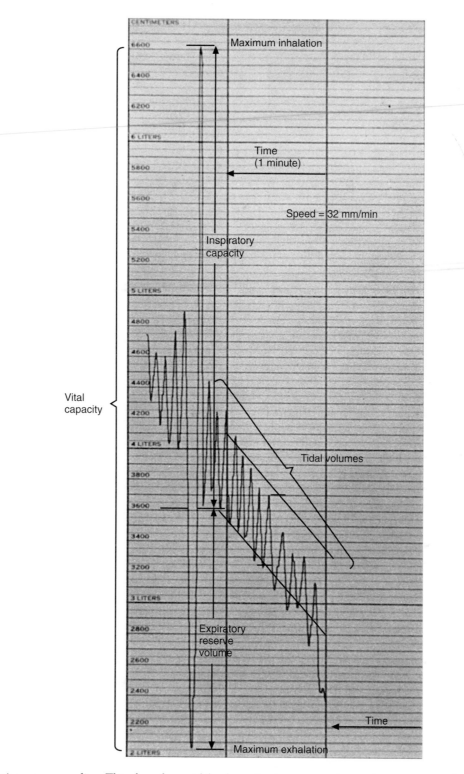

Figure 8.6 A spirometry recording. This chart shows tidal volume, inspiratory capacity, expiratory reserve volume, and vital capacity.

Figure 8.7 A close-up of tidal volume measurements from figure 8.6.

Example (from fig. 8.8)

Step 1   6,650 mL (maximum inhalation)
                 <u>– 3,650 mL</u> (normal exhalation)
                 3,000 mL

Step 2   3,000 mL (measured inspiratory capacity)
          ×<u>   1.1   </u> (BTPS factor)
                3,300 mL

Enter the corrected measured inspiratory capacity (IC) in the *Measured* column of the table in your laboratory report.

3. Obtain the measured expiratory reserve volume by subtracting the milliliters corresponding to the trough for maximum exhalation from the milliliters corresponding to the last normal exhalation before the vital capacity maneuver (same value used for step 2).

Figure 8.8 A close-up of the inspiratory capacity measurement from figure 8.6.

Example (from fig. 8.9)

Step 1   3,650 mL (normal exhalation)
                – 2,050 mL (maximum exhalation)
                   1,600 mL

Step 2   1,600 mL (measured expiratory reserve)
          ×   1.1   1,600 mL (volume)
            1,760 mL (BTPS factor)

Enter the corrected expiratory reserve volume (ERV) in the *Measured* column of the table in your laboratory report.

4. Obtain the measured vital capacity by either (1) adding the corrected inspiratory capacity (from step 2) and the corrected expiratory reserve volume (step 3) (since these values have already been BTPS standardized, an additional correction step is unnecessary); or (2) subtracting the milliliters corresponding to maximum exhalation from the milliliters corresponding to maximum inhalation. This value must then be multiplied by the BTPS factor.

Method 1

     3,300 mL (corrected inspiratory capacity)
     + 1,760 mL (corrected expiratory reserve volume)
     5,060 mL (corrected vital capacity)

Method 2

     6,650 mL (maximum inhalation)
     – 2,050 mL (maximum exhalation)
     4,600 mL

     4,600 mL (measured vital capacity)
     ×   1.1   (BTPS factor)
     5,060 mL (corrected capacity)

Enter the corrected vital capacity (VC) in the *Measured* column of the table in your laboratory report.

5. Obtain the *predicted vital capacity* for the subject's sex, age, and height from tables 8.3 and 8.4. The height in centimeters can be most conveniently obtained by referring to the height conversion scale in figure 8.10.

Example

Sex: male
Age: 34
Height: 174 cm
Predicted vital capacity: 4,140 mL

Enter the subject's predicted vital capacity from the tables of normal values in the *Predicted* column in your laboratory report.

6. To obtain an estimate of the predicted residual volume and the predicted total lung capacity, refer to table 8.1.

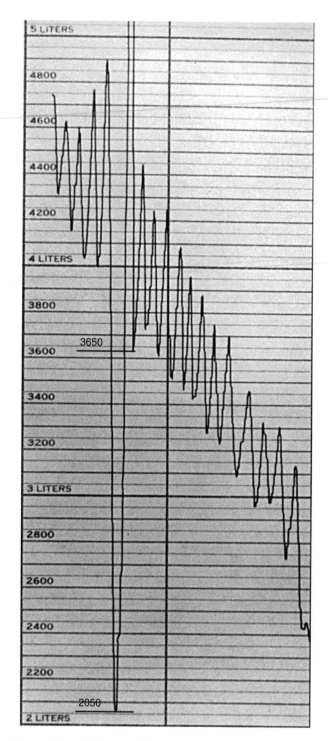

Figure 8.9   A close-up of the expiratory reserve volume measurement from figure 8.6.

Note: *These values cannot be measured by spirometry because residual volume cannot be exhaled; total lung capacity equals the vital capacity plus residual volume.*

7. Obtain the percent predicted value for all the measurements in the following way:

$$\text{Percent predicted} = \frac{\text{corrected measured value}}{\text{predicted value}} \times 100\%$$

Figure 8.10 A scale for converting height in feet and inches to height in centimeters.

Table 8.1 Factors for Obtaining the Predicted Residual Volume and Total Lung Capacity

Age	Residual Volume: Vital Capacity × Factor	Total Lung Capacity: Vital Capacity × Factor
16–34	0.250	1,250
35–49	0.305	1,305
50–69	0.445	1,445

Example

Measured vital capacity 5,060 mL (corrected to BTPS)
Predicted vital capacity 4,140 mL (from table 8.3 or 8.4)

$$\% \text{ predicted} = \frac{5,060 \text{ mL}}{4,140 \text{ mL}} \times 100\%$$
$$= 122\%$$

Enter the percent predicted values in the appropriate places in the table in your laboratory report.

Measurements of vital capacity that are consistently below 80% of the predicted value on repeated tests suggest the presence of a restrictive lung disease, such as emphysema.

B. Measurement of Forced Expiratory Volume

The ability to ventilate the lungs in a given amount of time is often of greater diagnostic value than measurements of simple lung volumes and capacities. One measurement that considers time intervals is the **forced expiratory volume (FEV),** otherwise known as the *timed vital capacity.*

In the forced expiratory volume test, the subject performs a vital capacity maneuver by inhaling maximally, holding, and then exhaling maximally. While holding at the point of peak inhalation, the respirometer kymograph drum speed is set to its fastest setting (1,920 mm/min); then the subject is instructed to exhale forcefully and maximally. This fast speed stretches out the exhalation tracing and the distance between heavy vertical lines is now traversed in 1 sec. From the recording, the percentage of the total vital capacity that is exhaled in the *first* second (FEV_1), the *second* second (FEV_2), and the *third* second (FEV_3) can be determined. A sample record of the forced expiratory volume is shown in figure 8.11.

Pulmonary Disorders

Chronic (long-term) pulmonary dysfunctions can be divided into two general categories: obstructive disorders and restrictive disorders. These two categories can be distinguished, in part, by the use of the spirometry tests performed in this exercise.

Since the flow of air through a tube is proportional to the fourth power of its radius (r^4), a small obstruction in the pulmonary airways results in a greatly magnified resistance to airflow. **Obstructive disorders** of the bronchioles, for example, are characteristic of *emphysema, bronchitis,* and *asthma.* This obstruction can result from

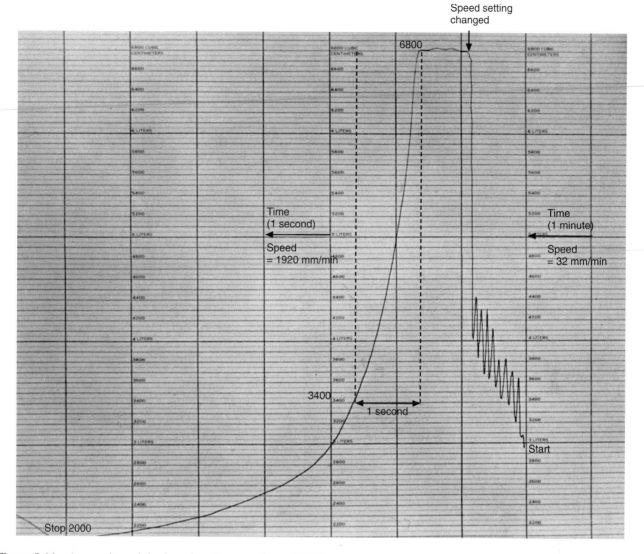

Figure 8.11 A recording of the forced expiratory volume (FEV) measurement.

bronchiolar secretions, inflammation and edema, or contraction of bronchiolar smooth muscle. These conditions make it difficult for sufferers to move air rapidly into and out of the lungs. The bronchioles may be weakened to such a degree that they collapse during exhalation before all the air has been emptied from the lungs. This condition, known as *air trapping*, is revealed by an increase in the functional residual capacity.

In **restrictive disorders,** actual damage to the lung tissue results in an abnormal vital capacity test. However, if the disease is purely restrictive (as in *pulmonary fibrosis*), the airways may be clear, resulting in a normal forced expiratory volume (FEV) test. The vital capacity is reduced, but it can be quickly exhaled. This is not true of emphysema, which is both a restrictive and an obstructive disease.

Caused primarily by cigarette smoking and aggravated by air pollution, emphysema reduces the number of

alveoli in the lungs. This results in an abnormally low vital capacity, indicative of restrictive disease. However, since the elastic alveolar tissue normally helps to keep the thin-walled bronchioles open during exhalation, the loss of alveoli in emphysema also results in a reduction in the elastic support of the bronchioles. During exhalation, such bronchioles may narrow (increasing the resistance to airflow) and even collapse. This adds an obstructive component to the disease, resulting in an abnormal forced expiratory volume test.

The FEV_1 test detects increased airway resistance, as occurs in emphysema, bronchitis, and asthma. Also used preoperatively, this test is used extensively to predict a patient's response to general anesthesia and to estimate the length of time the patient must be kept on a respirator postoperatively. The FEV_1 test is also valuable in research on the effects of air pollutants—such as cigarette smoke and ozone—on pulmonary function.

PROCEDURE

Note: *If the Spirocomp was used, the FEV₁ has already been computed and was displayed on the computer screen when the "V" test in part A was performed. The following procedure is for the Collins respirometer.*

1. Start the kymograph slowly rotating (32 mm/min), and have the subject breathe normally into the respirometer for a few breaths, until comfortable.
2. After a normal (unforced) exhalation, instruct the subject to take a deep, forceful inhalation and to hold this inhalation momentarily.
3. Switch the kymograph to the fast speed (1,920 mm/min), and instruct the subject to exhale as rapidly and as forcefully as possible.

Note: *At a speed of 1,920 mm/min, the time interval between two heavy vertical lines on the chart is 1 sec.*

Calculations: Forced Expiratory Volume (FEV₁)

1. Measure the vital capacity from the chart by subtracting the exhalation trough (which is flat, because all air has been expelled) from the inhalation peak (also flat, because the subject's breath is held). Do not multiply this value by the BTPS factor.

Example (from fig. 8.11)

6,800 mL (maximum inhalation)
− 2,000 mL (maximum exhalation)
4,800 mL (vital capacity)

Enter the *uncorrected* vital capacity below:
_____ mL

2. Measure the amount of air exhaled in the first second by subtracting the milliliters corresponding to the exhalation line after 1 sec (3,400 mL, in fig. 8.11) from the milliliters of the inhalation peak at the moment of exhale (6,800 mL). Remember that here the distance between heavy vertical lines is passed in 1 sec. If the subject does not begin to exhale exactly on a vertical line, use a ruler to measure 3.2 cm horizontally from the start of exhalation. (This is the distance between vertical lines and is equivalent to 1 sec at this chart speed.)

Example (from fig. 8.11)

6,800 mL (maximum inhalation)
− 3,400 mL (exhalation line after 1 sec)
3,400 mL (amount exhaled in first second)

Enter the amount exhaled in the first second below:
_____ mL

3. Calculate the percentage of the vital capacity exhaled in the first second (the FEV₁).

Example

$$FEV_1 = \frac{3,400 \text{ mL (step 2)}}{4,800 \text{ mL (step 1)}} \times 100\%$$
$$= 70.8\%$$

Enter the FEV₁ in the laboratory report. Refer to table 8.2, and enter the predicted percentage for the FEV₁ in the laboratory report.

Table 8.2 Predicted Percentage of the Vital Capacity (VC) Exhaled during the First Second (FEV₁)

Age	Predicted Percent VC (FEV₁)
18–29	82–80 %
30–39	78–77 %
40–44	75.5 %
45–49	74.5 %
50–54	73.5 %
55–64	72–70 %

From *Archives of Environmental Health,* Volume 12, p. 146, 1966. Reprinted with permission of the Helen Dwight Reid Educational Foundation. Published by Heidref Publications, 1319 Eighteenth St., N.W., Washington, D.C. 20036-1802. Copyright © 1966.

Table 8.3 Predicted Vital Capacities, Females (mL)

Height in Centimeters

Age	146	148	150	152	154	156	158	160	162	164	166	168	170	172	174	176	178	180	182	184	186	188	190	192	194
16	2950	2990	3030	3070	3110	3150	3190	3230	3270	3310	3350	3390	3430	3470	3510	3550	3590	3630	3670	3715	3755	3800	3840	3880	3920
17	2935	2975	3015	3055	3095	3135	3175	3215	3255	3295	3335	3375	3415	3455	3495	3535	3575	3615	3655	3695	3740	3780	3820	3860	3900
18	2920	2960	3000	3040	3080	3120	3160	3200	3240	3280	3320	3360	3400	3440	3480	3520	3560	3600	3640	3680	3720	3760	3800	3840	3880
20	2890	2930	2970	3010	3050	3090	3130	3170	3210	3250	3290	3330	3370	3410	3450	3490	3525	3565	3605	3645	3695	3720	3760	3800	3840
22	2860	2900	2940	2980	3020	3060	3095	3135	3175	3215	3255	3290	3330	3370	3410	3450	3490	3530	3570	3610	3650	3685	3725	3765	3800
24	2830	2870	2910	2950	2985	3025	3065	3100	3140	3180	3220	3260	3300	3335	3375	3415	3455	3490	3530	3570	3610	3650	3685	3725	3765
26	2800	2840	2880	2920	2960	3000	3035	3070	3110	3150	3190	3230	3265	3300	3340	3380	3420	3455	3495	3530	3570	3610	3650	3685	3725
28	2775	2810	2850	2890	2930	2965	3000	3040	3070	3115	3155	3190	3230	3270	3305	3345	3380	3420	3460	3495	3535	3570	3610	3650	3685
30	2745	2780	2820	2860	2895	2935	2970	3010	3045	3085	3120	3160	3195	3235	3270	3310	3345	3385	3420	3460	3495	3535	3570	3610	3645
32	2715	2750	2790	2825	2865	2900	2940	2975	3015	3050	3090	3125	3160	3200	3235	3275	3310	3350	3385	3425	3460	3495	3535	3570	3610
34	2685	2725	2760	2795	2835	2870	2910	2945	2980	3020	3055	3090	3130	3165	3200	3240	3275	3310	3350	3385	3425	3460	3495	3535	3570
36	2655	2695	2730	2765	2805	2840	2875	2910	2950	2985	3020	3060	3095	3130	3165	3205	3240	3275	3310	3350	3385	3420	3460	3495	3530
38	2630	2665	2700	2735	2770	2810	2845	2880	2915	2950	2990	3025	3060	3095	3130	3170	3205	3240	3275	3310	3350	3385	3420	3455	3490
40	2600	2635	2670	2705	2740	2775	2810	2850	2885	2920	2955	2990	3025	3060	3095	3135	3170	3205	3240	3275	3310	3345	3380	3420	3455
42	2570	2605	2640	2675	2710	2745	2780	2815	2850	2885	2920	2955	2990	3025	3060	3100	3135	3170	3205	3240	3275	3310	3345	3380	3415
44	2540	2575	2610	2645	2680	2715	2750	2785	2820	2855	2890	2925	2960	2995	3030	3060	3095	3130	3165	3200	3235	3270	3305	3340	3375
46	2510	2545	2580	2615	2650	2685	2715	2750	2785	2820	2855	2890	2925	2960	2995	3030	3060	3095	3130	3165	3200	3235	3270	3305	3340
48	2480	2515	2550	2585	2620	2650	2685	2715	2750	2785	2820	2855	2890	2925	2960	2995	3030	3060	3095	3130	3160	3195	3230	3265	3300
50	2455	2485	2520	2555	2590	2625	2655	2690	2720	2755	2785	2820	2855	2890	2925	2955	2990	3025	3060	3090	3125	3155	3190	3225	3260
52	2425	2455	2490	2525	2555	2590	2625	2655	2690	2720	2755	2790	2820	2855	2890	2925	2955	2990	3020	3055	3090	3125	3155	3190	3220
54	2395	2425	2460	2495	2530	2560	2590	2625	2655	2690	2720	2755	2790	2820	2855	2885	2920	2950	2985	3020	3050	3085	3115	3150	3180
56	2365	2400	2430	2460	2495	2525	2560	2590	2625	2655	2690	2720	2755	2790	2820	2855	2885	2920	2950	2980	3015	3045	3080	3110	3145
58	2335	2370	2400	2430	2460	2495	2525	2560	2590	2625	2655	2690	2720	2750	2785	2815	2850	2880	2920	2945	2975	3010	3040	3075	3105
60	2305	2340	2370	2400	2430	2460	2495	2525	2560	2590	2625	2655	2685	2720	2750	2780	2810	2845	2875	2915	2940	2970	3000	3035	3065
62	2280	2310	2340	2370	2405	2435	2465	2495	2525	2560	2590	2620	2655	2685	2715	2745	2775	2810	2840	2870	2900	2935	2965	2995	3025
64	2250	2280	2310	2340	2370	2400	2430	2465	2495	2525	2555	2585	2620	2650	2680	2710	2740	2770	2805	2835	2865	2895	2925	2955	2990
66	2220	2250	2280	2310	2340	2370	2400	2430	2460	2495	2525	2555	2585	2615	2645	2675	2705	2735	2765	2800	2825	2860	2890	2920	2950
68	2190	2220	2250	2280	2310	2340	2370	2400	2430	2460	2490	2520	2550	2580	2610	2640	2670	2700	2730	2760	2795	2820	2850	2880	2910
70	2160	2190	2220	2250	2280	2310	2340	2370	2400	2425	2455	2485	2515	2545	2575	2605	2635	2665	2695	2725	2755	2780	2810	2840	2870
72	2130	2160	2190	2220	2250	2280	2310	2335	2365	2395	2425	2455	2480	2510	2540	2570	2600	2630	2660	2685	2715	2745	2775	2805	2830
74	2100	2130	2160	2190	2220	2245	2275	2305	2335	2360	2390	2420	2450	2475	2505	2535	2565	2590	2620	2650	2680	2710	2740	2765	2795

Courtesy of Warren E. Collins, Inc., Braintree, MA.

Table 8.4 Predicted Vital Capacities, Males (mL)

Height in Centimeters

Age	146	148	150	152	154	156	158	160	162	164	166	168	170	172	174	176	178	180	182	184	186	188	190	192	194
16	3765	3820	3870	3920	3975	4025	4075	4130	4180	4230	4285	4335	4385	4440	4490	4540	4590	4645	4695	4745	4800	4850	4900	4955	5005
18	3740	3790	3840	3890	3940	3995	4045	4095	4145	4200	4250	4300	4350	4405	4455	4505	4555	4610	4660	4710	4760	4815	4865	4915	4965
20	3710	3760	3810	3860	3910	3960	4015	4065	4115	4165	4215	4265	4320	4370	4420	4470	4520	4570	4625	4675	4725	4775	4825	4875	4930
22	3680	3730	3780	3830	3880	3930	3980	4030	4080	4135	4185	4235	4285	4335	4385	4435	4485	4535	4585	4635	4685	4735	4790	4840	4890
24	3635	3685	3735	3785	3835	3885	3935	3985	4035	4085	4135	4185	4235	4285	4330	4380	4430	4480	4530	4580	4630	4680	4730	4780	4830
26	3605	3655	3705	3755	3805	3855	3905	3955	4000	4050	4100	4150	4200	4250	4300	4350	4395	4445	4495	4545	4595	4645	4695	4740	4790
28	3575	3625	3675	3725	3775	3820	3870	3920	3970	4020	4070	4115	4165	4215	4265	4310	4360	4410	4460	4510	4555	4605	4655	4705	4755
30	3550	3595	3645	3695	3740	3790	3840	3890	3935	3985	4035	4080	4130	4180	4230	4275	4325	4375	4425	4470	4520	4570	4615	4665	4715
32	3520	3565	3615	3665	3710	3760	3810	3855	3905	3950	4000	4050	4095	4145	4195	4240	4290	4340	4385	4435	4485	4530	4580	4625	4675
34	3475	3525	3570	3620	3665	3715	3760	3810	3855	3905	3950	4000	4045	4095	4140	4190	4225	4285	4330	4380	4425	4475	4520	4570	4615
36	3445	3495	3540	3585	3635	3680	3730	3775	3825	3870	3920	3965	4010	4060	4105	4155	4200	4250	4295	4340	4390	4435	4485	4530	4580
38	3415	3465	3510	3555	3605	3650	3695	3745	3790	3840	3885	3930	3980	4025	4070	4120	4165	4210	4260	4305	4350	4400	4445	4495	4540
40	3385	3435	3480	3525	3575	3620	3665	3710	3760	3805	3850	3900	3945	3990	4035	4085	4130	4175	4220	4270	4315	4360	4410	4455	4500
42	3360	3405	3450	3495	3540	3590	3635	3680	3725	3770	3820	3865	3910	3955	4000	4050	4095	4140	4185	4230	4280	4325	4370	4415	4460
44	3315	3360	3405	3450	3495	3540	3585	3630	3675	3725	3770	3815	3860	3905	3950	3995	4040	4085	4130	4175	4220	4270	4315	4360	4405
46	3285	3330	3375	3420	3465	3510	3555	3600	3645	3690	3735	3780	3825	3870	3915	3960	4005	4050	4095	4140	4185	4230	4275	4320	4365
48	3255	3300	3345	3390	3435	3480	3525	3570	3615	3655	3700	3745	3790	3835	3880	3925	3970	4015	4060	4105	4150	4190	4235	4280	4325
50	3210	3255	3300	3345	3390	3430	3475	3520	3565	3610	3650	3695	3740	3785	3830	3870	3915	3960	4005	4050	4090	4135	4180	4225	4270
52	3185	3225	3270	3315	3355	3400	3445	3490	3530	3575	3620	3660	3705	3750	3795	3835	3880	3925	3970	4010	4055	4100	4140	4185	4230
54	3155	3195	3240	3285	3325	3370	3415	3455	3500	3540	3585	3630	3670	3715	3760	3800	3845	3890	3930	3975	4020	4060	4105	4145	4190
56	3125	3165	3210	3255	3295	3340	3380	3425	3465	3510	3550	3595	3640	3680	3725	3765	3810	3850	3895	3940	3980	4025	4065	4110	4150
58	3080	3125	3165	3210	3250	3290	3335	3375	3420	3460	3500	3545	3585	3630	3670	3715	3755	3800	3840	3880	3925	3965	4010	4050	4095
60	3050	3095	3135	3175	3220	3260	3300	3345	3385	3430	3470	3500	3555	3595	3635	3680	3720	3760	3805	3845	3885	3930	3970	4015	4055
62	3020	3060	3110	3150	3190	3230	3270	3310	3350	3390	3440	3480	3520	3560	3600	3640	3680	3730	3770	3810	3850	3890	3930	3970	4020
64	2990	3030	3080	3120	3160	3200	3240	3280	3320	3360	3400	3440	3490	3530	3570	3610	3650	3690	3730	3770	3810	3850	3900	3940	3980
66	2950	2990	3030	3070	3110	3150	3190	3230	3270	3310	3350	3390	3430	3470	3510	3550	3600	3640	3680	3720	3760	3800	3840	3880	3920
68	2920	2960	3000	3040	3080	3120	3160	3200	3240	3280	3320	3360	3400	3440	3480	3520	3560	3600	3640	3680	3720	3760	3800	3840	3880
70	2890	2930	2970	3010	3050	3090	3130	3170	3210	3250	3290	3330	3370	3410	3450	3480	3520	3560	3600	3640	3680	3720	3760	3800	3840
72	2860	2900	2940	2980	3020	3060	3100	3140	3180	3210	3250	3290	3330	3370	3410	3450	3490	3530	3570	3610	3650	3680	3720	3760	3800
74	2820	2860	2900	2930	2970	3010	3050	3090	3130	3170	3200	3240	3280	3320	3360	3400	3440	3470	3510	3550	3590	3630	3670	3710	3740

Courtesy of Warren E. Collins, Inc., Braintree, MA..

Laboratory Report 8.1

Name _____

Date _____

Section _____

DATA FROM EXERCISE 8.1

A. Measurement of Simple Lung Volumes and Capacities

Enter your data (corrected to BTPS) under the *Measured* column, and enter your calculated *Percent Predicted,* in the table below.

Volume/Capacity	Measured	Predicted	Percent Predicted
TV		500 mL (avg. normal)	
IC		2,800 mL (avg. normal)	
ERV		1,200 mL (avg. normal)	
VC		(from tables)	
RV	not measured	(VC × factor)	cannot calculate
TLC	not measured	(VC × factor)	cannot calculate

B. Measurement of Forced Expiratory Volume

1. Enter your measured and predicted FEV_1 in the spaces below.

   Measured = _____%   Predicted range = _____%

2. Compare your values to the normal range and enter your conclusions in the space below.

REVIEW ACTIVITIES FOR EXERCISE 8.1

Test Your Knowledge of Terms and Facts

Identify the following lung volumes and capacities:
1. Maximum amount of air that can be expired after a maximum inspiration: _____ .
2. Maximum amount of air that can be expired after a normal expiration:_____ .
3. Maximum amount of air that can be inspired after a normal expiration: _____ .
4. Amount of air left in the lungs after a maximum expiration: _____ .
5. Category of pulmonary disorders in which the alveoli are normal but there is an abnormally high resistance to air flow: _____ .
6. An example of a disorder in the category described above is _____ .
7. A pulmonary function test for the category of disorders named in question 5 is the _____ test.

Test Your Understanding of Concepts

8. Calculate the following values for the spirogram shown below. Be sure to correct your values to BTPS (use a BTPS value of 1.1).

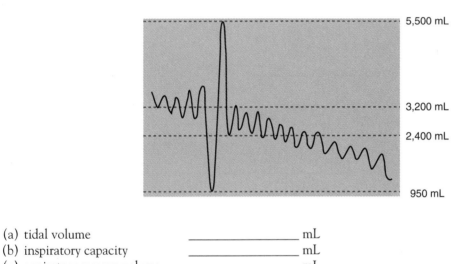

(a) tidal volume   _____ mL
(b) inspiratory capacity   _____ mL
(c) expiratory reserve volume   _____ mL
(d) vital capacity   _____ mL

9. Calculate the FEV_1 value for the spirogram shown below.

FEV_1 = _____ %

10. Distinguish between obstructive and restrictive disorders. How does spirometry aid in their diagnosis?

Test Your Ability to Analyze and Apply Your Knowledge

11. Does your chest expand because your lungs inflate, or do your lungs inflate because your chest expands? Explain.

12. When a person's lungs are ventilated by a tank of gas, the technique is known as *intermittent positive pressure breathing (IPPB)*. Analyzing this term, explain how this technique inflates the lungs for inspiration, and deflates the lungs for expiration.

Effect of Exercise
on the Respiratory System

MATERIALS

1. Collins 9-L respirometer or *Spirocomp* program and equipment
2. Nose clamps and disposable mouthpieces
3. Alternatively, the *Biopac* system may be used with the set up for *Biopac* lesson 12.

Total minute volume is the product of the rate and depth of breathing per minute. Oxygen consumption per minute is a measure of the metabolic rate. Total minute volume is adjusted by physiological mechanisms to compensate for changes in metabolic rate.

OBJECTIVES

1. Define the term *total minute volume* and explain how this measurement is obtained.
2. Describe how the rate of oxygen consumption is measured and explain how it is used as a measure of the metabolic rate.
3. Describe the relationship between the total minute volume and the rate of oxygen consumption; and explain how, and why, these measurements are changed during exercise.
4. Explain why oxygen consumption and total minute volume remain elevated after exercise has ceased.

Textbook Correlations*

Before performing this exercise, you may want to consult the following references in *Human Physiology*, seventh edition, by Stuart I. Fox:

- *Gas Exchange in the Lungs.* Chapter 16, pp. 496–501.
- *Regulation of Breathing.* Chapter 16, pp. 502–506.
- *Ventilation During Exercise.* Chapter 16, pp. 515–516.

   Those using different physiology textbooks may want to consult the corresponding information in those books.

The volume of air exhaled in a minute of resting breathing is known as the **total minute volume** and is equal to the product of tidal volume (milliliters per breath) and the frequency of breathing (breaths per minute). Only about two-thirds of this volume actually reaches the alveoli (this is known as the *alveolar minute volume*). The remaining one-third stays within the dead space of the lungs and is not involved in gas exchange.

Gas exchange occurs in the alveoli of the lungs. Oxygen diffuses from the alveolar air into the blood, while carbon dioxide diffuses from the blood into the alveolar air. Thus, the blood leaving the lungs is rich in oxygen and reduced in carbon dioxide. When the blood reaches the tissue capillaries, oxygen diffuses from the blood into the tissues, where it can be used by the cells in aerobic respiration. Meanwhile carbon dioxide, formed as a waste product of aerobic respiration, diffuses from the tissues into the capillary blood.

Since oxygen is consumed by the body's cells in aerobic respiration, the initial volume of air within the oxygen bell of the respirometer decreases as the subject breathes through the mouthpiece. (In addition, the exhaled carbon dioxide is removed by soda lime within the respirometer.) Thus, oxygen consumption results in the removal of air from the bell and on a spirogram is seen as an *upward slope* of the tidal volume tracing. The amount of oxygen consumed per minute can be calculated as the difference between milliliter levels before and after 1 minute of resting ventilation (fig. 8.12).

---

*See Appendix 3 for correlations to the A.D.A.M. *InterActive PHYSIOLOGY Modules*.

See Appendix 3 for correlations to the *Virtual Physiology Laboratory CD-ROM* by McGraw-Hill and Cypris Publishing, Inc.

See Appendix 3 for correlations to the *Intelitool Physiology Laboratory Exercises.* See Appendix 3 for correlations to the *Biopac Student Lab Exercises.*

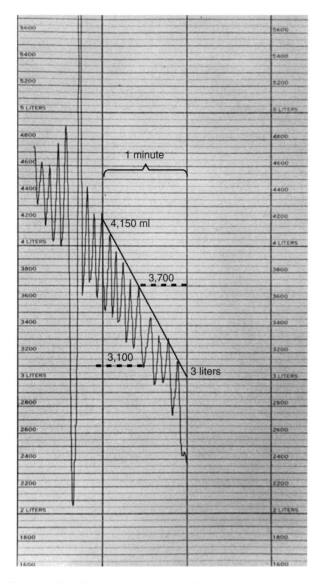

Figure 8.12 Spirogram showing tidal volume measurements over a 1 minute interval at rest. The rising slope of the tidal volume measurements indicates oxygen consumption.

The rate at which oxygen is consumed and carbon dioxide is produced by the body cells during aerobic respiration is related to the *metabolic rate* of the person. When the individual is relaxed and comfortable and has not eaten for 12 to 15 hours, the metabolic rate is lowest. This rate is referred to as the **basal metabolic rate (BMR).** Under these conditions, the metabolic rate is set primarily by the activity of the thyroid gland; in the past, measurements of BMR were used to assess thyroid function.

When a person exercises, however, the metabolic rate increases greatly (the metabolism of muscles can increase as much as sixtyfold during strenuous exercise). As a result, oxygen is consumed and carbon dioxide produced

at much more rapid rates than during resting conditions. The respiratory system keeps pace with this increased demand by increasing the total minute volume.

The **respiratory control center** in the *medulla oblongata* (fig. 8.13) contains inspiratory and expiratory neurons that regulate breathing via motor nerves to the respiratory muscles. The respiratory control center's activity, however, is influenced by neurons in the pons, and also by chemical changes in the blood. Specifically, an increase in plasma CO_2, acting through its lowering of blood pH and cerebrospinal fluid pH, stimulates **chemoreceptors.** The *peripheral chemoreceptors* are located in the **aortic** and **carotid bodies;** the *central chemoreceptors* are in the medulla oblongata. These chemoreceptors modify the activity of the respiratory control center in the medulla, such that a rise in plasma CO_2 (and consequent fall in pH) will stimulate breathing.

The increase in total minute volume that occurs during exercise may be due in part to an increase in CO_2 production, although concentrations of arterial CO_2 during exercise are not usually increased. Anticipation and excitement coming from conscious brain areas and sensory feedback from the exercising muscles may also contribute to the **hyperpnea** (increased breathing) of exercise.

Oxygen consumption and the total minute volume remain elevated immediately after exercise. This extra oxygen consumption (over resting levels) following exercise is called the **oxygen debt.** The extra oxygen is used to oxidize lactic acid produced as a result of anaerobic respiration in the exercising muscles and to support an increased metabolism within the warmed muscles.

Hypoventilation occurs when the alveoli are inadequately ventilated (alveolar minute volume is reduced). Due to a reduction in the total minute volume where either the tidal volume or the frequency of breathing is depressed, hypoventilation can also result from a pathological increase in dead air space in the lungs. The latter may be caused by any condition that affects lung tissue (such as emphysema) or by inadequate blood flow to well-ventilated alveoli. Such ventilated alveoli that are lacking the appropriate perfusion of capillary blood (producing an abnormal *ventilation/perfusion ratio*) are incapable of fully oxygenating blood. As a result of hypoventilation, there is inadequate elimination of carbon dioxide from the blood. Since the plasma carbon dioxide levels are directly affected by ventilation, hypoventilation may be operationally defined as an *abnormally increased plasma carbon dioxide level* (above 40 Torr, or mm Hg).

Midbrain

Pons

Brain stem respiratory centers
- Pneumotaxic area
- Apneustic area
- Rhythmicity area

Reticular formation

Medulla oblongata

Figure 8.13 Location of brain stem respiratory control centers.

PROCEDURE

1. Set the Collins respirometer to the slow speed of 32 mm/min. (At this speed the distance between two heavy vertical lines is traversed in 1 minute.) To avoid air leakage, position the mouthpiece securely in the mouth with the lips tightly sealed around it, and clamp the nostrils closed.

2. Under resting conditions, breathe normally into the respirometer for 1 minute (that is, perform the procedure for measuring tidal volume—see exercise 8.1).

Note: The tidal volume measurements must have an upward slope, indicating that the oxygen being consumed is from the air trapped in the oxygen bell. If the slope is downward, air is leaking into the system, usually through the corners of the mouth or the nose. In this event, reposition the mouthpiece, check the nose clamp, and begin the measurements again.

3. After 1 minute, stop the respirometer, remove the chart from the kymograph drum and determine the frequency of ventilation (number of breaths per minute) and the tidal volume (corrected to BTPS).

Example (from fig. 8.12)

```
   3,700 mL (inhalation peak of tidal volume)
 − 3,100 mL (exhalation trough of tidal volume)
     600 mL (uncorrected tidal volume)
```

```
       600 mL
 ×       1.1    (BTPS factor)
       660 mL (corrected tidal volume)
```

From figure 8.12, frequency = 10 breaths/min

Enter the subject's corrected tidal volume in the following space:
_____ mL

Enter the subject's frequency of ventilation in the following space:
_____ breaths/min

The average ventilation frequency is 14 breaths/min.

4. Determine the subject's total minute volume at rest by multiplying the frequency of ventilation by the tidal volume.

Enter this value in the data table in your laboratory report.

The average total minute volume is 6,750mL/min.

301

Example (from fig. 8.12)

$$660 \text{ mL/breath} \times 10 \text{ breaths/min}$$
$$= 6{,}600 \text{ mL/min total minute volume}$$

5. Use a straight edge to draw a line that touches either the peaks or the troughs of the tidal volume measurements. Determine the oxygen consumption per minute by subtracting the milliliters where this straight line intersects the heavy vertical chart line at the beginning of one minute from the milliliters where the straight line intersects the next heavy vertical line at the end of one minute.

Example (from fig. 8.12)

Using a line that averages the peaks:

$$
\begin{array}{ll}
4{,}150 \text{ mL} & \text{(at end of 1 minute)} \\
\underline{-\ 3{,}000 \text{ mL}} & \text{(at beginning of 1 minute)} \\
1{,}150 \text{ mL} & \text{(oxygen consumption)}
\end{array}
$$

Enter the resting oxygen consumption in the data table in your laboratory report.

6. Now, have the subject perform light exercise, such as 5 to 10 jumping jacks, and then repeat the respirometer measurements and data calculation described in steps 1–5. Alternatively, if a bicycle ergometer is available, the total minute volume may be determined while the student pedals lightly for one minute.

⚠ **Caution:** If breathing into the respirometer becomes difficult after exercise, the subject should stop the procedure. Results obtained in less than a minute can then be extrapolated to 1 minute. Alternatively, the bell can be filled with 100% oxygen to prevent the possible occurrence of hypoxia.

Enter the corrected tidal volume after exercise in the following space:
_____ mL

Enter the frequency of ventilation after exercise in the following space:
_____ breaths/min

Enter the total minute volume and the oxygen consumption per minute after exercise in the data table of the laboratory report.

7. Calculate the percent increase after exercise for total minute volume and for oxygen consumption. This is the difference between the exercise and resting measurements, divided by the resting measurement and multiplied by 100%. Enter these values in the data table in your laboratory report.

Name _____

Date _____

Section _____

DATA FROM EXERCISE 8.2

1. Enter the total minute volume and rate of oxygen consumption during rest and after exercise in the table below.
2. Calculate and enter the percent increases in your measurements after exercise, and enter these values in the table below.

Measurement	Resting	During or After Exercise	Percent Increase
Total minute volume			
Oxygen consumption			

REVIEW ACTIVITIES FOR EXERCISE 8.2

Test Your Knowledge of Terms and Facts

1. The net diffusion of oxygen and carbon dioxide in opposite directions across the wall of alveoli is known as _____ .
2. Define the term *tidal volume*. _____
3. The total minute volume is obtained by multiplying the _____ times the _____ .
4. The peripheral chemoreceptors include the _____ and the _____ _____ .
5. The stimulus that directly activates the peripheral chemoreceptors is _____ .
6. Define the term *hyperpnea*. _____
7. The term *hypoventilation* may be operationally defined as _____ .

Test Your Understanding of Concepts

8. Why did the rate of oxygen consumption increase during exercise? What happened to the plasma levels of carbon dioxide during exercise? Explain.

9. What happened to the total minute volume during exercise? What happened to the total minute volume right after exercise was completed? What physiological mechanisms might account for these changes?

10. Define the oxygen debt, and explain the functions of the oxygen debt.

11. High total minute volume during mild to moderate exercise is more accurately termed hyperpnea than hyperventilation. Distinguish between these two terms and explain why this statement is true?

12. How might the plasma CO_2 levels differ during the performance of a particular exercise in a trained versus an untrained person? Explain your answer.

Oxyhemoglobin Saturation

MATERIALS

1. Graduated cylinder and a 1-cc syringe
2. Test tube and distilled water
3. Colorimeter and cuvettes
4. Sodium dithionite (hydrosulfite), 1.0 g per 100 mL
5. Alcohol swabs and lancets

Textbook Correlations*

Before performing this exercise, you may want to consult the following references in *Human Physiology*, seventh edition, by Stuart I. Fox:

- *Partial Pressure of Gases in Blood* Chapter 16, pp. 497–498.
- *Hemoglobin and Oxygen Transport*, Chapter 16, pp. 506–512.

Those using different physiology textbooks may want to consult the corresponding information in those books.

The iron atoms within the heme groups of hemoglobin may be free (unbound), or they may be bonded to oxygen or carbon monoxide gas. Each of these different forms of hemoglobin has a slightly different color, which allows the percentage of each type in a mixture to be measured in an array known as an absorption spectrum. Absorption spectral analysis is used clinically to assess lung function and the capacity of blood to transport oxygen.

OBJECTIVES

1. Define the term percent saturation.
2. Explain the clinical significance of the percent oxyhemoglobin measurement.
3. Explain the clinical significance of the percent carboxyhemoglobin measurement.
4. Describe how an absorption spectrum is obtained, and explain how the absorption spectra of the different forms of hemoglobin are used to determine the percent saturation.

The ability of the blood to carry oxygen depends on (1) ventilation; (2) gas exchange across the alveoli of the lungs; (3) the red blood cell count and hemoglobin concentration; and (3) the chemical form of the hemoglobin.

There are two chemical forms of normal hemoglobin. Normal hemoglobin without oxygen is called **deoxyhemoglobin;** after it binds to oxygen, it is called **oxyhemoglobin.** If a person suffers from *carbon monoxide poisoning,* however, the abnormal hemoglobin form called

carboxyhemoglobin (hemoglobin bound to carbon monoxide) causes the blood to carry a lower amount of oxygen, because the carbon monoxide displaces oxygen and binds to hemoglobin with a higher affinity than does oxygen. Therefore, when health professionals need to determine the oxygen carrying capacity of the blood, they need to learn the relative proportion of each hemoglobin type as well as the total hemoglobin concentration.

The relative proportion of each type of hemoglobin is given as its **percent saturation.** The percent oxyhemoglobin saturation, for example, is the proportion of hemoglobin bound to oxygen. Normally, this value is approximately 97% in arterial blood and 75% in venous blood.

$$\frac{\% \text{ oxyhemoglobin}}{\text{saturation}} = \frac{\text{oxyhemoglobin}}{\text{total blood hemoglobin}} \times 100$$

The determination of percent oxyhemoglobin saturation is a very sensitive means of assessing the effectiveness of pulmonary function. When pulmonary function and blood hemoglobin are normal, the arterial blood has a percent oxyhemoglobin saturation of about 97%. Even when pulmonary function is normal, the percent oxyhemoglobin saturation can be decreased by carbon monoxide poisoning. When a person's blood has a carboxyhemoglobin

---

*See Appendix 3 for correlations to the A.D.A.M. InterActive PHYSIOLOGY Modules.

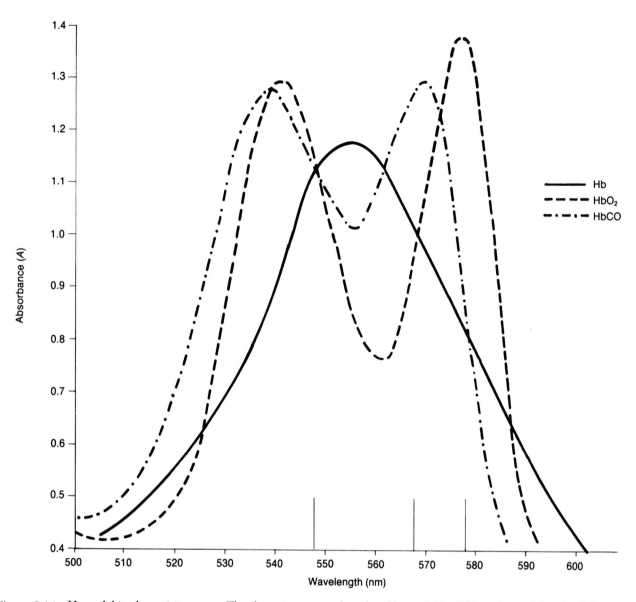

Figure 8.14 Hemoglobin absorption spectra. The absorption spectra for reduced hemoglobin (Hb), oxyhemoglobin (HbO$_2$), and carboxyhemoglobin (HbCO) are shown. Source: *Instrumentation Lab, Inc.*

saturation of 6.9%, for example (obtained in one study using cigarette-smoking New York taxicab drivers), the percent oxyhemoglobin saturation is decreased accordingly. Another abnormal form of hemoglobin is **methemoglobin,** which is oxidized hemoglobin that lacks the electron needed to bond with oxygen, and therefore cannot participate in oxygen transport. Notice that these impairments in oxygen transport cannot be detected by standard measurement of red blood cell count, hematocrit, or total blood hemoglobin.

The percent saturation of the different types of hemoglobins is measured by comparing the absorption spectrum of an unknown sample of blood with the absorption spectra of pure oxyhemoglobin, pure reduced hemoglobin, and pure carboxyhemoglobin. Since these hemoglobins have different colors, they absorb different amounts of light at each wavelength. A graph of absorbance versus

wavelength (where the concentration is constant) is called an **absorption spectrum** (fig. 8.14).

The absorption spectrum of an unknown sample of blood will display some combination of these three absorption spectra, since the blood contains all three types of hemoglobins. The relative contribution of each hemoglobin type to the absorption spectrum is proportional to the relative amount of each type in the blood. This analysis, which is obviously complex, is usually performed by a laboratory instrument specifically manufactured for this purpose.

In this exercise, you will construct an absorption spectrum for 100% oxyhemoglobin and 100% reduced hemoglobin. By bubbling air into a flask containing blood until the blood is in equilibrium with the air, 100% oxyhemoglobin may be obtained. This process essentially duplicates the process that occurs in the capillaries surrounding the lung alveoli. A simpler (but less accurate) method is to obtain a sample of blood from the fingertip, which has a high percent oxyhemoglobin saturation. The sample of 100% reduced hemoglobin may be obtained by adding sodium hydrosulfite ($Na_2S_2O_4$) to a second sample of blood. The sodium hydrosulfite (also called sodium dithionite) removes oxygen from oxyhemoglobin.

A person may have normal hemoglobin, hematocrit, and red blood cell counts and still not be delivering adequate amounts of oxygen to the body cells. This may be due to inadequate lung function, resulting in poor oxygenation of the blood, or due to an abnormally high blood concentration of carboxyhemoglobin or methemoglobin. In this event, the hemoglobin cannot become fully saturated with oxygen, and therefore the percent oxyhemoglobin saturation of arterial blood may drop below normal. These effects are similar to the effects of anemia, since in both cases the amount of oxygen carried by the blood is reduced. In this sense, carbon monoxide poisoning may be thought of as a functional anemia. The hemoglobin and red blood cell counts are normal, but the red blood cells are not transporting the normal amount of oxygen.

PROCEDURE

1. Add 8.0 mL of distilled water to a test tube. Obtain a large drop of blood by wiping the fingertip with 70% alcohol, let dry, and puncturing it with a sterile lancet. Then, mix this blood with the distilled water by inverting the test tube over the punctured finger.
2. Transfer half the contents of the test tube (4.0 mL) to a second tube.
3. Add 0.20 mL of 1.0% sodium dithionite solution to the second test tube and mix thoroughly.

Note: *The dithionite solution should be freshly prepared just prior to use, and the absorbance values of the two tubes should be determined within 5 minutes of the time the dithionite is added to the second tube.*

4. Transfer the two solutions to two cuvettes. Fill a third cuvette with distilled water and use it as a blank to standardize the spectrophotometer at 500 nm. (See exercise 2.1 for a description of the spectrophotometer, standardizing procedures, and Beer's law.)
5. Record the absorbances of solutions 1 and 2. Continue using the blank to standardize the spectrophotometer at each of the successive wavelengths from 510 to 600 nm, and then record the absorbances of the two solutions in the laboratory report.
6. Graph the absorption spectra of oxyhemoglobin and reduced hemoglobin on the graph provided in the laboratory report.

Name _____

Date _____

Section _____

DATA FROM EXERCISE 8.3

Wavelength	Oxyhemoglobin Absorbance	Reduced Hemoglobin Absorbance	Wavelength	Oxyhemoglobin Absorbance	Reduced Hemoglobin Absorbance
500 nm			560 mn		
510 nm			570 mn		
520 mn			580 mn		
530 mn			590 mn		
540 mn			600 mn		
550 mn					

REVIEW ACTIVITIES FOR EXERCISE 8.3

Test Your Knowledge of Terms and Facts

1. In the lungs, normal hemoglobin without oxygen, or _____ , binds to oxygen to become _____ .
2. The type of hemoglobin bound to carbon monoxide: _____ .
3. The type of hemoglobin where the heme iron is in the oxidized Fe^{2+} state: _____ .
4. What is the normal percent oxyhemoglobin saturation of arterial blood? _____ %
5. A graph of the absorbance of light as a function of the wavelength of light is called a(n) _____ _____ .
6. The different forms of hemoglobin can be distinguished visually because they have different _____ _____ .

Test Your Understanding of Concepts

7. What blood measurement would be abnormally increased in a person with carbon monoxide poisoning? What are the dangers of carbon monoxide poisoning?

8. In what way are carbon monoxide poisoning and anemia different? In what way are they similar? Explain.

9. Suppose a person's percent oxyhemoglobin saturation in the venous blood was 75% at rest, but decreased to 35% during moderate exercise. (a) Explain why the venous percent oxyhemoglobin saturation decreased during the exercise; and (b) predict how the arterial percent oxyhemoglobin saturation might change during exercise. Explain your answer.

10. The percent oxyhemoglobin saturation is measured in babies under treatment for *respiratory distress syndrome (RDS)* and in patients under general anesthesia. What information would these measurements provide under these conditions? Explain.

Respiration and Acid-Base Balance

MATERIALS

1. pH meter, droppers, beakers, straws
2. Buffer, pH = 7 (made from purchased concentrate); concentrated HCl
3. concentrated NaOH
4. phenolphthalein solution (saturated)

Carbon dioxide in plasma can combine with water to produce carbonic acid, which in turn dissociates to produce protons (H$^+$) and bicarbonate ions (HCO$_3^-$). Ventilation regulates the carbon dioxide concentration of the plasma and has an important role in acid-base balance.

OBJECTIVES

1. Describe the pH scale and define the terms acid and base.
2. Explain how carbonic acid and bicarbonate are formed in the blood and describe their functions.
3. Define the terms acidosis and alkalosis and explain how these conditions relate to hypoventilation and hyperventilation.
4. Explain how ventilation is adjusted to help maintain acid-base balance.

Textbooks Correlations*

Before performing this exercise, you may want to consult the following references in *Human Physiology*, seventh edition, by Stuart I. Fox:

- *Regulation of Breathing.* Chapter 16, pp. 502–506.
- *Carbon Dioxide Transport and Acid-Base Balance.* Chapter 16, pp. 513–515.
- *Ventilation During Exercise.* Chapter 16, pp. 515–516.

Those using different physiology textbooks may want to consult the corresponding information in those books.

Ventilation has two different but related functions: (1) oxygenation of the blood, accomplished by bringing new air into the alveoli during the inhalation phase, and (2) elimination of carbon dioxide from the blood, accomplished by the diffusion of CO_2 from the blood into the alveoli and the extrusion of this CO_2 air by exhalation. The first function serves to maintain aerobic cell respiration; the second serves to maintain the normal pH of the blood.

The pH (see Appendix 1) indicates the concentration of H$^+$ (hydrogen ion) in a solution and is defined by the following formula:

$$pH = \log \frac{1}{\left[H^+\right]}$$

where H$^+$ is the concentration of H$^+$ in moles (atomic weight in grams) per liter. Some water molecules ionize to produce equal amounts of H$^+$ and OH$^-$ (hydroxyl ion). In pure water, the H$^+$ concentration is 10^{-7} moles/L. (Because hydrogen has an atomic weight of 1, this is the same as 10^{-7} g/L.) This is equal to a pH of 7.0 and is called a *neutral solution*. An *acidic solution* has a higher H$^+$ concentration and a lower pH; a *basic solution* has a lower H$^+$ concentration and a higher pH (table 8.5).

An **acid** is a molecule that can donate free H$^+$ to a solution and lower its pH. *Carbonic acid* (H_2CO_3) is formed from the combination of CO_2 and water within

---

* See Appendix 3 for correlations to the A.D.A.M. *InterActive PHYSIOLOGY Modules.*

the red blood cells. This reaction is catalyzed by an enzyme called *carbonic anhydrase* (fig. 8.15).

$$CO_2 + H_2O \xrightarrow{\text{carbonic anhydrase}} H_2CO_3$$

Some of the carbonic acid formed can immediately dissociate to yield H^+ and *bicarbonate ion* (HCO_3^-). The H^+ derived from carbonic acid and other acids in the blood gives normal arterial blood a pH of 7.40 ± 0.05 (table 8.6 and fig. 8.15).

$$H_2CO_3 \longrightarrow HCO_3^- + H^+$$

A. ABILITY OF BUFFERS TO STABILIZE THE pH OF SOLUTIONS

Plasma has a particular concentration of bicarbonate as a result of the dissociation of carbonic acid. Bicarbonate serves as the major *buffer* of the blood, helping to stabilize the pH of plasma despite the continuous influx of H^+ from molecules of lactic acid, fatty acids, ketone bodies,

and other metabolic products. The H^+ released by these acids is prevented from lowering the blood pH because it is combined with bicarbonate. Although a new acid molecule (carbonic acid) is formed, this reaction prevents a rise in the free H^+ concentration (fig. 8.15).

$$H^+ + HCO_3^- \longrightarrow H_2CO_3$$

Carbonic acid formed in this way can provide a source of new H^+ if the blood pH should begin to rise (from a loss of blood H^+) beyond normal levels. The carbonic acid/bicarbonate buffer system helps to stabilize the blood pH under normal conditions. Disease states, however, may cause the blood pH to fall below 7.35 or to rise above 7.45. These conditions are called *acidosis* and *alkalosis*, respectively.

Normally, the rate of ventilation is matched to the rate of CO_2 production by the tissues, so that the carbonic acid, bicarbonate, and H^+ concentrations in the blood remain within the normal range. If **hypoventilation** occurs, however, the carbonic acid levels will rise above normal and the pH will fall below 7.35. This condition is called **respiratory acidosis** (table 8.6). **Hyperventilation,** conversely, causes an abnormal decrease in carbonic acid and a corresponding rise in blood pH. This condition is called **respiratory alkalosis.** Thus, respiratory acidosis or alkalosis occurs when the blood CO_2 level (as measured by its partial pressure or P_{CO_2}, in millimeters of mercury) is different from the normal value (40 mm Hg) as a result of abnormal breathing patterns.

PROCEDURE

1. Allow the pH meter to warm up by setting the selector switch to the *standby* position. Be sure that the pH electrodes are immersed in buffer and are not allowed to dry. Verify that the temperature selector switch is set at the current room temperature.
2. Turn the selector switch to *pH* and take a reading of the buffer. Use the calibration knob to set the pH meter to the correct pH of the buffer (7.000). Now, turn the selector switch back to the *standby* position

Table 8.5 The pH Scale

	H^+ Concentration (Molar)	pH	OH^- Concentration (Molar)
	1.0	0	10^{-14}
	0.1	1	10^{-13}
	0.01	2	10^{-12}
Acids	0.001	3	10^{-11}
	0.0001	4	10^{-10}
	10^{-5}	5	10^{-9}
	10^{-6}	6	10^{-8}
Neutral	10^{-7}	7	10^{-7}
	10^{-8}	8	10^{-6}
	10^{-9}	9	10^{-5}
	10^{-10}	10	0.0001
Bases	10^{-11}	11	0.001
	10^{-12}	12	0.01
	10^{-13}	13	0.1
	10^{-14}	14	1.0

Table 8.6 The Effect of Respiration on Blood pH

P_{CO_2} (mm Hg)	H_2CO_c (mEq/L)[1]	HCO_3^- (mEq/L)[1]	HCO_3^-/H_2CO_3 Ratio	Blood pH	Condition
20	0.6	24	40/1	7.70	Respiratory alkalosis
30	0.9	24	26.7/1	7.53	Respiratory alkalosis
40	1.2	24	20/1	7.40	Normal
50	1.5	24	16/1	7.30	Respiratory alkalosis
60	1.8	24	13.3/1	7.22	Respiratory alkalosis

1. Ion concentrations are commonly measured in milliequivalents (mEq) per liter. This measurement is equal to the millimolar concentration of the ion multiplied by its number of charges.

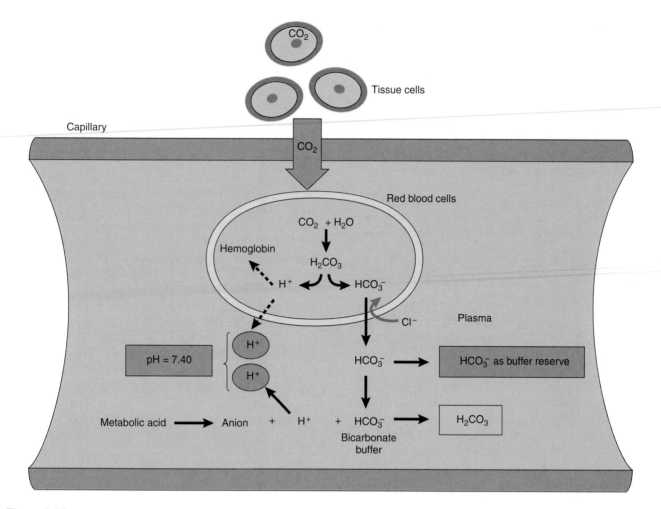

Figure 8.15 Maintentance of acid-base balance. Carbon dioxide produced by tissue cells forms carbonic acid, which adds both H^+ and bicarbonate (HCO_3^-) to the plasma. The bicarbonate released from red blood cells buffers the H^+ produced by ionization of metabolic (nonvolatile)acids, such as lactic acid and ketone bodies.

and transfer the pH electrodes to a beaker of distilled water. Turn the selector switch to *pH* and record the pH of distilled water. Return the selector switch to the *standby* position and the electrodes back to the buffer.

3. Add one drop of concentrated hydrochloric acid (HCl) to the beaker of distilled water and mix thoroughly. Transfer the electrodes to this solution, turn the selector switch to the *pH* position, and record the pH of the solution in your laboratory report.

Note: *After recording the pH of a solution, always turn the selector switch to standby and use a squeeze bottle of distilled water to rinse the electrodes thoroughly. Wipe the electrodes with lint-free paper and return them to the buffer solution. Check the pH of the buffer solution after the cleaning procedure to be sure you have adequately cleaned the electrodes.*

4. Add one drop of concentrated NaOH to a fresh beaker of distilled water and record the pH of the water before and after adding the NaOH.
5. Add one drop of concentrated HCl to a beaker containing standard buffer solution (pH = 7.000).
6. Add one drop of concentrated NaOH to a fresh beaker of standard buffer solution (pH = 7.000).
7. Add two drops of concentrated HCl to a beaker of fresh standard buffer solution (pH = 7.000).
8. Add two drops of concentrated NaOH to a beaker of fresh standard buffer solution (pH = 7.000).

B. EFFECT OF EXERCISE ON THE RATE OF CO_2 PRODUCTION

Increased muscle metabolism during exercise results in an increase in CO_2 production. Despite this, the CO_2 levels and pH of arterial blood do not normally change significantly during exercise. This is because the increased rate of CO_2 production is matched by an increase in the rate of its elimination through ventilation. The mechanisms responsible for exercise *hyperpnea* (increased breathing) are complex and incompletely understood.

 Hypoventilation results in the retention of carbon dioxide and in the excessive accumulation of carbonic acid; this produces a fall in blood pH called **respiratory acidosis.** *Hyperventilation* results in the excessive elimination of CO_2, lowered carbonic acid, and a rise in pH, causing **respiratory alkalosis.** This differs from the normal hyperpnea (increased total minute volume) that occurs during exercise, where increased respiration matches increased CO_2 production so that the arterial CO_2 levels and pH remain in the normal range.

PROCEDURE

1. Fill a beaker with 200 mL of distilled water and add 5.0 mL of 0.10N NaOH and a few drops of phenolphthalein indicator. This indicator is pink in alkaline solutions and clear in neutral or acidic solutions. Divide this solution into two beakers.
2. While sitting quietly, exhale through a glass tube or straw (or double straws) into the solution in the first beaker. Carefully record the time required to turn the solution from pink to clear in your laboratory report.
3. Exercise vigorously for 2 to 5 minutes by running up and down stairs or by doing jumping jacks. Exhale through a glass tube or straw (or double straws) into the second beaker, and again record the time it takes to clear the pink solution.

C. ROLE OF CARBON DIOXIDE IN THE REGULATION OF VENTILATION

The carbon dioxide concentration of the blood reflects a balance between the rate of its production (by aerobic cell respiration) and the rate of its elimination through the lungs. When a person consciously holds his or her breath for a sufficiently long time, the carbon dioxide level rises (and the pH falls) to such an extent that reflex breathing occurs. On the other hand, during hyperventilation, the pH of the blood can rise to the point that the desire to breathe is eliminated until the amount of carbon dioxide in the blood again rises above the critical point.

PROCEDURE

1. Count the number of breaths you take in 1 minute of relaxed, unforced breathing. Enter this number in your laboratory report.
2. Force yourself to hyperventilate for about 10 seconds; stop if you begin to feel dizzy.
3. Immediately after hyperventilation, count the number of breaths you take in 1 minute of relaxed, unforced breathing.

Laboratory Report 8.4

Name _____

Date _____

Section _____

DATA FROM EXERCISE 8.4

A. Ability of Buffers to Stabilize the pH of Solutions

1. Enter your data in the spaces below:

   pH of distilled water: _____

   pH of water + 1 drop HCl: _____

   pH of water + 1 drop NaOH: _____

   pH of buffer: 7.000

   pH of buffer + 1 drop HCl: _____

   pH of buffer + 1 drop NaOH: _____

   pH of buffer + 3 drops HCl: _____

   pH of buffer + 3 drops NaOH: _____

2. Does your data support the statement that "buffers help to stabilize the pH of solutions"? Explain your answer.

B. Effect of Exercise on the Rate of CO_2 Production

1. Enter your data in the spaces below:

   Time for color change at rest: _____

   Time for color change after exercise: _____
2. Explain your results in the space below:

C. Role of Carbon Dioxide in the Regulation of Ventilation

1. Enter your data in the spaces below:

   Rate of breathing at rest: _____ breaths/min

   Rate of breathing after hyperventilation: _____ breaths/min
2. Explain your results in the space below.

REVIEW ACTIVITIES FOR EXERCISE 8.4

Test Your Knowledge of Terms and Facts

1. A solution with a H^+ concentration of 10^{-9} molar has a pH of _____ .
2. Hypoventilation produces a condition called respiratory _____ ; hyperventilation produces a condition called respiratory _____ .
3. Define the following terms:
   (a) *acid* _____
   (b) *base* _____
   (c) *acidosis* _____
   (d) *alkalosis* _____
4. What is the normal measurement of arterial carbon dioxide levels? _____ mm Hg
5. The free bicarbonate in the plasma serves as the major _____ of the blood.
6. The enzyme in red blood cells that catalyzes the formation of carbonic acid is _____ .

Test Your Understanding of Concepts

7. Draw equations to show how carbon dioxide affects the blood concentration of H^+ (and thus the pH) and the blood concentration of HCO_3^-. Indicate the directions of change in these values if blood carbon dioxide levels were to rise.

8. Use the equations shown in question 7 to explain how hyperventilation and hypoventilation affect the blood pH.

9. Describe how your breathing rate changed following 10 seconds of hyperventilation. Explain the physiological mechanisms responsible for the changed breathing pattern.

Test Your Ability to Analyze and Apply Your Knowledge

10. People who hyperventilate may get dizzy (due to cerebral vasoconstriction), causing anxiety and further hyperventilation. Such people are sometimes urged to breathe into a paper bag. What good would this do? Explain the physiological mechanisms involved.

11. Intravenous infusions of sodium bicarbonate are often given to acidotic patients to correct the acidosis and relieve the strain of rapid breathing. Explain why bicarbonate is helpful in this situation. What would happen if too much bicarbonate were given? Explain.

Renal Function and Homeostasis

The kidneys are responsible for the elimination of most of the waste products of metabolism. These wastes include urea and creatinine (derived from protein catabolism) and ketone bodies (derived from fat catabolism). The kidney must also retain (or *reabsorb*) molecules essential for normal body function, such as glucose, amino acids, and bicarbonate.

Through these actions, the kidneys are involved in maintaining a constant internal environment (homeostasis), including the regulation of *electrolyte concentrations, fluid balance,* and *acid-base balance.* The fluid volume of the blood is maintained by the reabsorption of 98% to 99% of the water that leaves the blood in the initial step of urine formation, while the electrolyte and pH balance of the blood is maintained by the selective reabsorption of such ions as Na^+, K^+, and HCO_3^-.

The kidneys contain approximately 2 million functional units called **nephrons** (fig. 9.1*c*). Each nephron is composed of two parts: (1) the *glomerulus,* which is a tightly woven, highly permeable capillary bed at the end of an arteriole, and (2) the *renal tubule,* which is a bent and convoluted tubule lined by epithelial cells. The mouth of each renal tubule (Bowman's capsule) envelops a glomerulus. The last part of the renal tubule (the collecting duct) empties its contents into the renal pelvis as urine, which is then funneled down the ureters (fig. 9.1).

The formation of urine in the nephron occurs in two stages. (1) The hydrostatic pressure of the blood squeezes fluid out of the capillary wall of the glomerulus, producing an *ultrafiltrate* of blood. Except for proteins, which are usually too large to leave the capillaries, the glomerular filtrate contains the same solute molecules as plasma and is isotonic to plasma. (2) As the glomerular filtrate passes through the renal tubules, the cells of the tubules selectively reabsorb and secrete solute molecules and ions. The solution that emerges at the end of the collecting duct is urine, and thus very different in composition and concentration from the glomerular filtrate that enters the tubule.

Exercise 9.1   Renal Regulation of Fluid and Electrolyte Balance

Exercise 9.2   Renal Plasma Clearance of Urea

Exercise 9.3   Clinical Examination of Urine

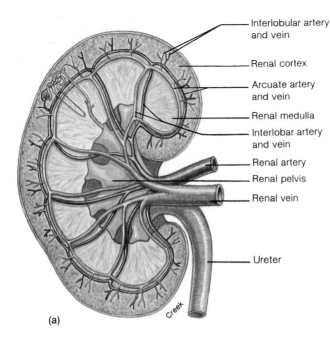

(a)

- Interlobular artery and vein
- Renal cortex
- Arcuate artery and vein
- Renal medulla
- Interlobar artery and vein
- Renal artery
- Renal pelvis
- Renal vein
- Ureter

Creek

(b)

- Convolut tubule
- Glomerul

- Glomerulus
- Glomerular capsule
- Efferent arteriole
- Afferent arteriole
- Interlobular artery
- Proximal convoluted tubule
- Arcuate artery and vein
- Interlobar artery and vein
- Nephron loop
  - Descending limb
  - Ascending limb
- Peritubular capillaries
- Distal convoluted tubule
- Renal cortex
- Interlobular vein
- Renal medulla
- Vasa recta
- Collecting duct

(c)

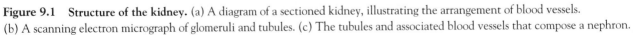

Figure 9.1 **Structure of the kidney.** (a) A diagram of a sectioned kidney, illustrating the arrangement of blood vessels. (b) A scanning electron micrograph of glomeruli and tubules. (c) The tubules and associated blood vessels that compose a nephron.

Renal Regulation of Fluid and Electrolyte Balance

MATERIALS

1. Urine collection cups
2. Urinometers and droppers

Note: *Specific gravity can also be measured visually using disposable urine dip strips (Miles Inc., Curtin Matheson Scientific Inc.)*

3. pH paper (pH range 3–9), potassium chromate (20 g per 100 mL), silver nitrate (2.9 g per 100 mL)
4. NaCl crystals or salt tablets
5. Alternatively, normal and abnormal artificial urine is available (Wards Biology). However, modifications must be made by the instructor to simulate the conditions in this exercise.

Textbook Correlations*

Before performing this exercise, you may want to consult the following references in *Human Physiology*, seventh edition, by Stuart I. Fox:

- *Reabsorption of Salt and Water.* Chapter 17, pp. 534–541.
- *Renal Control of Electrolyte and Acid-Base Balance.* Chapter 17, pp. 546–552.

Those using different physiology textbooks may want to consult the corresponding information in those books.

Urine volume, solute concentration, and electrolyte content are adjusted by the kidneys to maintain homeostasis of the blood. Drinking excess water or eating salty foods results in a rising blood volume, which is followed by compensatory increases in the urinary excretion of the salt and water.

OBJECTIVES

1. Describe the roles of ADH and aldosterone in the regulation of fluid and electrolyte balance.
2. Calculate the concentration of ions in solution (in milliequivalents per liter).
3. Demonstrate and explain how the kidneys respond to water and salt loading by changes in urinary volume, specific gravity, pH, and electrolyte composition.

The reabsorption of fluid and electrolytes (ions) into the blood from the renal tubular filtrate is adjusted to meet the needs of the body by the action of hormones. The major hormones involved in this process are **antidiuretic hormone (ADH),** released by the posterior pituitary gland, and **aldosterone,** secreted by the cortex of the adrenal gland (fig. 9.2).

The release of antidiuretic hormone by the posterior pituitary is regulated by osmoreceptors in the hypothalamus. These receptors are stimulated by an increase in the osmotic pressure of the blood, as might occur in dehydration. The ADH released in response to this stimulus promotes the reabsorption of water from the renal tubules, resulting in (1) the retention of water and therefore a decrease in the osmotic pressure of the blood back to the normal level, and (2) the excretion of a small volume of highly concentrated (hypertonic) urine.

The secretion of aldosterone by the adrenal cortex may be stimulated by an increase in blood K^+ or by a decrease in blood Na^+ or blood volume. An increase in blood K^+ directly stimulates the adrenal cortex to secrete aldosterone. A decrease in blood Na^+ or blood volume indirectly affects aldosterone secretion by stimulating the kidneys to secrete the enzyme **renin** into the blood (fig. 9.2). Renin catalyzes the reaction that leads to the formation of a polypeptide known as **angiotensin II,** which has two major effects. Angiotensin II: (1) stimulates vasoconstriction, thus increasing the blood pressure, and (2) stimulates the secretion of aldosterone from the adrenal cortex. This released aldosterone then promotes the reabsorption of Na^+ from the glomerular filtrate into

---

*See Appendix 3 for correlations of the A.D.A.M. InterActive PHYSIOLOGY Modules.

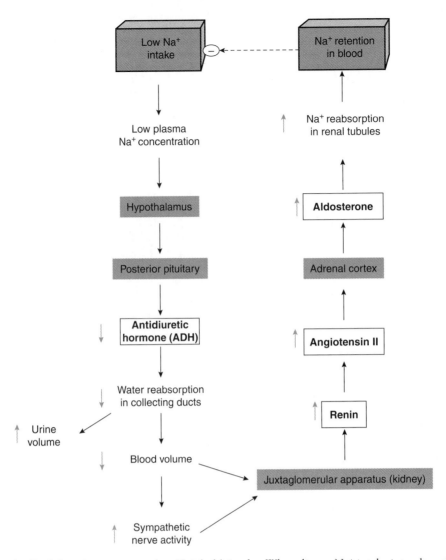

Figure 9.2 Negative feedback loop in response to low Na⁺ (salt) intake. When dietary Na⁺ intake is too low, there is a decrease in ADH secretion and consequent fall in blood volume. In response to the fall in blood flow through the kidneys, the renin-angiotensin-aldosterone system is activated. Increased aldosterone stimulates increased renal reabsorption of Na⁺.

the blood, in exchange for K⁺, which is secreted from the blood into the renal tubules. When Na⁺ ions are reabsorbed, water follows passively owing to the osmotic gradient that is created. In this way, aldosterone prompts a rise in blood Na⁺ and volume, correcting the original deviation and maintaining homeostasis.

MILLIEQUIVALENTS

The concentrations of ions in body fluids (electrolytes) are usually given in terms of milliequivalents (mEq) per liter. To convey the meaning and significance of this unit of measurement, consider the chloride (Cl⁻) concentration of the urine.

Suppose a urine sample had a chloride concentration of 610 mg per 100 mL. How does this number of ions and this number of charges compare with the number of other ions and charges that are present in the urine? To

determine this, we must first convert the chloride concentration from milli*grams* per 100 mL to milli*moles* per liter.

Example

The atomic weight of chloride is 35.5. Therefore,

$$\frac{610 \text{ mg of Cl}^-}{100 \text{ mL}} \times \frac{1 \text{ g}}{1,000 \text{ mg}} \times \frac{1,000 \text{ mL}}{1 \text{ L}} \times \frac{1 \text{ mole}}{35.5 \text{ g}}$$

$$= 0.171 \text{ M} \times \frac{1,000 \text{ mM}}{1 \text{ M}} = \mathbf{171 \text{ mM}}$$

One mole of chloride has the same number of ions as 1 mole of Na⁺ or 1 mole of Ca²⁺ or 1 mole of anything else. One mole of Ca²⁺, however, has twice the number of charges (*valence*) as 1 mole of Cl⁻. Therefore, 2 moles of Cl⁻ are required to neutralize 1 mole of Ca²⁺. If charges are taken into account by multiplying the moles by the valence, the product is termed the **equivalent weight** of an ion. One-thousandth of the equivalent weight dissolved in 1 liter of solution gives a concentration of *milliequivalents per liter (mEq/L)*.

Example

$$171 \text{ mM} \times 1 \text{ (the valence of Cl}^-)$$
$$= 171 \text{ mEq/L of Cl}^-$$

The major advantage of expressing the concentrations of ions in milliequivalents per liter is that the total concentration of anions can be easily compared with the total concentration of cations. In an average sample of venous plasma, for example, the total anions and the total cations are each equal to 156 mEq/L. Chloride, the major anion, has a plasma concentration of 103 mEq/L, whereas the chloride concentration in the urine is highly variable, ranging from 61 to 310 mEq/L.

PROCEDURE

1. The students void their urine into collection cups at the beginning of the laboratory session. In the analyses done in step 4, this sample will serve as the control (time zero).
2. The students drink 500 mL of water. One group just drinks the water; another group ingests NaCl (salt tablets are easiest to take) in addition to drinking the water. Most people can tolerate up to 4.5 g of salt, but the amount ingested should not be so great as to cause nausea.

Due in large part to the effects of ADH and aldosterone, the kidneys can vary their excretion of water and electrolytes to maintain homeostasis of the blood volume and composition. Abnormally low blood volume can produce **hypotension** (low blood pressure) and may result in circulatory shock; abnormally high blood volume contributes to **hypertension.** Renal regulation of Na$^+$ balance is also critical for health. Changes in blood Na$^+$ cause secondary changes in blood volume, as water follows sodium by osmosis. Changes in blood K$^+$ affect the bioelectrical properties of all cells, but the effects on the heart are particularly serious. **Hyperkalemia** (high blood K$^+$) is usually fatal when the K$^+$ concentration rises from 4 mEq/L (normal) to over 10 mEq/L. This may be caused by a variety of conditions, including inadequate aldosterone secretion (in Addison's disease), or by an excessive intake of potassium.

Note: *Students with hypertension or on sodium-restricted diets should not perform the salt-loading exercise. Interestingly, 4.5 g/500 mL of NaCl is a 0.9 g/100 mL saline solution that is isotonic to plasma (normal saline). How did this solution taste to you?*

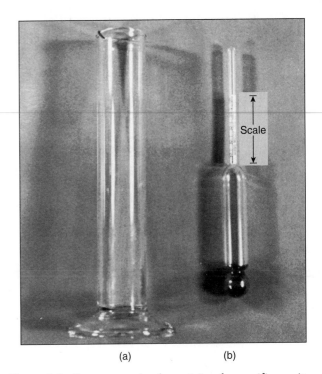

(a)   (b)

Figure 9.3 Instruments for determining the specific gravity of urine. (a) A glass cylinder and (b) a urinometer float.

3. After drinking the solutions described in step 2, the students void their urine every 30 minutes for 2 hours. The urine samples are analyzed as described in step 4.
4. Each of the five urine samples collected are analyzed for pH, specific gravity, and chloride content as follows:
   (a) **Volume (mL).** Measure the approximate volume of urine obtained and enter the data in the table of the laboratory report.
   (b) **pH.** Determine the pH of the urine samples by dipping a strip of pH paper into the urine and matching the color developed with a color chart. The urine normally has a pH between 5.0 and 7.5.
   (c) **Specific gravity.** Determine the specific gravity of the urine samples by floating a urinometer in a cylinder (fig. 9.3) nearly filled with the specimen. Read the specific gravity at the meniscus on the urinometer scale, making sure that the urinometer float is not touching the bottom or the sides of the cylinder. The specific gravity is directly related to the amount of solutes in the urine and ranges from 1.010 to 1.025. (Pure water should have a specific gravity of 1.000.)
   (d) **Chloride concentration.** When Na$^+$ is reabsorbed by the renal tubules, Cl$^-$ follows passively by electrostatic attraction. Follow the steps below to determine the chloride concentration of the urine samples.

(1) Measure 10 drops of urine into a test tube (1 drop is approximately 0.05 mL).

(2) To this tube add 1 drop of 20% potassium chromate solution with a second dropper.

(3) Add 2.9% silver nitrate solution one drop at a time using a third dropper, while shaking the test tube continuously. Count the number of full drops required to cause a permanent change in the color of the solution from yellow to brown.

(4) Determine the chloride concentration of the urine sample. Since each drop of 2.9% silver nitrate added in step 3 is equivalent to 61 mg of Cl⁻ per 100 mL of urine, simply multiply the number of drops by 61 to obtain the chloride concentration of the urine in milligrams per 100 mL.

Example

If 10 drops of 2.9% silver nitrate were required,

$$10 \times 61 \text{ mg of Cl}^-/100 \text{ mL} = 610 \text{ mg Cl}^-/100 \text{ mL}$$

5. Convert the chloride concentration to mEq/L and enter your data in the appropriate table in the laboratory report.

Laboratory Report 9.1

DATA FROM EXERCISE 9.1

Enter your data in the appropriate table below.

1. Ingestion of water only

Time	Volume (mL)	pH	Specific Gravity	Chloride (mEq/L)
0				
30				
60				
90				
120				

2. Ingestion of water and NaCl

Time	Volume (mL)	pH	Specific Gravity	Chloride (mEq/L)
0				
30				
60				
90				
120				

REVIEW ACTIVITIES FOR EXERCISE 9.1

Test Your Knowledge of Terms and Facts

1. Osmoreceptors in the hypothalamus of the brain are stimulated by a(n) _____ (increase/decrease) in the plasma osmotic pressure.
2. As a result of stimulation, the osmoreceptors stimulate the secretion of _____ from the _____ gland.
3. The hormone named in question 2 specifically stimulates the kidneys to _____ .
4. The hormone that stimulates the reabsorption of Na^+ from the nephron tubules, and also stimulates the secretion of K^+ into the tubules, is _____ .
5. The substance that stimulates vasoconstriction and also stimulates the secretion of the hormone named in question 4: _____ .
6. The measurement that is 1.000 for pure water and that increases in proportion to the general solute concentration of a solution: _____ .

Test Your Understanding of Concepts

7. Calcium is normally present at a concentration of about 0.1 g/L. Calculate the mEq/L of Ca^{2+} in the plasma (the atomic weight of calcium is 40).

8. Describe how the measurement of urine specific gravity changed after drinking water, and explain the physiological mechanisms responsible for this change.

Test Your Ability to Analyze and Apply Your Knowledge

9. Imagine a dehydrated desert prospector and a champagne-quaffing partygoer, each of whom drinks a liter of water at time zero and voids urine over a period of 3 hours. Using their urine samples, compare the probable differences in volume and composition. (*Hint:* Alcohol inhibits ADH secretion.) Use relative terms, such as *increased* or *decreased*.

Observation	Prospector	Partygoer
Urine volume		
Specific gravity		
Na^+ and Cl^- Content		

Explain the answers you provided in the table above.

10. Many diuretic drugs used clinically inhibit Na^+ reabsorption in the loop of Henle. Predict the effect of these drugs on the urinary excretion of Cl^- and K^+, and explain your answer.

Renal Plasma Clearance of Urea

MATERIALS

1. Pipettes with mechanical pipettors (or Repipettes), mechanical microliter pipettes (capacity μL), disposable tips
2. Colorimeter and cuvettes
3. BUN reagents and standard (Stanbio, through Curtin Matheson Scientific, Inc.)
4. Sterile lancets and 70% alcohol
5. File
6. Microhematocrit centrifuge and heparinized capillary tubes
7. Container for the disposal of blood-containing objects
8. As an alternative, normal and abnormal artificial urine is available (Wards Biology). However, modifications must be made by the instructor to simulate the conditions in this exercise.

Urea and other waste products in the plasma are filtered by the kidneys and excreted in the urine. The efficiency of the kidneys in performing these processes for each solute excreted is measured by its renal plasma clearance.

OBJECTIVES

1. Describe the chemical nature of urea and explain its physiological significance.
2. Define renal plasma clearance and explain how this value is calculated.
3. Perform a renal plasma clearance measurement for urea and explain the physiological significance of this measurement.
4. Explain how the renal plasma clearance for a solute is affected by filtration, reabsorption, and secretion.

Textbook Correlations*

Before performing this exercise, you may want to consult the following references in *Human Physiology*, seventh edition, by Stuart I. Fox:

- *Renal Plasma Clearance.* Chapter 17, pp. 541–546.

Those using different physiology textbooks may want to consult the corresponding information in those books.

When amino acids are broken down in the process of cellular respiration, or when they are converted to glucose (a process known as gluconeogenesis), the amino (—NH_2) groups are removed and secreted into the blood in the form of **urea.** This function is performed by the liver.

$$NH_2 - \underset{\underset{O}{\|}}{C} - NH_2$$

Urea

The urea is filtered by the glomeruli and enters the renal tubules. Although urea is a waste product of amino acid metabolism, some of it is transported (through facilitative diffusion) out of the nephron tubules by specific carriers. This increases the urea concentration in the interstitial fluid of the renal medulla, thus contributing to the hypertonicity of the renal medulla. Because urea is reabsorbed after filtration, only 60% of the blood filtered by the glomeruli is cleared of urea. Since the average glomerular filtration rate (GFR) is 125 mL/min (for both kidneys), this amounts to an average of 75 mL of plasma cleared of urea per minute. This value is termed the **renal plasma clearance** of urea (fig. 9.4).

---

*See Appendix 3 for correlations to the *A.D.A.M. InterActive PHYSIOLOGY Modules.*

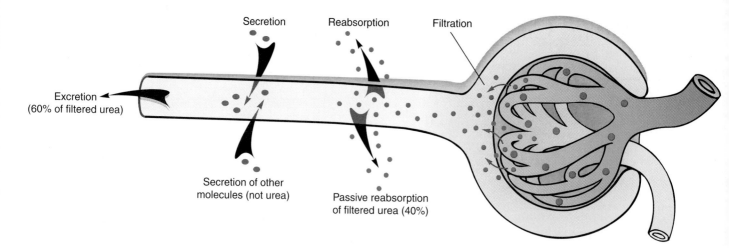

Figure 9.4 Filtration, reabsorption, and secretion by the nephron. Urea is filtered and about 40% of the amount filtered is passively reabsorbed (there is no secretion of urea). Therefore, approximately 60% of the amount filtered is excreted in the urine.

The clearance rate of a substance depends on the size of the kidneys and the rate of urine production, as well as the other factors discussed. In this exercise, it will be assumed that the kidneys are of average size and that urine production is equal to or greater than 2.0 mL/min. Under these conditions, the clearance of urea can be calculated using the formula

$$\text{Clearance} = \frac{U \times V}{P}$$

where,

> U is the concentration of urea in the urine, in milligrams per 100 mL
>
> P is the concentration of urea in the plasma, in milligrams per 100 mL
>
> V is the urine excreted in milliliters per minute

The concentration of a substance in urine, U, multiplied by the volume, V, of urine produced per minute gives the milligrams of the substance excreted in the urine per minute. When this figure is divided by the concentration, P, of that substance in the plasma, the result indicates the volume of plasma that contained the amount of the substance excreted per minute. This is the amount of plasma "cleared" by passage through the kidneys per minute.

If the substance is filtered from the glomeruli but is not reabsorbed or secreted (as with the polysaccharide *inulin*), the renal clearance rate equals the glomerular filtration rate (GFR) (fig. 9.5). If a substance is filtered but then reabsorbed into the blood (as with glucose, amino acids, urea, and many other substances), the renal plasma clearance must be less than the glomerular filtration rate. (How much less depends on the degree of reabsorption.) If a substance enters the renal tubules both by filtration and by active transport from the capillaries into the nephron tubules (a process called **secretion**), the renal plasma clearance is greater than the glomerular filtration rate.

 The plasma concentration of **blood urea nitrogen (BUN)** reflects both the rate of urea formation from protein in the liver and the rate of urea excretion by filtration through the glomeruli of the kidneys. In the absence of liver malfunction and abnormal protein metabolism, a rise in BUN indicates a kidney disorder such as nephritis, pyelonephritis, or kidney stones.

Since the renal plasma clearance for urea is substantially less than the glomerular filtration rate (GFR) due to its reabsorption, the urea clearance is not a particularly good indicator of kidney function. More useful clinically are the renal plasma clearances for exogenously administered *inulin* (a large polysaccharide) and endogenous *creatinine* (a byproduct of creatine, a molecule found primarily in muscle). Since inulin is neither reabsorbed nor secreted by the nephron, its clearance equals the GFR. Creatinine is secreted to a slight degree by the renal nephron, so its clearance rate is 20–25% greater than the true GFR (as defined by the inulin clearance test).

This is the case with the substance *para-aminohippuric acid* (PAH), which is almost entirely removed or "cleared" from the blood in a single passage through the kidneys.

PROCEDURE

Collection of Plasma and Urine Samples

1. Empty the urinary bladder. Then, have the subject drink 500 mL of water as quickly as is comfortable and record the time (time zero).

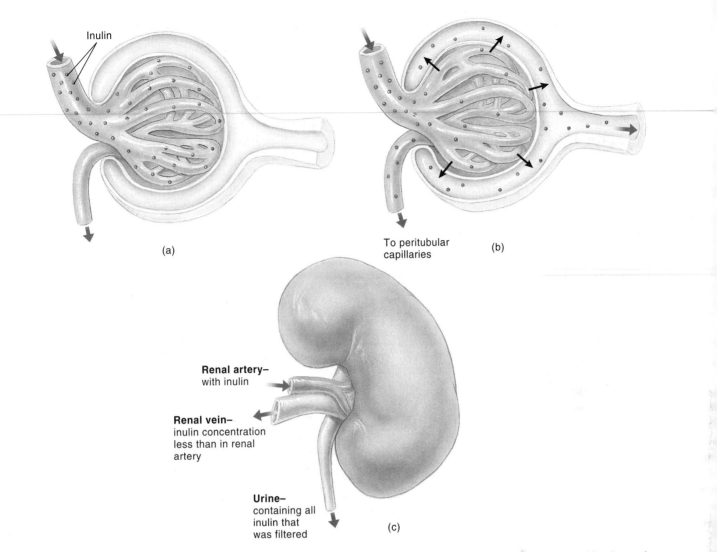

Inulin

To peritubular
capillaries

(a)   (b)

Renal artery–
with inulin

Renal vein–
inulin concentration
less than in renal
artery

Urine–
containing all
inulin that
was filtered   (c)

Figure 9.5   Renal clearance of inulin. (a) Inulin is present in the blood entering the glomeruli, and (b) some of this blood, together with its dissolved inulin, is filtered. All of this filtered inulin enters the urine, whereas most of the filtered water is returned to the vascular system (is reabsorbed). (c) The blood leaving the kidneys in the renal vein, therefore, contains less inulin than the blood that entered the kidneys in the renal artery. Since inulin is filtered but neither reabsorbed nor secreted, the inulin clearance rate equals the glomerular filtration rate (GFR).

2. About 20 minutes later, cleanse the fingertip with 70% alcohol and, using a sterile lancet, obtain a drop of blood from the subject's fingertip. Fill a heparinized capillary tube at least halfway with blood, plug one end, and centrifuge in a microhematocrit centrifuge for 3 minutes (as in performing a hematocrit measurement—see exercise 6.1).

⚠ **Caution:** *Always handle only your own blood and place all objects that have been in contact with blood in the container indicated by the instructor.*

3. Score the capillary tube lightly with a file at the plasma-cell junction and break the tube at the scored mark. Expel the plasma into a labeled small beaker or test tube.

4. Collect a complete urine sample 30 minutes after drinking the 500 mL of water. Measure the milliliters of water produced in 30 minutes, divide by 30, and enter the volume of urine produced per minute in the laboratory report.

5. Dilute the urine 1:20 with water. This can be done by adding 19 mL of water to 1 mL of urine, or by adding 1.9 mL of water to 0.10 mL (100 μL) of urine. Mix the diluted urine solution.

MEASUREMENT OF UREA

6. Label four test tubes or cuvettes: *B* (blank), *S* (standard), *P* (plasma), and *U* (urine).

7. Pipette 2.5 mL of *Reagent A* into each tube; then pipette 20 μL of the following into the indicated tubes:

Tube B: 20 mL of distilled water
Tube S: 20 mL of urea standard (28 mg/dL)
Tube P: 20 mL of plasma from the capillary tube
Tube U: 20 mL of the 30-minute urine sample that was previously diluted to 1/20 of its original concentration

Note: *Since 20 μL is a very small volume, pipetting must be carefully performed to be accurate. If an automatic microliter pipette is used, depress the plunger several times to wet the disposable tip thoroughly in the solution before withdrawing the 20 μL sample. When delivering the sample, slowly depress the plunger to the first stop with the disposable tip against the wall of the test tube. Then pause several seconds before depressing the plunger to the second stop. Be sure to mix the test tube well to wash the sample from the test tube wall.*

8. Mix and incubate the tubes at room temperature for 10 minutes.
9. Pipette 2.5 mL of *Reagent B* into each tube and incubate for an additional 10 minutes.
10. Set the colorimeter at 600 nm, standardize with the reagent blank, and record the absorbance values of the standard, plasma, and urine samples in the data table in your laboratory report.

11. Calculate the urea concentration of the plasma (using Beer's law, as described in exercise 2.1) and enter this value (P) in your laboratory report.
12. Calculate the urea concentration of the diluted urine sample in the manner described in step 11. Multiply this answer by the dilution factor (20), and enter this corrected value for the urea concentration of the urine (U) in your laboratory report.
13. Using your values for the volume of urine produced per minute (V), plasma urea concentration (P), and urine urea concentration (U), calculate your renal plasma clearance rate for urea. Enter this value in your laboratory report.

> The normal range of urea in plasma is 5–25 mg/dL

Example

$$\text{Suppose, } V \text{ (from step 4)} = 2.0 \text{ mL/min}$$
$$P \text{ (from step 11)} = 10 \text{ mg/dL}$$
$$U \text{ (from step 12)} = 375 \text{ mg/dL, then}$$

$$\text{Renal plasma clearance} = \frac{375 \text{ mg/dL} \times 2.0 \text{ mL/min}}{10 \text{ mg/dL}}$$

> The normal range for the urea renal plasma clearance rate is 64–99 mL/min.

Laboratory Report 9.2

Name _____

Date _____

Section _____

DATA FROM EXERCISE 9.2

1. Enter the *volume* of urine produced per minute in the space below.

   V = _____ ml/min

2. Enter your absorbance values in the table below.

Tube	Contents	Absorbance
P	Plasma	
U	Diluted urine	
S	Standard (28 mg/dL)	

3. Enter the urea concentration of the:

   plasma (P): _____ mg/dL

   urine (U): _____ mg/dL

4. Calculate your renal plasma clearance rate for urea and enter this value in the space below.

   Clearance = _____ mL/min

REVIEW ACTIVITIES FOR EXERCISE 9.2

Test Your Knowledge of Terms and Facts

Identify a molecule in the plasma that is:
1. filtered, but neither reabsorbed nor secreted _____ .
2. filtered and partially reabsorbed _____ .
3. filtered and completely secreted _____ .
4. not filtered to a significant degree _____ .

Identify the substance from the description of the renal plasma clearance.
5. The clearance is greater than zero but less than the GFR for _____ .
6. The clearance is equal to the GFR for _____ .
7. The clearance is only slightly greater than the GFR for _____ .
8. The clearance is equal to the total plasma flow rate to the kidneys for _____ .

Test Your Understanding of Concepts

9. Explain what is meant by the *renal plasma clearance* and describe how it is measured.

10. What is the significance of the glomerular filtration rate (GFR) measurement? Describe two ways that this measurement may be obtained.

Test Your Ability to Analyze and Apply Your Knowledge

11. Suppose substance "A" is reabsorbed about 30%, whereas substance "B" is secreted about 30%. Explain how their renal plasma clearance values relate to each other and to the glomerular filtration rate (GFR).

12. Explain the mechanism by which each of the following conditions might produce an increase in the plasma concentration of urea: (a) increased protein catabolism; (b) decreased blood pressure, as in circulatory shock; and (c) kidney failure.

Clinical Examination of Urine

MATERIALS

1. Microscopes
2. Urine collection cups, test tubes, microscope slides, and coverslips
3. Albustix, Clinitest tablets, Ketostix, Hemastix, Ictotest tablets, or Multistix (all from Ames Laboratories)
4. Sediment stain (such as Sternheimer-Mablin stain)
5. Centrifuge and centrifuge tubes
6. Transfer pipettes (droppers)
7. Alternatively, normal and abnormal artificial urine is available with a few test strips (Wards Biology).

The presence of abnormally large amounts of proteins and casts in the urine can indicate damage to the glomeruli of the kidney. The presence of bacteria and a large number of white blood cells in the urine sediment indicates urinary tract infection. Abnormal concentrations of glucose, ketone bodies, bilirubin, and other plasma solutes in the urine may reflect abnormally high concentrations in the plasma.

OBJECTIVES

1. Describe the physiological processes responsible for normal urinary concentrations of protein, glucose, ketone bodies, and bilirubin in the urine.
2. Describe the pathological processes that may produce abnormal solute concentrations and explain the clinical significance of this information.
3. Describe the normal constituents of urine sediment and explain how the microscopic examination of urine sediment can be clinically useful.

Textbook Correlations*

Before performing this exercise, you may want to consult the following references in *Human Physiology,* seventh edition, by Stuart I. Fox:

- *Glomerular Filtration.* Chapter 17, pp. 531–534.
- *Reabsorption of Glucose.* Chapter 17, pp. 545–546.

Those using different physiology textbooks may want to consult the corresponding information in those books.

Table 9.1 Appearance of Urine and Cause

Color	Cause
Yellow-orange to brownish green	Bilirubin from obstructive jaundice
Red to red-brown	Hemoglobinuria
Smoky red	Unhemolyzed RBCs from urinary tract
Dark wine color	Hemolytic jaundice
Brown-black	Melanin pigment from melanoma
Dark brown	Liver infections, pernicious anemia, malaria
Green	Bacterial infection (*Pseudomonas aeruginosa*)

Note: Certain foods (e.g., beets, rhubarb) and some commonly prescribed drugs and vitamin supplements may also alter the color of urine.

A clinical examination of urine may provide evidence of urinary tract infection (UTI) or kidney disease. Additionally, since urine is derived from plasma, an examination of the urine provides a convenient, nonintrusive means of assessing the composition of plasma and of detecting a variety of systemic diseases. A clinical examination of the urine includes an observation of its appearance (table 9.1), tests of its chemical composition, and a microscopic examination of urine sediment.

*See Appendix 3 for correlations to the A.D.A.M. InterActive PHYSIOLOGY Modules.

When the kidneys are inflamed the permeability of the glomerular capillaries may be increased, resulting in the leakage of proteins into the urine (**proteinuria**) and the appearance of casts in the urine sediment. Since this leakage represents a continuous loss of the protein solutes responsible for the colloid osmotic pressure of plasma, fluid may accumulate in the tissues, resulting in *edema*.

The appearance of glucose (**glycosuria**) in the urine suggests the presence of *diabetes mellitus*. If diabetes is suspected, however, the test for glucose in the urine alone is not sufficient because a person may have hyperglycemia without glycosuria. In such a case, the glucose concentration in the plasma at the time of the test may not be high enough to exceed the ability of the nephron tubules to completely reabsorb glucose from the filtrate (known as the *transport maximum* (T_m) for glucose). Because tubular reabsorption is so efficient (the T_m is high), the urine will be free of glucose until the concentrations of glucose in the blood exceed its *renal plasma threshold*. A more conclusive test for diabetes is the *oral glucose tolerance test*.

When the glucose concentration both in the plasma and in the glomerular ultrafiltrate is within the normal limits (70–110 mg per 100 mL), there is a sufficient number of carrier molecules along the renal tubules to transport all the glucose back into the blood. However, if the blood glucose level exceeds a certain limit, called the **renal plasma threshold** for glucose, (about 180 mg per 100 mL), the number of glucose molecules in the glomerular ultrafiltrate will be greater than the number of available carrier molecules, and the nontransported glucose will "spill over" into the urine.

The chief cause of glycosuria is diabetes mellitus, although other conditions, such as hyperthyroidism, hyperpituitarism, and liver disease may also have this effect. Glycosuria, therefore, is not a renal disease but a symptom of other systemic diseases that raise the blood sugar level.

PROCEDURE

1. Place 10 drops of water and 5 drops of urine in a test tube.
2. Add a Clinitest tablet.
3. Wait 15 seconds and compare the color developed with the color chart provided.
   Or: Dip the end of a disposable Clinistix strip into a urine sample, wait the required amount of time, then compare the color developed with the chart provided on the bottle. Record the glucose concentration in the data table of the laboratory report.

C. TEST FOR KETONURIA

When there is carbohydrate deprivation, such as in starvation or high-protein diets, the body relies increasingly on the metabolism of fats for energy. This pattern is also seen in people with diabetes mellitus, where lack of the hormone insulin prevents the body cells from utilizing the large amounts of glucose available in the blood. This occurs because insulin is necessary for the transport of glucose from the blood into the body cells.

The metabolism of fat proceeds in a stepwise manner:

1. triglycerides are hydrolyzed to fatty acids and glycerol;
2. fatty acids are converted into smaller intermediate compounds—*acetoacetic acid*, β-*hydroxybutyric acid*, and *acetone* (collectively known as **ketone bodies**);
3. the ketone bodies are broken down in aerobic cellular respiration, releasing energy. When the production of ketone bodies from fatty acid metabolism exceeds the ability of the body to metabolize these compounds, they accumulate in the blood (*ketonemia*) and spill over into the urine (*ketonuria*).

A. TEST FOR PROTEINURIA

Since proteins are very large molecules (macromolecules), they are not normally present in measurable amounts in the glomerular ultrafiltrate or the urine. The presence of proteins in the urine may therefore indicate an abnormal increase in the permeability of the kidney glomeruli. Such permeability changes may be caused by renal infections (glomerulonephritis) or by other diseases that have secondarily affected the kidneys, such as diabetes mellitus, jaundice, or hyperthyroidism.

PROCEDURE

Dip the yellow end of a disposable Albustix strip into a urine sample and compare the color developed with the chart provided. Record the albumin (protein) concentration in the data table in your laboratory report.

B. TEST FOR GLYCOSURIA

Although glucose is easily filtered by the glomerulus, it is not normally present in the urine. All of the filtered glucose is normally reabsorbed from the renal tubules into the blood. This reabsorption process is carrier mediated—that is, the filtered glucose is transported across the wall of the renal tubule by a protein carrier.

PROCEDURE

Dip a disposable Ketostix strip into a urine sample; 15 seconds later, compare the color developed with the color chart. Record the ketone concentration in the data table of the laboratory report.

D. TEST FOR HEMOGLOBINURIA

Hemoglobin may appear in the urine in the event of hemolysis in the systemic blood vessels (e.g., in transfusion reactions), of rupture in the capillaries of the glomerulus, or of hemorrhage in the urinary system. In the latter condition, whole red blood cells may be found in the urine (*hematuria*), although the low osmotic pressure of the urine may cause hemolysis and the release of hemoglobin (*hemoglobinuria*) from these red cells. Hemoglobinuria is normally found in the samples from menstruating women.

PROCEDURE

Dip the test end of a disposable Hemastix strip into the urine sample, wait 30 seconds, and compare the color developed with the color chart. Enter the hemoglobin content in the data table of the laboratory report.

E. TEST FOR BILIRUBINURIA

The fixed phagocytic cells of the spleen and bone marrow (*reticuloendothelial system*) destroy old red blood cells and convert the *heme* groups of hemoglobin into the pigment **bilirubin.** The bilirubin is secreted into the blood and carried to the liver, where it is bonded to (*conjugated* with) glucuronic acid, a derivative of glucose. Most of this conjugated bilirubin is secreted into the bile as bile pigment; the rest is released into the blood.

The blood normally contains a small amount of free and conjugated bilirubin. An abnormally high level of blood bilirubin (*hyperbilirubinemia*) may result from (1) an increased rate of red blood cell destruction, seen in hemolytic anemia, for example; (2) liver damage, as in hepatitis and cirrhosis; or (3) obstruction of the common bile duct, as might occur because of a gallstone. The increase in blood bilirubin results in **jaundice,** a condition characterized by a brownish yellow pigmentation of the skin, sclera of the eye, and mucous membranes.

Normally, the kidneys can excrete only the more water-soluble bilirubin that is conjugated with glucuronic acid. An increase in the urine bilirubin (*bilirubinuria*), therefore, may be associated with jaundice due to liver disease or bile duct obstruction, but is not normally observed in jaundice due to hemolytic anemia. In the latter case, the excess bilirubin is in a free, nonpolar state where it binds to plasma proteins, and thus is not filtered into the nephron tubules.

PROCEDURE

1. Place 5 drops of urine on a square of the test mat.
2. Place an Ictotest tablet in the center of the mat.
3. Place 2 drops of water on the tablet.
4. Interpret the test as follows:
   (a) Negative: Mat has no color or a slight pink-to-red color.
   (b) Positive: Mat turns blue to purple. The speed and intensity of color development is proportional to the amount of bilirubin present.
5. Record your observations in the data table in your laboratory report.

F. MICROSCOPIC EXAMINATION OF URINE SEDIMENT

Microscopic examination of the urine sediment may reveal the presence of various cells, crystals, bacteria, and casts (figs. 9.6, 9.7, and 9.8). **Casts** are cylindrical structures formed by the precipitation of protein molecules within the renal tubules. Although a small number of casts are found in normal urine, a large number indicate renal disease, such as *glomerulonephritis* or *nephrosis*. The casts may lack cells, or contain cells such as leukocytes, erythrocytes, or epithelial cells (fig. 9.6). The presence of large numbers of erythrocytes, leukocytes, or certain epithelial cells in the urine is indicative of renal disease.

Like casts, a small number of crystals are present in normal urine. Their presence in large numbers, however, may suggest a tendency to form kidney stones. Additionally, large numbers of *uric acid* crystals occur in *gout*, a form of arthritis (fig. 9.7).

PROCEDURE

1. Fill a conical centrifuge tube three-quarters full of urine, and centrifuge at a moderate speed for 5 minutes.
2. Discard the supernatant and place one drop of Sternheimer-Mablin stain (or similar sediment stain) on the sediment. Mix by aspiration with a transfer pipette.
3. Place 1 drop of the stained sediment on a clean slide and cover with a coverslip.
4. Scan the slide with the low-power objective under reduced illumination (close the diaphragm) and identify the components of the sediment.

Red blood cells (erythrocytes)

White blood cells (leukocytes)

Epithelial cells of the renal tubules

Epithelial cells of the bladder

Epithelial cells of the urethra

Bacteria

Hyaline cast

Waxy cast

Figure 9.6   Components of urine sediment.

Hippuric acid

Ammonium-magnesium
phosphate

Cholesterol

Calcium oxalate

Calcium phosphate

Uric acid

Tyrosine

Cystine

Figure 9.7 Crystals found in urine sediment.

Hyaline casts; phase contrast. (400×)

Oval fat bodies; "mulberry cell." Urine from a diabetic rat. (400×)

"Drug" crystal (sulfa); polarized light. Note sharp needle-like structure that may cause damage to the tubules. (400×)

Leucine crystals; note size variation. (400×)

Tyrosine crystals (rosette needles). (100×)

Uric acid crystals (plates); polarized light. (400×)

Triple phosphate crystals; "coffin-lid" form. (400×)

Calcium Oxalate crystals. (400×)

Figure 9.8   Photographs of urine sediment.

340

Laboratory Report 9.3

DATA FROM EXERCISE 9.3

Record your data and the interpretations of your data in the table below.

Urine Test	Result of Exercise (Positive or Negative)	Physiological Reason for Negative Result	Clinical Significance of Positive Result
Proteinuria			
Glycosuria			
Ketonuria			
Hemoglobinuria			
Bilirubinuria			

1. Describe the appearance (such as color, intensity, turbidity) of the urine sample and explain the possible causes of these observations.

2. List the casts, cells, or crystals of the urine sediment identified under low-power; and suggest the possible clinical significance if these components were found in abnormally high concentrations.

REVIEW ACTIVITIES FOR EXERCISE 9.3

Test Your Knowledge of Terms and Facts

Match the following molecules with their descriptions:

_____1. normally not filtered
_____2. filtered, then normally completely reabsorbed
_____3. derived from fat breakdown
_____4. derived from heme groups of hemoglobin

(a) ketone bodies
(b) bilirubin
(c) glucose
(d) protein

5. Common cells found in normal urine sediment _____ .
6. An abnormal component of urine sediment, formed from protein _____ .
7. Indicate a possible cause for each of the following conditions:
   (a) proteinuria _____
   (b) glycosuria _____
   (c) ketonuria _____
   (d) bilirubinuria _____

Test Your Understanding of Concepts

8. Describe the composition of urinary casts and explain how they can get into the urine.

9. Which component of the urine would be raised if a person were on a very low carbohydrate weight-reducing diet? Explain the process involved.

10. Is it possible for someone to have an abnormally high plasma glucose concentration and yet not have glycosuria? Explain your answer. Relate this answer to a person who eats a couple of sugar-coated doughnuts compared to a person with uncontrolled diabetes mellitus.

11. Proteinuria and the presence of numerous casts in the urine sediment are often accompanied by edema. What is the relationship between these symptoms? (*Hint:* Review exercise 2.1— particularly the clinical applications box regarding plasma protein concentration.)

Digestion and Nutrition

The digestive tract is a continuous tube that is open to the external environment at both ends, by way of the mouth and the anus (fig. 10.1). Material inside this digestive tract is outside the body in the sense that it can contact only the epithelial cells that line the tract. For this material to reach the inner cells of the body, it must pass through the epithelial cells of the tract (a process known as **absorption**) into the blood. Before nutritive material can be absorbed, however, it must first be broken down by physical processes, such as chewing (mastication), and by enzymatic hydrolysis into its monomers. The process of hydrolyzing larger food molecules (polymers) into absorbable monomers is known as **digestion.**

The embryonic digestive system consists of a hollow tube only one cell layer thick. As the embryo develops, different regions of the digestive tract become specialized for different functions. Some epithelial cells that line the tract become secretory, forming exocrine glands that secrete mucus, HCl, or particular hydrolytic enzymes characteristic of a certain region of the digestive tract. The liver, which produces bile, and the pancreas, which produces the many digestive enzymes found in "pancreatic juice," develop as outpouchings (*diverticula*) of the embryonic small intestine and maintain their connections with the intestine by means of the hepatic and pancreatic ducts. Other regions of the small intestine become specialized for absorption through an increase in surface area. This increased surface area is produced by epithelial folds known as **villi** and minute foldings of the epithelial cell membranes called **microvilli.**

The digestive tract may be visualized as a "disassembly" line, where the food is conveyed, by means of muscular movements of the tract (*peristalsis*) and the opening and closing of sphincter muscles, from one stage of processing to the next. Coordination of these processes is achieved by neural reflexes and by hormones secreted by the gastrointestinal tract (table 10.1).

Diet maintains the consistent supply of nutrients to the body cells as energy sources for fuel, for growth, and for the replacement of cellular parts. *Carbohydrates, fats, proteins, vitamins, minerals,* and *water* are six nutrient classes recommended for daily consumption. Only carbohydrates, fats, and proteins can provide energy, measured in **kilocalories** (*kcals*). The **basal metabolic rate (BMR)** is the minimum amount of energy required by the body at rest.

Exercise 10.1   Histology of the Gastrointestinal Tract, Liver, and Pancreas

Exercise 10.2   Digestion of Carbohydrate, Protein, and Fat

Exercise 10.3   Nutrient Assessment, BMR, and Body Composition

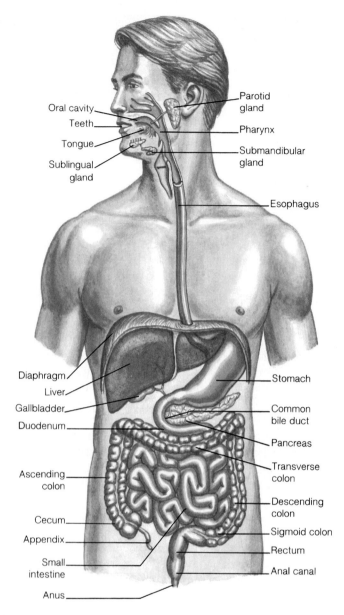

Figure 10.1   Organs of the digestive system.

Table 10.1   Physiological Effects of Gastrointestinal Hormones

Secreted by	Hormone	Effects
Stomach	Gastrin	Stimulates parietal cells to secrete HCl
		Stimulates chief cells to secrete pepsinogen
		Maintains structure of gastric mucosa
Small intestine	Secretin	Stimulates water and bicarbonate secretion in pancreatic juice
		Potentiates actions of cholecystokinin on pancreas
Small intestine	Cholecystokinin (CCK)	Stimulates contraction of gallbladder
		Stimulates secretion of pancreatic juice enzymes
		Inhibits gastric motility and secretion
		Maintains structure of exocrine pancreas (acini)
Small intestine	Gastric inhibitory peptide (GIP)	Inhibits gastric motility and secretion
		Stimulates secretion of insulin from pancreatic islets
Ileum and colon	Glucagon-like peptide-1 (GLP-1)	Inhibits gastric motility and secretion
		Stimulates secretion of insulin from pancreatic islets
	Guanylin	Stimulates intestinal secretion of Cl^-, causing excretion of NaCl and water in the feces

Histology of the Gastrointestinal Tract, Liver, and Pancreas

MATERIALS

1. Microscopes
2. Prepared tissue slides of the digestive system

All regions of the gastrointestinal tract have a mucosa, submucosa, muscularis, and serosa, but these layers of the wall display different specializations in different regions of the tract. The histology of the liver and pancreas provides insights into the functions of these organs.

OBJECTIVES

1. Identify the mucosa, submucosa, muscularis, and serosa layers of different regions of the gastrointestinal tract.
2. Describe the structure and function of the layers of the esophagus, stomach, small intestine, and large intestine.
3. Describe the microscopic anatomy of the liver and explain its functional significance.
4. Describe the microscopic anatomy of the pancreas and distinguish the parts involved in the endocrine and exocrine functions of the pancreas.

Textbook Correlations

Before performing this exercise, you may want to consult the following references in *Human Physiology*, seventh edition, by Stuart I. Fox:

- *Esophagus and Stomach*. Chapter 18, pp. 565–571.
- *Small Intestine*. Chapter 18, pp. 571–574.
- *Large Intestine*. Chapter 18, pp. 575–577.
- *Liver, Gallbladder, and Pancreas*. Chapter 18, pp. 578–586.

Those using different physiology textbooks may want to consult the corresponding information in those books.

The tubular digestive tract, including the esophagus, stomach, small intestine, and large intestine, consists of four major layers, or *tunics* (fig. 10.2). From the innermost layer outward, they are as follows:

1. The **mucosa,** or mucous membrane, consists of an inner epithelium spread over a thin layer of connective tissue, the *lamina propria*, which is bordered by a ribbon of smooth muscle, the *muscularis mucosa*. The epithelium is stratified squamous in the esophagus and anal canal and simple columnar in the stomach, small intestine, and large intestine.
2. The **submucosa** is connective tissue and therefore has abundant extracellular space for blood vessels, nerves, and mucus-secreting glands. Parasympathetic fibers and ganglia can be seen as the *submucosal (Meissner's) plexus* in the submucosa.
3. The **muscularis externa** consists of smooth muscle, arranged in an inner circular and outer longitudinal layer throughout most of the digestive tract. Parasympathetic fibers and ganglia can be seen as the *myenteric (Auerbach's) plexus* in this layer.
4. The **serosa** consists of a *simple squamous* epithelium and connective tissue, and is the outermost covering of the digestive tract.

A. ESOPHAGUS AND STOMACH

The mucosa layer of the esophagus is lined with a stratified squamous epithelium (fig. 10.3). The muscles of the first third of the esophagus, like those of the pharynx and mouth, are striated to provide voluntary control of swallowing. The middle third contains a mixture of striated and smooth muscle, and the last third of the esophagus contains only involuntary smooth muscle.

The submucosa of the stomach is thrown into large folds, or *rugae*, which can be seen with the unaided eye. Microscopic examination of the mucosa shows that it, too, is folded. The openings of these folds into the stomach lumen are called *gastric pits*. The cells that line the folds of mucosa are secretory and form the **gastric glands** (figs. 10.4 and 10.5).

The gastric glands include: (1) *goblet cells*, which secrete mucus; (2) *parietal cells*, which secrete hydrochloric

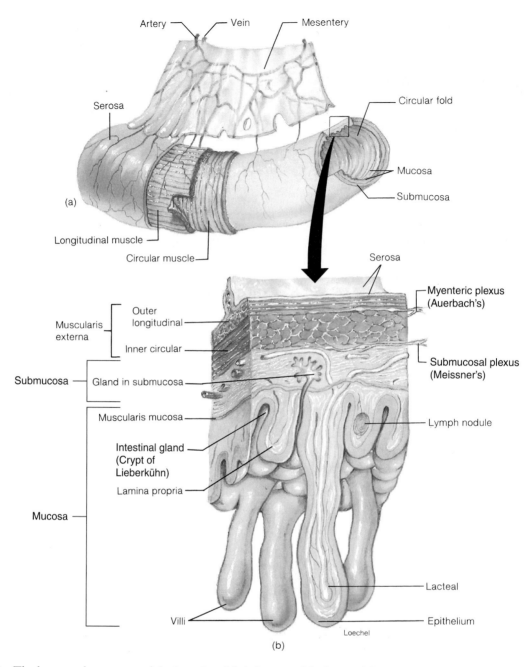

(a)

Artery — Vein — Mesentery

Serosa

Circular fold

Mucosa

Submucosa

Longitudinal muscle

Circular muscle

Serosa

Muscularis externa
— Outer longitudinal
— Inner circular

Submucosa
— Gland in submucosa

Muscularis mucosa

Intestinal gland (Crypt of Lieberkühn)

Lamina propria

Mucosa

Villi

Myenteric plexus (Auerbach's)

Submucosal plexus (Meissner's)

Lymph nodule

Lacteal

Epithelium

Loechel

(b)

Figure 10.2   The layers and structures of the intestine. (a) A diagram of the layers of the intestine in longitudinal and cross section. (b) An enlarged cross section of the intestinal wall.

Serosa

Muscularis

Submucosa

Mucosa

Lumen

Figure 10.3   The histology of the esophagus (low power).

Figure 10.4 The microscopic structure of the stomach.

Labels on figure 10.4:
Lumen
Columnar epithelium of mucosa
Gastric pit
Gastric glands with chief and parietal cells
Lamina propria

acid (HCl); (3) *chief cells*, which secrete pepsinogen (the inactive precursor of pepsin, a protein-digesting enzyme); (4) *enterochromaffin-like cells (ECL)* which secrete histamine; (5) *G cells*, which secrete the hormone gastrin into the blood; and (6) *D cells*, which secrete the hormone somatostatin. The gastric mucosa also secretes a polypeptide called *intrinsic factor*, which aids in the absorption of vitamin B_{12} in the intestine.

PROCEDURE

1. Observe a cross section of the esophagus under 100× (using the 10× objective lens) and note the four major layers, or tunics.

2. Hold a slide of a stomach section up to a light source and observe a fold, or ruga. Now, place the slide on a microscope and, under 100×, observe the gastric pits and glands in the mucosa, the submucosa, and the muscularis externa.

3. Using the high-dry objective lens (45×), observe the gastric glands in the mucosa under a total magnification of 450×. Identify goblet cells near the surface of the gastric pits. These mucus-secreting cells are numerous and clear in appearance. Near the base of the glands, identify parietal cells (with red-staining cytoplasm) and chief cells (with blue-staining cytoplasm). Argentaffin and G cells cannot be easily identified without specially stained slides.

Figure 10.5 Gastric pits and gastric glands of the mucosa. (a) Gastric pits are the openings of the gastric glands. (b) Gastric glands consist of mucous cells, chief cells, and parietal cells, each of which produces a specific secretion.

349

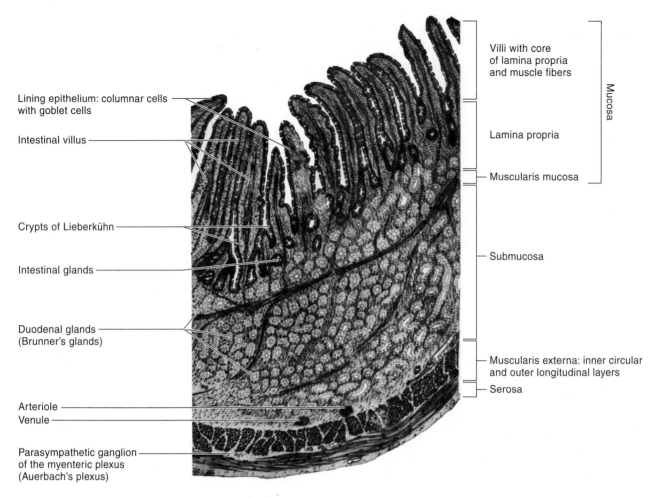

Figure 10.6 The microscopic structure of the duodenum.

Labels (clockwise from left):
- Lining epithelium: columnar cells with goblet cells
- Intestinal villus
- Crypts of Lieberkühn
- Intestinal glands
- Duodenal glands (Brunner's glands)
- Arteriole
- Venule
- Parasympathetic ganglion of the myenteric plexus (Auerbach's plexus)
- Villi with core of lamina propria and muscle fibers
- Lamina propria
- Muscularis mucosa
- Mucosa
- Submucosa
- Muscularis externa: inner circular and outer longitudinal layers
- Serosa

B. Small Intestine and Large Intestine

The small intestine is approximately 21 feet long (in a cadaver) and divided into three regions. The first region, approximately 12 inches long, is called the **duodenum.** The next region, the **jejunum,** is about 8 feet long and constitutes two-fifths of the entire length of the intestine. The **ileum,** about 12 feet long (constituting three-fifths of the intestine), is the terminal region.

The mucosa and submucosa of the small intestine form large folds called the *plicae circulares*. The surface area of the mucosa is further increased by microscopic folds that form fingerlike projections called *villi* (fig. 10.6). Each villus has a core of connective tissue (the lamina propria) covered with a simple columnar epithelium. The apical surface (facing the lumen) of each epithelial cell has a slightly blurred, "brush border" appearance because of numerous projections of its cell membrane in the form of *microvilli*. Microvilli can be clearly seen only with an electron microscope.

The microvilli, villi, and plicae circulares increase the surface area of the small intestine tremendously, thus maximizing the rate at which the products of digestion can be absorbed by transport through the epithelium into the blood. Various digestive enzymes—called **brush-border enzymes**—are fixed to the cell membranes of the microvilli and act together with enzymes from pancreatic juice to catalyze hydrolysis reactions of food molecules. The small intestine epithelium is also the source of important gastrointestinal hormones (table 10.2).

The epithelium at the base of the villi invaginates to form pouches called *crypts of Lieberkühn*, or intestinal crypts. Although somewhat similar in appearance to gastric glands of the stomach, these intestinal crypts are not secretory. Instead, the cells within the crypts seem to undergo mitotic division and push upward to replace those cells that are continuously shed into the lumen from the tips of the villi. Within the submucosa of the duodenum are alkaline mucus-secreting *Brunner's glands* (fig. 10.6). The jejunum and ileum lack these glands but are otherwise similar in structure (fig. 10.7).

Waste products from the small intestine pass into the **colon** of the large intestine where water, sodium, and potassium are absorbed. The mucosa of the large intestine contains crypts of Lieberkühn but not villi, so its surface has a flat appearance. As with the small intestine, numerous lymphocytes can be seen in the lamina propria, and

Table 10.2 Characteristics of the Major Digestive Enzymes

Enzyme	Site of Production	Source	Substrate	Optimum pH	Product(s)
Salivary amylase	Mouth	Saliva	Starch	6.7	Maltose
Pepsin	Stomach	Gastric glands	Protein	1.6–2.4	Shorter polypeptides
Pancreatic amylase	Duodenum	Pancreatic juice	Starch	6.7–7.0	Maltose, maltriose, and oligosaccharides
Trypsin, chymotrypsin, carboxypeptidase			Polypeptides	8.0	Amino acids, dipeptides, and tripeptides
Pancreatic lipase			Triglycerides	8.0	Fatty acids and monoglycerides
Maltase		Epithelial membranes	Maltose	5.0–7.0	Glucose
Sucrase			Sucrose	5.0–7.0	Glucose + fructose
Lactase			Lactose	5.8–6.2	Glucose + galactose
Aminopeptidase			Polypeptides	8.0	Amino acids, dipeptides, tripeptides

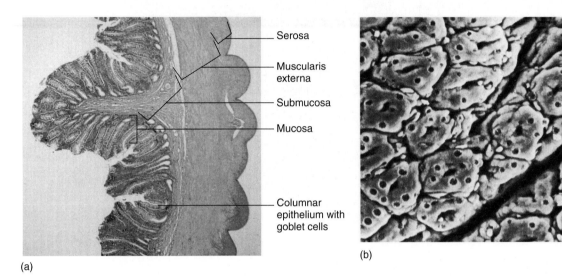

(a)

(b)

Figure 10.7   The microscopic structure of the ileum and large intestine. (a) A photomicrograph of a cross section of the ileum. (b) A scanning electron micrograph of the luminal surface of the large intestine (the arrow points to the opening of a goblet cell). (b is from R. G. Kessel and R. H. Kardon, *Tissues and Organs: A Text-Atlas of Scanning Electron Microscopy* © 1979 W. H. Freeman and Company).

large lymphatic nodules appear at the junction of the mucosa and submucosa. Lymphatic nodules are clearly evident in a section of the *appendix,* a short outpouching from the cecum.

PROCEDURE

1. Use the lowest power available on the microscope to observe the layers of a section of small intestine. Identify the villi, submucosa, and muscularis externa.
2. Observe the villi using the 45× objective lens. Identify the goblet cells in the epithelium and the numerous lymphocytes (small, blue-staining cells) in the lamina propria within each villus. Also within the lamina propria, note the *central lacteal*—a lymphatic vessel that transports absorbed fat from the intestine.
3. Observe a slide of the large intestine using the 10× objective lens (fig. 10.7). Note the four layers, absence of villi, and goblet cells in the columnar epithelium of the mucosa.

C. LIVER

The liver aids digestion by producing and secreting **bile,** which emulsifies fat. Bile leaves the liver in the *common hepatic duct,* which branches to form the *cystic duct* and the *common bile duct.* The cystic duct channels bile to the gallbladder where it is stored and concentrated. The common bile duct joins the pancreatic duct and, together, they empty into the duodenum.

The liver also serves to modify the composition of the blood that drains from capillaries of the intestine through the *hepatic portal vein.* Before this rich venous blood can return to the heart and circulation, it must pass through *sinusoids* in the liver tissue (figs. 10.8 and 10.9). Liver sinusoids, however, are wider than most types of capillaries and are lined with phagocytic cells called *Kupffer cells.* Blood is drained from these sinusoids by small *central veins* that ultimately merge to form the large *hepatic vein,* which carries blood away from the liver. Liver sinusoids also receive arterial blood from

Figure 10.8 The microscopic structure of the liver. (a) A cross section of a liver lobule. (b) A longitudinal section of a liver lobule, showing the arrangement of blood vessels, hepatic plates, and hepatic sinusoids. (c) A photomicrograph of a liver lobule in cross section.

(a)

Portal vein

Hepatic artery

Bile duct

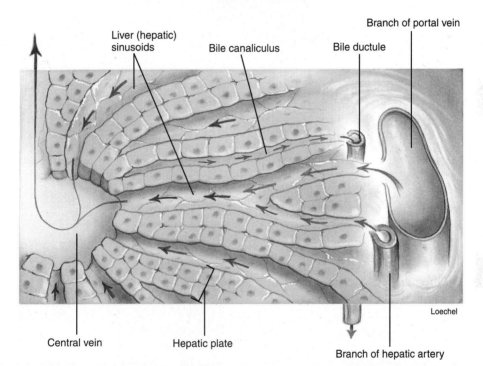

Liver (hepatic) sinusoids

Bile canaliculus

Bile ductule

Branch of portal vein

Central vein

Hepatic plate

Branch of hepatic artery

Loechel

(b)

Figure 10.9 **The structure of a liver lobule.** (a) A scanning electron micrograph and (b) a diagram showing the flow of blood and bile in the lobule.

(Photo from R. G. Kessel and R. H. Kardon, *Tissue and Organs: A Text-Atlas of Scanning Electron Microscopy.* © 1979 W. H. Freeman and Company.)

branches of the *hepatic artery*. Arterial blood mixes with blood from the hepatic portal vein, and together they pass through the sinusoids to the central vein (fig. 10.9).

Bile is produced and secreted by the liver cells (*hepatocytes*), but does not mix with blood because bile never enters the sinusoids. Instead, the hepatocytes secrete bile into *bile canaliculi* that are located between adjacent hepatocytes (fig. 10.9). Bile is drained from the canaliculi into *bile ducts*, located near the entry of the portal vein and hepatic artery into the sinusoid. The grouping of the por-

tal vein, hepatic artery, and bile duct that one sees in a microscopic view of the liver is called a *portal*, or *hepatic*, *triad* (fig. 10.8).

PROCEDURE

1. Examine a slide of the liver under low magnification and locate a central vein and a portal triad. The central vein appears as a large "hole" into which a number of sinusoids empty. The portal vein may also

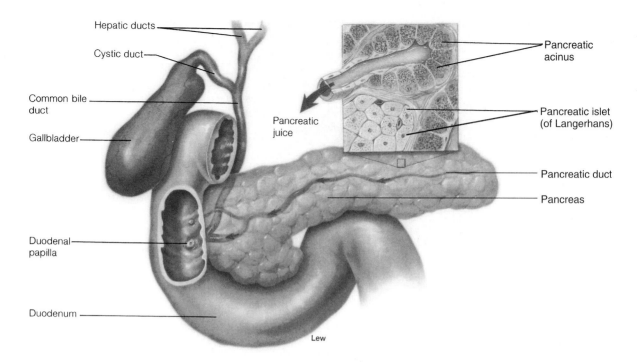

Figure 10.10 **Exocrine and endocrine structures of the pancreas.** The pancreatic acini secrete pancreatic juice into the pancreatic duct, which carries this exocrine secretion into the duodenum. The pancreatic islets (of Langerhans) secrete hormones (insulin and glucagon) into the blood.

appear as a large hole, but it can easily be distinguished from the central vein because the hepatic artery and bile duct are in close proximity to the portal vein.

2. Observe a portal triad under high magnification and distinguish between the portal vein, hepatic artery, and bile duct (fig. 10.9). The portal vein is the largest of the three. The hepatic artery can be identified by its layer of smooth muscle and lining of simple squamous endothelium; by contrast, the bile duct lacks smooth muscle and is lined with simple columnar epithelium.

D. PANCREAS

The pancreas is both an exocrine and an endocrine gland. Most of the pancreas is composed of exocrine tissue, with secretory cells arranged in clusters, or **acini** (fig. 10.10). Each pancreatic acinus secretes **pancreatic juice** into an opening of the *pancreatic duct,* which carries these secretions to the duodenum. Pancreatic juice contains water, bicarbonate, and a variety of digestive enzymes, including *trypsin, lipase,* and *amylase* (for the digestion of protein, fat, and carbohydrates, respectively).

Scattered among the exocrine acini are islands of endocrine cells. These are the pancreatic **islets of Langerhans** (fig. 10.11). The islets contain **alpha cells,** which secrete the hormone *glucagon,* and **beta cells,** which secrete the hormone *insulin.* These hormones are secreted into surrounding blood capillaries rather than into the pancreatic duct.

Figure 10.11 **A photomicrograph of the pancreas.** The exocrine acini compose most of the pancreas; the endocrine islets of Langerhans can be seen scattered within the exocrine tissue.

Inflammation of the liver, or **hepatitis,** may be caused by bacterial or viral infections, alcohol abuse, allergy, or drugs. This condition is usually reversible. In **cirrhosis,** however, large areas of liver tissue are destroyed and replaced with permanent connective tissue and "regenerative nodules" of hepatocytes that lack the plate-like structure of normal liver tissue. Since, among its many functions, the liver produces plasma *albumin* and converts ammonia to *urea,* liver disease may be accompanied by a decrease in the plasma albumin concentration and in the appearance of ammonia in the blood.

Inflammation of the pancreas, or **pancreatitis,** can result from the action of digestive enzymes on pancreatic tissue. The digestive enzymes produced within the pancreas are normally inactive until they enter the duodenum, but activated enzymes may reflux from the duodenum into the pancreatic duct. This produces an inflammation reaction accompanied by a "leakage" of enzymes into the blood. This may be detected clinically by a rise in the concentration of pancreatic amylase in the plasma.

PROCEDURE

1. Examine the pancreas under low magnification. The numerous small clusters of cells are the pancreatic acini. Occasional larger groupings of cells less intensely stained are pancreatic islets of Langerhans (fig. 10.11).
2. Scan the slide for a pancreatic duct in cross section. When one has been found, change to high magnification and observe its simple columnar epithelium.

Laboratory Report 10.1

REVIEW ACTIVITIES FOR EXERCISE 10.1

Test Your Knowledge of Terms and Facts

1. Identify the cells of the gastric mucosa that secrete:
   (a) HCl _____
   (b) pepsinogen _____
   (c) histamine _____
2. The three components of the mucosa layer of the digestive tract are the _____ ,
   the _____ , and the _____ .
3. Microscopic fingerlike projections of mucosa in the small intestine are called _____ .
4. The foldings of the plasma membrane of intestinal epithelial cells that produce the "brush border" are
   called _____ .
5. Blood is transported from the intestine to the liver in a large vessel known as the _____ .
6. Once blood has reached the liver, it travels through large capillaries called _____ .
7. The microscopic exocrine units of the pancreas are called _____ ;
   the endocrine structures are known as the _____ .
8. The glands in the duodenum that secrete an alkaline mucus: _____ .

Test Your Understanding of Concepts

9. Describe the structural adaptations of the small intestine that help increase the surface area and the rate at which
   digestion products can be absorbed.

10. Describe the composition of pancreatic juice and explain its functions.

11. Describe the microscopic anatomy of the liver, explain how this structure (a) allows the liver to modify the chemical composition of the blood and (b) keeps blood separate from bile.

Test Your Ability to Analyze and Apply Your Knowledge

12. Describe the location of the digestive enzymes produced by the small intestine, and explain how these cooperate with pancreatic juice enzymes.

13. Suppose you were to grind up and homogenate a pancreas. Do you think it would be possible to isolate insulin from this homogenate? (*Hint:* Remember that the pancreas is also an exocrine gland, producing pancreatic juice). Explain your answer.

Digestion of Carbohydrate, Protein, and Fat

MATERIALS

1. Water bath (set at 37°C), Bunsen burners, test tubes, test-tube clamp, graduated cylinders, droppers
2. Starch solution: dissolve 1.0 g per 100 mL water, over heat
3. Iodine solution (Lugol's reagent): dissolve 1.0 g iodine and 2.0 g potassium iodide in 300 mL of water
4. Benedict's reagent: dissolve 50.0 g sodium carbonate, 85.0 g sodium citrate, and 8.5 g copper sulfate in 5.0 L of water
5. As an alternative to saliva, α-amylase (type VI-B from porcine or human pancreas) may be used. Dilute to a concentration of about 300 units per mL, equivalent to saliva. Amylase may be purchased as a solid with 500,000 units of activity (order A-3176) from Sigma Chemical Company.
6. Freezer or ice bath; white of hard-boiled eggs
7. Pepsin (5 g per 100 mL), 2 N HCl, 10 N NaOH
8. pH meter or short-range pH paper
9. Pancreatin solution (1 g per 100 mL); cream or vegetable oil

Textbook Correlations*

Before performing this exercise, you may want to consult the following references in *Human Physiology,* seventh edition, by Stuart I. Fox:

• *Digestion and Absorption of Carbohydrates, Lipids, and Proteins,* Chapter 18, pp. 591–595.

Those using different physiology textbooks may want to consult the corresponding information in those books.

*See Appendix 3 for correlations to the *Virtual Physiology Laboratory* CD-ROM by McGraw-Hill and Cypris Publishing, Inc.

Saliva contains salivary amylase, an enzyme that digests starch into sugars. Acidic stomach secretions contain pepsin, an enzyme that hydrolyzes specific peptide bonds of large proteins into smaller peptides. Pancreatic juice contains lipase, an enzyme that digests fats emulsified by bile salts into fatty acids and glycerol. The digestive activity of these enzymes can be assessed by measuring the disappearance of substrates or the appearance of the products of the enzyme.

OBJECTIVES

1. Explain the digestive action of salivary amylase, pepsin, and pancreatic lipase. Describe the procedures demonstrating the presence of these enzymes.
2. Explain how the activity of salivary amylase and pepsin is influenced by changes in pH and temperature.
3. Describe the ability of pepsin to digest large proteins.
4. Discuss why the stomach does not normally digest itself; and explain how peptic ulcers may be formed.
5. Define the term *emulsification,* and explain the individual roles of bile and pancreatic lipase in the digestion of fat.
6. Following fat digestion, describe how fatty acids and glycerol are absorbed and transported in the body.

A. DIGESTION OF CARBOHYDRATE (STARCH) BY SALIVARY AMYLASE

The digestion of a carbohydrate such as starch begins in the mouth, where it is mixed with saliva containing the enzyme **salivary amylase,** or *ptyalin*. Starch, a long chain of repeating glucose subunits, is hydrolyzed by amylase first into shorter polysaccharide chains and eventually into the disaccharide maltose, which consists of two glucose subunits (table 10.2). Maltose, glucose, and other monosaccharides are known as *reducing sugars.*

In this exercise, the effects of pH and temperature on the activity of amylase will be tested by checking for the disappearance of substrate (starch) and the appearance of product (maltose) at the end of an incubation period. The appearance of a reducing sugar (maltose) in the incubation medium will be determined by the *Benedict's test*, where an alkaline solution of cupric ions (Cu^{2+}) is reduced to cuprous ions (Cu^+), forming a yellow-colored precipitate of cuprous oxide (Cu_2O).

Although starch digestion begins in the mouth with the action of salivary amylase, this is usually of minor importance in digestion (unless one chews excessively). Most of the digestion of polysaccharides and complex sugars to monosaccharides occurs in the small intestine when exposed to pancreatic and fixed brush-border enzymes. The hydrolytic action of amylase in combination with the buffering action of saliva may help prevent the accumulation of fermentable carbohydrates between the gums (gingiva) and the teeth, thus serving to protect against the growth of harmful bacteria that result in dental cavities (**caries**).

PROCEDURE (SEE FIG. 10.12)

Step 1:
(1) Label four clean test tubes 1–4.
(2) Obtain 10 mL of saliva in a small, graduated cylinder. (Salivation can be aided by chewing a piece of paraffin.) If only 5 mL of saliva is obtained, dilute the saliva with an equal volume of distilled water.

Note: As an alternative to saliva, the instructor may provide a solution of amylase derived from the pig or human pancreas.

Step 2:
(1) Add 3.0 mL of distilled water to tube 1.
(2) Add 3.0 mL of saliva to tubes 2 and 3.
(3) Add 3 drops of concentrated HCl to tube 3.
(4) Boil the remaining saliva in a separate Pyrex test tube by passing the tube through the flame of a Bunsen burner. Use a test-tube clamp and keep the tube at an angle, pointed away from yourself and others. When cool, add 3.0 mL of this boiled saliva to tube 4.

Step 3: Add 5.0 mL of cooked starch (provided by the instructor) to each of the four tubes.
Step 4: Incubate all tubes for 1–1 1/2 hours in a 37°C water bath.
Step 5: Divide the contents of each sample by pouring half into four new test tubes.

Step 6: Test one set of four solutions for starch by adding a few drops of iodine solution (*Lugol's reagent*). A positive test is indicated by the development of a purplish black color.
Step 7: Test the other set of four solutions for reducing sugars in the following way:
(1) Add 5.0 mL of Benedict's reagent to each of the four test tubes and immerse them in a rapidly boiling water bath for 2 minutes.
(2) Remove the tubes from the boiling water with a test-tube clamp and rate the amount of reducing sugar present according to the following scale:

Blue (no maltose)	–
Green	+
Yellow	++
Orange	+++
Red (most maltose)	++++

Enter your results in the data table in your laboratory report.

B. DIGESTION OF PROTEIN (EGG ALBUMIN) BY PEPSIN

Although amylase is most active at the pH of saliva (pH 6–7), the enzyme **pepsin** has a pH optimum that is adapted to the normal pH of the stomach (pH less than 2—see table 10.2). The low pH of the stomach is due to the secretion of hydrochloric acid (HCl) by parietal cells in the gastric glands. The strong acidity of the stomach coagulates proteins, thus facilitating their digestion by pepsin and, later by other proteolytic enzymes in the small intestine.

Gastric ulcers are apparently not due to an increase in stomach acidity, but rather to a breakdown in the normal mucosal barriers to digestion. The barriers are believed to be (1) the tight junctions between adjacent epithelial cells that prevent hydrogen ions from entering the mucosa and (2) the rapid renewal of surface epithelial cells. (The stomach sheds half a million cells a minute, completely renewing the gastric mucosa every three days.) By itself, the thick layer of mucus that covers the gastric epithelium is not an effective barrier to self-digestion. The weakening of the mucosal barrier to hydrogen ions is promoted by alcohol and salicylates, such as aspirin. The bacterium *Helicobacter pylori* resides in the gastrointestinal tract of many people, and this bacterium may be a causative agent in peptic ulcers. Indeed, peptic ulcer can be treated and even cured in some people by antibiotic therapy.

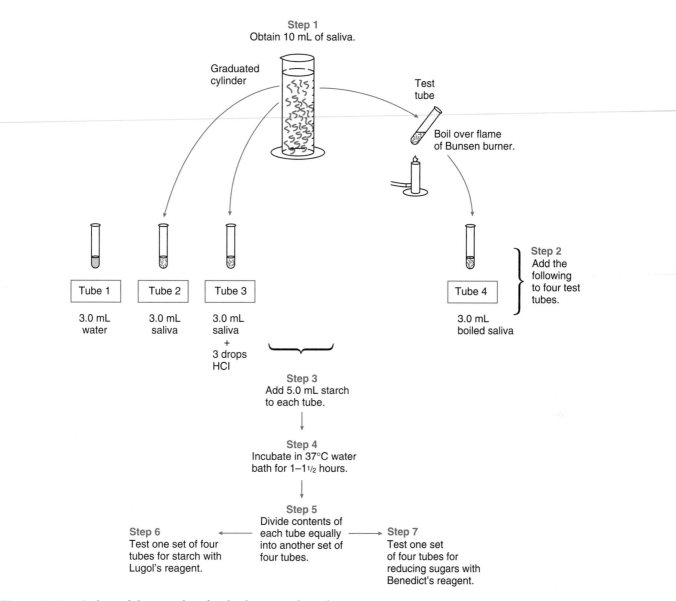

Figure 10.12 A chart of the procedure for the digestion of starch.

Pepsin, secreted by the *chief cells* of the gastric glands, is responsible for the digestion of less than 15% of ingested protein. Removal of the entire stomach (complete gastrectomy) thus has little effect on protein digestion. The major site of protein digestion is the small intestine, where the enzymes *trypsin* and *chymotrypsin* (secreted by the pancreas) and the *dipeptidases* (fixed in the intestinal brush-border mucosa) hydrolyze proteins and smaller polypeptides into absorbable amino acids (table 10.2).

The stomach does not normally digest itself. A *peptic ulcer* may form when the mucosa of the stomach (gastric ulcer) or duodenum (duodenal ulcer) is digested by the strongly acidic gastric juice. Although the etiology of peptic ulcers is not entirely known, it is believed that ul-

cers are caused by acid (H^+) from the gastric lumen eroding the mucosal surface rather than by the digestive action of pepsin on structural proteins within the mucosa. When the stomach produces an excess of acid, protective mechanisms may not be sufficient to protect the intestinal mucosa. Excessive stomach acid may be produced in susceptible individuals as a result of vagus stimulation and aggravate a duodenal ulcer.

When the acidic products of the stomach (called *chyme*) enter the duodenum, the intestine is stimulated to release the hormone *secretin*, which inhibits the gastric secretion of acid and stimulates the release of alkaline pancreatic juice (table 10.1). The acidic chyme, therefore, is diluted and neutralized in the small intestine.

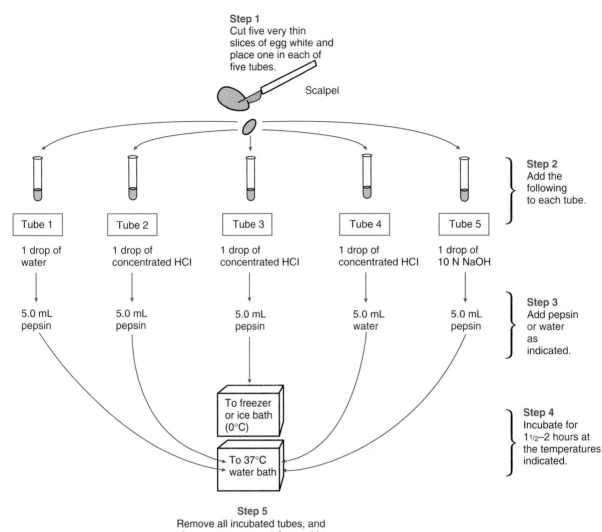

Step 1
Cut five very thin slices of egg white and place one in each of five tubes.

Scalpel

Step 2
Add the following to each tube.

| Tube 1 | Tube 2 | Tube 3 | Tube 4 | Tube 5 |

1 drop of water | 1 drop of concentrated HCl | 1 drop of concentrated HCl | 1 drop of concentrated HCl | 1 drop of 10 N NaOH

Step 3
Add pepsin or water as indicated.

5.0 mL pepsin | 5.0 mL pepsin | 5.0 mL pepsin | 5.0 mL water | 5.0 mL pepsin

To freezer or ice bath (0°C)

To 37°C water bath

Step 4
Incubate for 1½–2 hours at the temperatures indicated.

Step 5
Remove all incubated tubes, and observe appearance of egg white.

Figure 10.13   **A chart of the procedure for the digestion of protein.**

PROCEDURE (SEE FIG. 10.13)

Step 1: Label five clean test tubes 1–5. Using a sharp scalpel or razor blade, cut five slices of egg white about the size of a fingernail and as thin as possible. It is essential that the slices be very thin and uniform in size. Place a slice of egg white in each of the five test tubes.

Step 2:
(1) Add 1 drop of distilled water to tube 1.
(2) Add 1 drop of concentrated hydrochloric acid (HCl) to tubes 2, 3, and 4.
(3) Add 1 drop of concentrated (10N) NaOH to tube 5.

Step 3:
(1) Add 5.0 mL of pepsin solution to tubes 1, 2, 3, and 5.
(2) Add 5.0 mL of distilled water to tube 4.

Step 4:
(1) Place tubes 1, 2, 4, and 5 in a 37°C water bath. Place tube 3 in a freezer or ice bath.

(2) Incubate all tubes for 1–1 1/2 hours, remove the tubes (thaw the one that was frozen).

Step 5: Remove all incubated tubes and record the appearance of the egg white in the data table in your laboratory report.

C. DIGESTION OF FAT BY PANCREATIC JUICE AND BILE

Although the stomach produces a gastric lipase, the major digestion of fat occurs in the small intestine through the action of **pancreatic** and **intestinal lipase** (table 10.2). The digestion of fat in the small intestine is dependent upon the presence of *bile*, which is produced by the liver and transported to the duodenum via the bile duct (fig. 10.14). (The gallbladder serves only to store and concentrate the bile.)

Since fat is not soluble in water, dietary fat enters the duodenum in the form of large fat droplets containing the fat-soluble vitamins A, D, E, and K. The detergent action of bile salts lowers the surface tension of these large

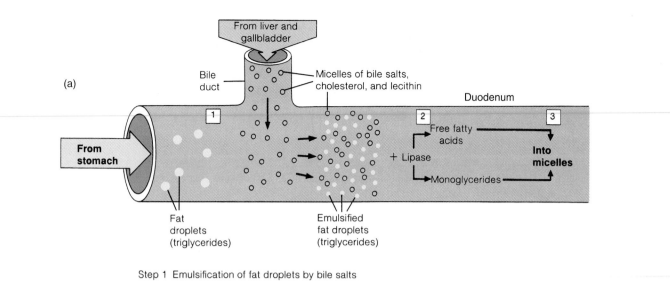

(a)

From liver and gallbladder

Bile duct

Micelles of bile salts, cholesterol, and lecithin

Duodenum

From stomach

1

2

3

Free fatty acids

+ Lipase

Into micelles

Monoglycerides

Fat droplets (triglycerides)

Emulsified fat droplets (triglycerides)

Step 1 Emulsification of fat droplets by bile salts

Step 2 Hydrolysis of triglycerides in emulsified fat droplets into fatty acid and monoglycerides

Step 3 Dissolving of fatty acids and monoglycerides into micelles to produce "mixed micelles"

(b)

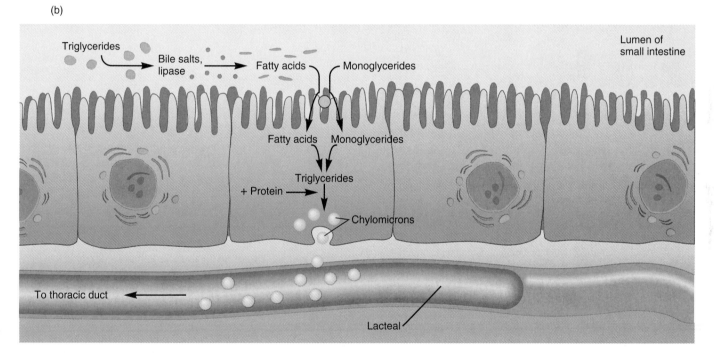

Lumen of small intestine

Triglycerides

Bile salts, lipase

Fatty acids

Monoglycerides

Fatty acids

Monoglycerides

Triglycerides

+ Protein

Chylomicrons

To thoracic duct

Lacteal

Figure 10.14 **Fat digestion and absorption.** (a) The steps in the digestion of fat. (b) The process of fat absorption into the intestinal epithelial cells and secretion into lymphatic capillaries (lacteals).

droplets, breaking them up into smaller droplets in a process called **emulsification.** Following this process, more surface area is presented to the lipase enzymes, promoting the digestion of fat into monoglycerides and fatty acids, and the release of the fat-soluble vitamins (fig. 10.14).

The absorption of fat is more complicated than that of the water-soluble monomers. The glycerol and fatty acids produced by lipase action aggregate to form spherical structures (*micelles*), which are absorbed by the intestinal epithelium. Once in the epithelial cells, the monomers are resynthesized to form tiny lipid droplets primarily composed of triglycerides (*chylomicrons*), which are then secreted into lymphatic vessels (lacteals) of intestinal villi (fig. 10.14). From there, chylomicrons are carried via lymph to veins. Unlike the other products of digestion, therefore, lipids enter the blood as polymers rather than monomers. It should be emphasized, however, that all foodstuffs, including fats, must be completely digested into their monomers before they can be absorbed by the digestive epithelium.

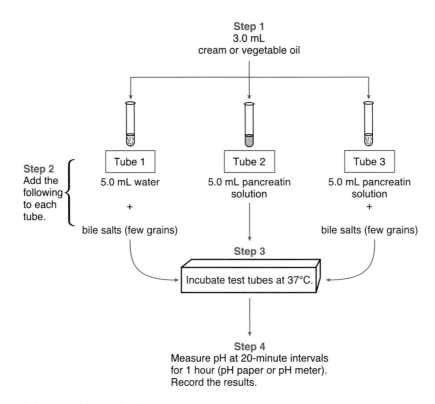

Step 1
3.0 mL
cream or vegetable oil

Tube 1	Tube 2	Tube 3

Step 2
Add the following to each tube.

5.0 mL water

+

bile salts (few grains)

5.0 mL pancreatin solution

5.0 mL pancreatin solution

+

bile salts (few grains)

Step 3
Incubate test tubes at 37°C.

Step 4
Measure pH at 20-minute intervals for 1 hour (pH paper or pH meter).
Record the results.

Figure 10.15   **A chart of the procedure for fat digestion.**

In this exercise, we will test the digestion of fat into glycerol and fatty acids by measuring the decrease in pH produced by the liberation of free fatty acids, as the digestion of triglycerides proceeds.

PROCEDURE (SEE FIG. 10.15)

Step 1: Add 3.0 mL of cream or vegetable oil to three test tubes, numbered 1–3.

Step 2: Add the following:
(1) To tube 1, add 5.0 mL of water and a few grains of bile salts
(2) To tube 2, add 5.0 mL of pancreatin solution
(3) To tube 3, add 5.0 mL of pancreatin solution and a few grains of bile salts

Step 3: Incubate the tubes at 37° C for 1 hour, checking the pH of the solutions at 20-minute intervals with a pH meter or with short-range pH paper.

Step 4: Record your data in the table in your laboratory report.

The formation of **gallstones** is believed to be due, in part, to an excessive concentration of cholesterol in the bile. Blockage of the bile duct with a gallstone can result in the inadequate flow of bile to the intestine, producing obstructive jaundice and steatorrhea. *Obstructive jaundice* is an elavetion in the blood levels of the bile pigment bilirubin due to blockage of the bile duct. High bilirubin levels produce a yellowish discoloration of the skin, the sclera of the eyes, and the mucous membrane. *Steatorrhea,* the appearance of fat in the feces due to the inadequate digestion and absorption of fat, is associated with a deficiency in the uptake of fat-soluble vitamins A, D, E, and K. Since vitamin K is necessary for normal blood clotting, this condition can be serious. Treatment for gallstones include surgery, dissolution by drugs, and fragmentation by ultrasound.

Laboratory Report 10.2

DATA FROM EXERCISE 10.2

A. Digestion of Carbohydrate (Starch) by Salivary Amylase

Enter your data in the table below using the rating method described in the procedure.

Contents Before Incubation	Starch After Incubation	Maltose After Incubation
Tube 1: Starch + distilled water		
Tube 2: Starch + saliva		
Tube 3: Starch + saliva + HCl		
Tube 4: Starch + boiled saliva		

B. Digestion of Protein (Egg Albumin) by Pepsin

Enter your observations in the data table below.

Incubation Condition	Appearance of Egg White After Incubation
Tube 1: Protein + pepsin at 37°C	
Tube 2: Protein + pepsin + HCl at 37°C	
Tube 3: Protein + pepsin + HCl at 0°C	
Tube 4: Protein + HCl at 37°C	
Tube 5: Protein + pepsin + NaOH at 37°C	

C. Digestion of Fat (Oil) by Pancreatic Juice and Bile

Record your data in the table below.

Time	pH		
	Tube 1: Fat + Bile Salts	Tube 2: Fat + Pancreatin	Tube 3: Fat + Bile Salts + Pancreatin
0 minutes			
20 minutes			
40 minutes			
60 minutes			

REVIEW ACTIVITIES FOR EXERCISE 10.2

Test Your Knowledge of Terms and Facts

1. Starch is partially digested into maltose by the action of _____ .
2. The enzyme in gastric juice that partially digests proteins is _____ ; this enzyme has a pH optimum of _____ .
3. Which food group—carbohydrates, lipids, or proteins—is not digested significantly until it reaches the small intestine? _____
4. Bile is produced by the _____ and stored in the _____ .
5. What is the function of bile salts? _____
6. The particles consisting of a combination of triglycerides and protein, which are secreted by intestinal epithelial cells into the central lacteals of the villi, are called _____ .

Test Your Understanding of Concepts

7. In exercise A, which tube(s) contained the most starch following incubation? Which tube(s) contained the most reducing sugars? What conclusion can you draw from these results?

8. In exercise B, which tube showed the most digestion of egg albumin? What can you conclude about the pH optimum of pepsin?

9. Compare the effects of HCl on protein digestion by pepsin and on starch digestion by salivary amylase. Explain the physiological significance of these effects.

10. In part C, which test tube displayed the most rapid fall in pH? Explain the reason for this, including an explanation of how the digestion of fat can cause a fall in the pH of the solution, and the function of bile salts.

11. How does the digestion and absorption of fat differ from the digestion and absorption of glucose and amino acids?

12. Why doesn't the stomach normally digest itself? Why doesn't gastric juice normally digest the duodenum?

Test Your Ability to Analyze the Apply Your Knowledge

13. Suppose, in performing part A, the test for starch and for reducing sugars both came out positive after a 1 hour incubation, but after the tubes incubated for 2 hours, the test for starch came out negative. How could you explain these results?

14. Reviewing your data from part A, what do you think happens to the digestion of a bite of bread after you swallow it?

15. Reviewing your data from part B, explain why frozen food keeps longer than food kept at room temperature.

16. A person with gallstones may have jaundice and an abnormally long clotting time. Explain the possible relationship between gallstones, jaundice, and blood clotting.

Nutrient Assessment, BMR, and Body Composition[1]

MATERIALS

1. Home scale or physicians height-weight scale
2. Tape measure, fat calipers (if available)
3. Calorie counting guide such as the U.S. Department of Agriculture Handbook, cookbooks, or popular diet books
4. Alternatively, caloric values of food; and the caloric expenditure of exercise, can be obtained from the Web. One good source is

www.caloriecontrol.org

Textbook Correlations

Before performing this exercise, you may want to consult the following references in *Human Physiology*, seventh edition, by Stuart I. Fox:

- *Nutritional Requirement.* Chapter 19, pp. 602–607.
- *Regulatory Functions of Adipose Tissue.* Chapter 19, pp. 609–611.

Those using different physiology textbooks may want to consult the corresponding information in those books.

Energy consumed in food and expended by the metabolic activities of the body is measured in kilocalories. Weight is gained when the energy consumption is greater than the energy expenditure, and weight is lost when the reverse is true.

OBJECTIVES

1. Describe the different nutrient classes; and list the calories per gram for carbohydrates, lipids, and fats.
2. Demonstrate how a dietary record is used to assess food and fluid consumption.
3. Define BMR and demonstrate two different methods for estimating BMR.
4. Define activity factor (AF) and demonstrate how to estimate the number of calories burned for various activities.
5. Calculate and balance the number of calories consumed in the diet and calories expended in activities over three days.
6. Describe how each pound of body weight gain or loss is related to calories consumed or expended.

A. BODY COMPOSITION ANALYSIS

Body composition refers to relative proportion of both the actively metabolizing tissues, or *lean body mass*, and fat tissue in the body. Although some adipose tissue is essential, an excess of fat is detrimental to health. An estimated 25% of the U.S. population is *obese* (more than 20% above the "ideal body weight"), with greater risk of health problems.

People with **apple-shaped bodies** (more fat around the abdominal area) seem to be more likely to develop cardiovascular disease, hypertension, and diabetes mellitus than those with **pear-shaped bodies** (more fat in the hips, buttocks, and thighs). Since males tend to become apple shaped, and females tend to become pear shaped, it appears that sex hormones help to direct the distribution of fat. For reasons still unknown, abdominal fat (stored deep in the body within the greater omentum) seems to pose a greater health risk than the subcutaneous fat stored under the skin in the hips and thighs. For example, the pear-to-apple shift in fat distribution seen in postmenopausal females is accompanied by an increase in the risk of diseases, such as cardiovascular disease (usually more common in males) and breast cancer.

---

1. Courtesy of Dr. Lawrence G. Thouin, Jr., Pierce College.

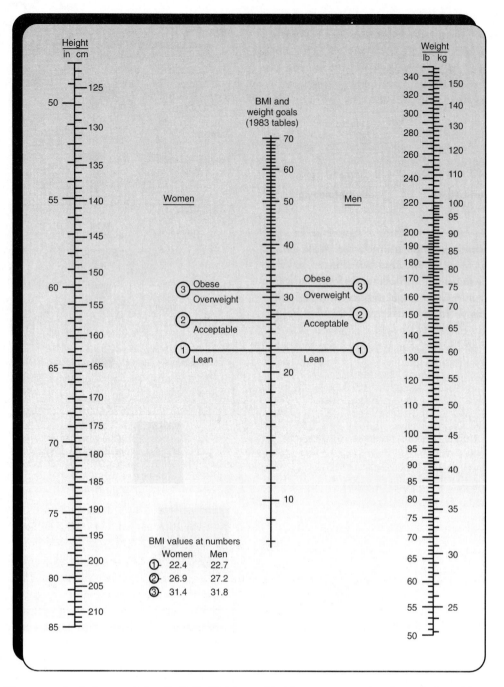

Figure 10.16 Nomogram for body mass index (kg/in²). Weights and heights are measured without clothing. The ratio weight/height² (metric units) is read from the central scale after a straight edge is placed between height and body weight.

Body composition can be estimated using a variety of techniques. One estimate of body fat is a height-to-weight ratio called the *body mass index* (*BMI*). BMI equals the body weight in kilograms divided by height in meters squared (the arithmetic can be avoided using the nomogram in fig. 10.16). In general, a person with a BMI of 27 or greater is obese, with an associated increased risk of health problems. *Skinfold calipers* measure the thickness of the subcutaneous fat layer at particular body sites. These measurements are then compared to norms for total body fat. Underwater weighing, one of the most accurate methods of determining body fat, is also the most expensive and inconvenient. *Bioelectric impedance* instruments send a small current through skin electrodes to measure the resistance in the electrolyte-rich body fluids. This method estimates body water, and then uses a computer data bank to derive body fat indirectly (lean body mass has more water than fat tissue). *Infrared wands* derive the percent body fat from density data taken over the biceps brachii muscle.

PROCEDURE (BODY MASS INDEX)

1. The body mass index, or BMI, is obtained by the following formula:

$$BMI = \frac{w}{h^2}$$

where,

w = weight in kilograms (pounds divided by 2.2)

h = height in meters (inches divided by 39.4)

2. Alternatively, the BMI may by obtained by use of the nomogram in figure 10.16. Simply draw a straight line between your height in the left column and your weight in the right column. Your BMI will be shown in the middle column.

3. Alternatively, you can simply enter your height and weight in the calculator provided at the following website:
   www.caloriecontrol.org/bmi.html

4. Record your BMI below:

   _____ .

> A BMI of over 30 places a person at high risk for the diseases of obesity. A BMI of under 27 is considered healthy, and a BMI in the range of 23–25 appears to be optimum for health.

PROCEDURE (WAIST-TO-HIP RATIO):

1. Stand and measure your waist at the navel. Record this value (in cm): _____ cm.

2. Measure your hips at the greatest circumference (buttocks). Record this value (in cm): _____ cm.

3. Divide the waist circumference by the hip circumference to get the waist-to-hip ratio. Record this waist-to-hip ratio: _____

> According to the American Heart Association, a waist-to-hip ratio above 1.0 for men and above 0.8 for women is associated with an increased risk of cardiovascular disease.

B. ENERGY INTAKE AND THE THREE-DAY DIETARY RECORD

Both the chemical energy consumed in foods and the metabolic energy expended by the cell are measured in **kilocalories (kcal),** or **Calories (C).** The major sources of food calories are carbohydrates, fats (lipids), and proteins. When allowances are made for inefficiency in the assimilation of each nutrient, one gram (*g*) of each of the three energy nutrients provides the body with approximately the following number of calories:

1 g carbohydrate = 4.0 kcal
1 g fat = 9.0 kcal
1 g protein = 4.0 kcal

The primary **carbohydrates** in food are the *sugars* (such as glucose, fructose, and sucrose) and *complex carbohydrates* (such as starches and dietary fiber). To meet the energy requirements of children and adults, recommended dietary allowances suggest that more than half (about 55%) of the calories consumed per day come from carbohydrate sources in the diet. Emphasis should be on the increased consumption of complex carbohydrates, especially dietary fibers found in fruits, vegetables, legumes, and whole-grain cereals. In addition to providing a source of calories, dietary fiber has been associated with improving overall health by promoting normal stool elimination, enhancing satiety, and lowering plasma cholesterol levels.

Lipids are generally divided into *triglycerides, phospholipids,* and *sterols* (steroids). The digestion, emulsification, and absorption of lipids also facilitates the absorption of *fat-soluble vitamins A, D, E,* and *K* and the *essential fatty acids.* Two primary unsaturated fatty acids are considered essential and must be present in the diet to maintain health: *linoleic acid* and α-linolenic acid.

Triglycerides are the major lipid components of foods and the most concentrated source of energy (9 kcal/g). The average American currently derives about 36% of the daily total dietary calories from fats. High dietary fat and cholesterol intakes have been associated with an increased risk of cardiovascular disease and cancer. As recommended by the Food and Nutrition Board's Committee on Diet and Health, the fat content of the U.S. diet should be lowered so as not to exceed 30% of the caloric intake (10% of fat calories from *saturated* fatty acids, 10% from *polyunsaturated* fatty acids, and 10% from *monounsaturated* fatty acids); and dietary cholesterol should be less than 300 mg/day (National Research Council, 1989).

According to U.S. Department of Agriculture (USDA) surveys, about 15% of the total food energy intake of the average American is derived from **protein.** Most of this protein, about 65%, is derived from animal sources, primarily meat and dairy products, with only about 20% from cereal grains. Despite increased protein requirements for certain populations, such as growing children, pregnant or lactating females, and the elderly, the typical American diet normally meets or exceeds the requirements. Interestingly, there is little evidence that physical exercise increases the need for protein, other than that required during the initial conditioning period. Therefore, individuals eating a typical American diet need make no adjustment to the recommended allowance for protein.

Nutrients that do not contribute energy to the body but that still are required to maintain body functions are vitamins, minerals, and water. The **fat-soluble vitamins** (A, D, E, and K) are absorbed from the small intestine with other food lipids, and are concentrated to some degree in adipose tissue. The **water-soluble vitamins** are vitamin C, thiamine (B1), riboflavin (B1), niacin (B1), pyridoxine (B6), folate, cyanocobalamin (B12), biotin, and pantothenic acid. Most water-soluble vitamins serve as coenzymes that assist enzymes in the regulation of metabolism. The **major minerals,** required in higher quantities, are calcium, phosphorus, magnesium, and the electrolytes (sodium, chloride, and potassium). The **minor minerals,** or **trace elements** are required in lesser quantities. They include iron, zinc, iodine, selenium, copper, manganese, fluoride, chromium, and molybdenum. Recommended quantities of these nutrients are normally met when a variety of foods are consumed. For the average person eating a typical American diet, therefore, no supplementation is recommended. Indeed, high intakes of certain nutrients in supplement form can be toxic.

Water is also an essential dietary nutrient. Although assessments of adequate water intake involve many complex factors, the general recommendation is a minimum of 1.0 mL of water per kilocalorie of energy expended per day. Therefore, an individual expending 2,000 kilocal per day should consume at least 2 liters of water.

PROCEDURE

1. Complete the 3-day dietary record (provided in the laboratory report). The 3 days must be consecutive (attempt to include at least one weekend day). At roughly the same time each day, try to weigh yourself under consistent conditions (comparable clothing, same scale etc.). Record your weight in pounds in the space provided for that day.
2. Record all foods and fluids consumed each day in the appropriate columns, noting the approximate time of day. Estimate food quantities by weight (such as ounces) and fluids by volume (e.g., cups, or liters), depending upon the units listed in your calorie guide. Record the total volume of fluids consumed at the bottom of the column.
3. Look up the estimated number of calories for each food or fluid noted in your diet record. The caloric values can be looked up in reference books or on the net, for example at *www.caloriescount.com/calculator.html.* Record the total calorie intake at the bottom of the column.
4. Comment on where the food or fluid was consumed. Do you always eat sitting down at a table, in a quiet,

relaxing environment? Or are you sometimes in the car, between classes, in front of the TV, or in bed studying?
5. Comment on why you ate or drank. Do you always eat because you are hungry? On occasion, do you eat because you are bored, or because the food is there, or because someone else is paying for it?

C. ENERGY OUTPUT: ESTIMATES OF THE BMR AND ACTIVITY

The total energy expended each day includes the energy required at rest and that expended during physical activity. For most people, the calories consumed at rest make up most of the total daily energy expenditure. This energy is used to pump blood, inflate the lungs, transport ions, and carry on the other functions of life. Measurement of this resting energy expenditure shortly after awakening and at least 12 hours after the last meal is known as the **basal metabolic rate (BMR).** With all other factors equal, BMR is influenced most by the amount of actively metabolizing tissues, or *lean body mass.* The BMR is higher in younger, more muscular people, and in males (who have a higher average muscle mass than females). The BMR is also influenced by **thyroxine.** People who are *hypothyroid* have a low BMR; those who are *hyperthyroid* have a high BMR.

While most of our caloric output is spent at rest, most people are physically active and expend calories beyond the BMR. The additional number of activity calories expended will vary with the individual, and with the duration, intensity, and types of activities performed. This increased caloric output can be estimated by multiplying an *activity factor* (AF) by the BMR. In general, the average sedentary person raises the total number of calories burned per day to about 130% (AF = 1.3) of the estimated BMR. Moderately active people may raise daily expenditures upwards of 150% (AF = 1.5) above the BMR estimates. Top athletes may double the BMR estimates, or more (AF ≥ 2.0). Aerobic activities, such as running, swimming, bicycling, and dancing, normally burn more calories than anaerobic activities, such as weight lifting.

Activity	Activity Factor (AF)
Lying in bed all day—equal to BMR	1.00–1.29
Mild activity—normal routine, no exercise	1.30–1.49
Moderate activity—1 hour of aerobic exercise	1.50–1.69
Heavy activity—2 to 4 hours of aerobic exercise	1.70–1.99
Rigorous athletic training	2.0 and above

PROCEDURE

Step 1. Estimate your basal metabolic rate (BMR) using two different methods:

Method 1: Your weight in kilograms (2.2 lb/kg)

_____ (kg):

Female BMR = 0.7 kcal/kg/hr

Male BMR = 1.0 kcal/kg/hr

Use the above conversion factors to calculate your kilocalories per hour. Then, multiply this figure by 24 (hours per day) and enter your answer in the space below.

In one day (24 hours), your BMR is approximately _____ kcal.

Method 2: Estimate your *Ideal Body Weight (IBW)* in pounds.

Female IBW = 100 lb for the first 5 feet in height + 5 lb per inch above 5 feet in height

Male IBW = 106 lb for the first 5 feet in height + 6 lb per inch above 5 feet in height

Your Ideal Body Weight is approximately _____ lb. Next, multiply your IBW by 10 for your daily estimated BMR.

In one day (24 hours), your BMR is approximately _____ kcal.

Note: *Since these BMRs are only estimates, a difference between the two values is to be expected. Select the one BMR estimate you feel is most accurate, write that number in the space provided on the dietary record in the laboratory report, and use it for calculations.*

Step 2. For each day in your 3-day dietary record, select one activity factor (AF) from the chart above that best reflects your total activity for that 24-hour period, with 1.30 typical for the casual college routine (without exercise); and write that decimal for each day in the AF box of the dietary record.

Note: *For variety (and more fun) attempt to vary your activities and the AF each day—another reason to include one weekend day in this report.*

Step 3. Calculate the total number of calories expended (output) each day as follows:

Total calories expended (kcal) =

Activity AF (from box) × BMR (from Step 1, Method 1 or 2)

Write this total in the line 3 of the dietary record.

Step 4. Subtract total calories expended from total calories consumed to determine the caloric balance lost or gained that day (in your record subtract line 3 from line 1). Write the number of excess calories in line 4 of the dietary record; and circle either calories "gained" or "lost."

Step 5. Assuming that 1 lb of body tissue (not just fat) that is gained or lost represents approximately 3,500 kcal, convert the excess calories from line 4 into body weight:

Body weight (lost/gained) = _____ (kcal) + 3,500 (kcal/lb) = _____ (lb)

Write the pounds lost/gained that day in line 5 of the dietary record.

Step 6. Complete the evaluation section in the laboratory report, which follows the 3-day dietary record.

3-Day Dietary Record

Day # _____ Day of Week _____ Your Weight (lb) _____

Time (A.M./P.M.)	Foods-Units (oz, g, . . .)	Fluid-Units (cup, mL, . . .)	Calories (kcals)	Where ?	Why ?	Activities
		Total fluid	Total kcals			

1. Total calories consumed (intake): _____ kcal
2. Estimated BMR at rest: _____ kcal (Step 1, Method 1 or 2)
3. Total calories expended (output): _____ kcal (Step 3)
4. Today's caloric balance: _____ kcal (Step 4)
   gained / lost (circle one)
5. Given that 1 lb of body weight (not just fat) is equal to approximately 3,500 kcal gained or lost, how many pounds of body weight were gained or lost today? _____ lb.
6. Enter values from today's report into the Dietary Record Evaluation form following these daily reports.

AF

3-Day Dietary Record

Day # _____   Day of Week _____   Your Weight (lb) _____

Time (A.M./P.M.)	Foods-Units (oz, g, . . .)	Fluid-Units (cup, mL, . . .)	Calories (kcals)	Where ?	Why ?	Activities
		Total fluid	Total kcals			

1. Total calories consumed (intake): _____ kcal
2. Estimated BMR at rest: _____ kcal (Step 1, Method 1 or 2)
3. Total calories expended (output): _____ kcal (Step 3)
4. Today's caloric balance: _____ kcal (Step 4)
   gained / lost (circle one)
5. Given that 1 lb of body weight (not just fat) is equal to approximately 3,500 kcal gained or lost, how many pounds of body weight were gained or lost today? _____ lb.
6. Enter values from today's report into the Dietary Record Evaluation form following these daily reports.

AF
☐

3-Day Dietary Record

Day # _____   Day of Week _____   Your Weight (lb) _____

Time (A.M./P.M.)	Foods-Units (oz, g, . . .)	Fluid-Units (cup, mL, . . .)	Calories (kcals)	Where ?	Why ?	Activities
		Total fluid	Total kcals			

1. Total calories consumed (intake): _____ kcal
2. Estimated BMR at rest: _____ kcal (Step 1, Method 1 or 2)
3. Total calories expended (output): _____ kcal (Step 3)   **AF**
4. Today's caloric balance: _____ kcal (Step 4)
   gained / lost (circle one)
5. Given that 1 lb of body weight (not just fat) is equal to approximately 3,500 kcal gained or lost, how many pounds of body weight were gained or lost today? _____ lb.
6. Enter values from today's report into the Dietary Record Evaluation form following these daily reports.

Laboratory Report 10.3

DATA FROM EXERCISE 10.5

3-Day Dietary Record Evaluation

1. Record the total number of calories consumed over 3 days:
    Day 1 _____ Day 2 _____ Day 3 _____ = _____ kcal
2. Record the total number of calories expended over 3 days:
    Day 1 _____ Day 2 _____ Day 3 _____ = _____ kcal
3. Subtract the total kilocalories expended from the total kilocalories consumed (line 2 from line 1), and indicate whether the difference represents a gain (+) or a loss (−).
    _____ total kcal gained/lost (circle one) over 3 days
4. Divide your total from step 3 by 3,500 kcal/lb to convert kilocalories to pounds. Indicate whether this represents a weight gain or a weight loss.
    _____ lb gained/lost (circle one) over 3 days
5. Determine your measured weight change by subtracting your weight on day 3 from your weight on day 1. Compare your calculated weight gain/loss with your actual weight gain/loss, and discuss many of the variables that cause the numbers to differ. (*Hint:* Explain individual differences as well as variables built into the exercise.)

6. Study your 3-day dietary record. Make a detailed list of at least five specific suggestions for improving this diet. (*Hint:* Discuss nutrient classes, except water; make food exchanges for improvements, and include where and why entries. Exercise is not dietary, and water is discussed next.)

    a.

    b.

    c.

    d.

    e.

7. Record the total volume of fluids consumed over 3 days:

Day 1 _____ Day 2 _____ Day 3 _____ = _____ Total fluid (cups, oz etc.)
Based on the recommendation of at least 1.0 mL of water per kilocalorie of energy expended each day, did you meet this requirement each day? If not, propose a plan that would allow you to meet this requirement. Were diuretics such as caffeine or alcohol included? If so, what adjustments should be made to your daily fluid requirements?

REVIEW ACTIVITIES FOR EXERCISE 10.3

Test Your Knowledge of Terms and Facts

1. Fat provides _____ kilocalories per gram, whereas carbohydrates and protein provide _____ kilocalories per gram.
2. For optimum health, most of the calories consumed in a diet should come from _____ (carbohydrates/fat/protein).
3. Vitamins A, D, E, and K are grouped together as the _____ vitamins.
4. The lowest rate of energy expenditure of the body is called the _____.
5. A measurement of body fat that involves a height-to-weight ratio is called the _____.

Test Your Understanding of Concepts

6. Define the *basal metabolic rate (BMR)* and explain how physical exercise can influence the body weight.

7. Is it healthier to eat a diet high in complex carbohydrates, or high in fat? Explain why this is so.

8. A college student who gained unwanted weight wishes to lose 20 pounds before the graduation ceremonies, which are 10 months away. By how many *kilocalories per week* would the student have to reduce intake or increase expenditures in order to lose the 20 pounds?

9. How many *kilocalories per day* would the student have to lose over the 10-month period? Is this a practical weight loss program? Explain fully.

10. If this student wants to lose all of this weight by only removing fat calories from foods consumed, how many *grams of fat per day* must be removed from the diet? Is this a practical weight loss program? Explain fully.

Unlike the physiology of other body systems, the function of the reproductive system is not to maintain homeostasis, but rather to ensure the continuity of the species. The study of the reproductive system includes the anatomy and physiology of many other body systems. Neural and endocrine regulatory mechanisms are very active and an integral part of reproductive physiology. The cardiovascular, respiratory, muscular, digestive, and other systems are likewise involved in the physiology of reproduction.

The primary sex organs are the **gonads:** *testes* in the male and *ovaries* in the female. The gonads produce **gametes** (*sperm* and *ova*) and **sex steroid hormones.** The functions of the gonads in both sexes are regulated by gonadotropic hormones (FSH and LH) secreted by the anterior pituitary gland. Secretion of these gonadotropic hormones, in turn, is regulated in part by hormones produced in the hypothalamus and by feedback effects from the sex steroids. The regulation of ovarian function by the FSH and LH follows a cyclical pattern in females, producing the menstrual cycle, whereas the stimulation of the testes is noncyclical.

Associated with the functions of the gonads are the **accessory sex organs.** These include the uterus, uterine (fallopian) tubes, vagina, labia, and clitoris in a female (fig. 11.1), and the epididymis, vas (ductus) deferens, seminal vesicles, prostate, ejaculatory ducts, and penis in a male (fig. 11.2).

The uterine tubes in a female serve to convey the ovum toward the uterus. **Fertilization** normally occurs in the uterine tube, whereas **implantation** of the embryo occurs a week later in the inner lining, or *endometrium,* of the uterus. The endometrium undergoes monthly cycles of shedding (in *menstruation*) and growth as a result of the cyclical changes in the secretion of sex steroid from the ovary.

Sperm produced in the seminiferous tubules of the testes within the scrotum are conveyed to a tubular structure, the *epididymis,* where they undergo maturation. The vas deferens transports the sperm out of the epididymis and into the body cavity, where fluid from the seminal vesicles and prostate is added to form *semen,* or *seminal fluid.* The ejaculatory duct fuses with the urethra so that semen can be ejaculated out of the urethra at the tip of the penis. Although millions of sperm are ejaculated into the female vagina, only about 100 survive to reach a uterine tube and only one sperm cell, or *spermatozoon,* is allowed to fertilize the ovum ovulated by a single follicle in the ovary. This fertilized ovum, or *zygote,* given the proper environment within the uterus, can develop into a new and genetically unique human being.

Exercise 11.1   Ovarian Cycle as Studied Using a Vaginal Smear of the Rat
Exercise 11.2   Human Chorionic Gonadotropin and the Pregnancy Test
Exercise 11.3   Patterns of Heredity

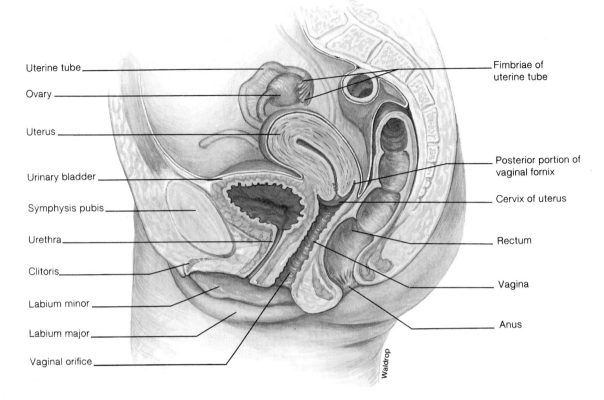

Figure 11.1 Organs of the female reproductive system.

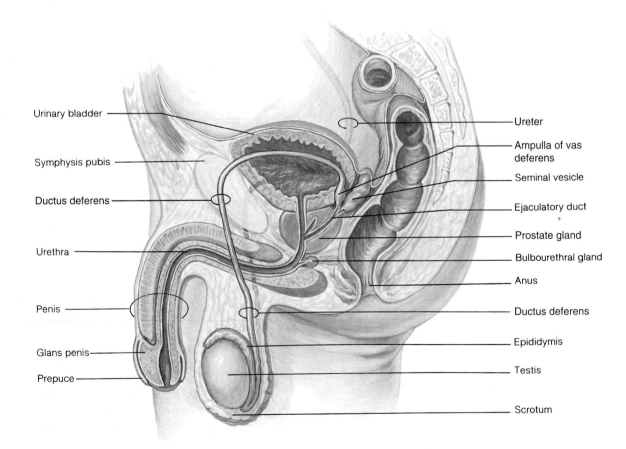

Figure 11.2 Organs of the male reproductive system.

Ovarian Cycle as Studied Using a Vaginal Smear of the Rat

11.1

MATERIALS

1. Young female rats
2. Ether jar (large, widemouth jar with close-fitting lid) and ether
3. Isotonic saline and cotton swabs
4. Giemsa's stain (dilute concentrate 1:50) and absolute methyl alcohol in staining jars
5. Microscopes and microscope slides

The cyclic changes in ovarian hormone secretion cause cyclic changes in the epithelium of the female reproductive tract. By observing exfoliated epithelial cells, the stage of the ovarian cycle and the level of ovarian hormone secretion can be determined.

OBJECTIVES

1. Identify the phases of the ovarian cycle.
2. Describe the changes that occur in the endometrium and correlate these changes with the stages of the ovarian cycle.
3. Describe the appearance of a vaginal smear at different stages of the cycle and explain the clinical usefulness of a vaginal smear.

Textbook Correlations

Before performing this exercise, you may want to consult the following references in *Human Physiology,* seventh edition, by Stuart I. Fox:

• *Menstrual Cycle.* Chapter 20, pp. 661–667.

Those using different physiology textbooks may want to consult the corresponding information in those books.

The amount of gonadotropic hormones (FSH and LH) secreted by the anterior pituitary of females increases and decreases in a cyclical fashion. The secretion of estrogen and progesterone by the ovary will follow the same cycle. In most mammals, sexual receptivity (heat, or estrus) occurs during a specific part of the cycle. This pattern is called an **estrous cycle.** In human and subhuman primates, sexual receptivity occurs throughout the cycle, with monthly bleeding occurring at the beginning of each cycle. These cycles are called **menstrual cycles** (*menses* means "monthly") (fig. 11.3).

The uterus is one of the target organs of the ovarian hormones. As the secretions of estrogen and progesterone increase during the cycle, the inner lining of the uterus (the endometrium) increases in thickness (fig. 11.3). The ovarian hormones are preparing the uterus for the possible implantation of the developing embryo should fertilization occur. If fertilization does not occur, the cyclical decrease in plasma levels of estrogen and progesterone causes the necrosis (cellular death) and sloughing off of the upper two-thirds of the endometrium. The cycle is ready to begin anew (fig. 11.3).

The different stages of an estrous cycle or a menstrual cycle may be followed by observing the cyclical changes in the cells of the endometrium. However, since these changes are correlated with the types of cells found in the vaginal lumen, the stage of the cycle can be determined more conveniently by taking a *vaginal smear.*

The estrous cycle of a rat is usually completed in 4 to 5 days. The cycle is roughly divisible into four stages (fig. 11.4).

1. **Proestrus.** Proestrus is the beginning of a new cycle. The follicles of the ovary start to mature under the influence of the gonadotropic hormones, and the ovary starts to increase its secretion of estrogen. *Vaginal smear.* Nucleated epithelial cells (fig. 11.4*a,b*).
2. **Estrus.** The uterus is enlarged and distended because of the accumulation of fluid. Estrogen secretion is at its height. The rat thus becomes sexually receptive as ovulation occurs. *Vaginal smear.* Squamous cornified cells (fig. 11.4*c*).

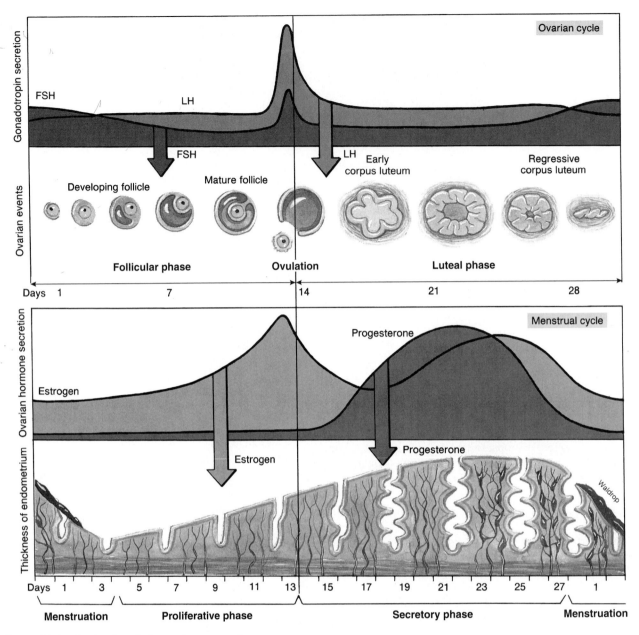

Figure 11.3   Cycle of ovulation and menstruation.

Figure 11.4 **The vaginal smear of a rat.** (a,b) Nucleated epithelial cells and leukocytes; (c) nucleated epithelial cells and cornified cells.

3. **Metestrus.** Metestrus is the stage that follows ovulation. The ovary contains functioning corpora lutea secreting progesterone, preparing the uterus for implantation.
Vaginal smear. Many leukocytes, some cornified epithelial cells.

4. **Diestrus.** The corpora lutea regress, and the declining secretion of estrogen and progesterone causes regression of the uterine endometrium.
Vaginal smear. Entirely leukocytes.

The stages of the estrous cycle correspond to the phases of the menstrual cycle as follows (fig. 11.3):

Proestrus corresponds to the *follicular phase*.
Estrus corresponds to the *ovulatory phase*.
Metestrus corresponds to the *luteal phase*.
Diestrus corresponds to the *menstrual phase*.

Like the endometrium of the uterus, the vaginal epithelium undergoes cycle changes during the menstrual cycle. Estrogen stimulates growth and development of the vaginal epithelium, and under conditions of high estrogen four layers can be distinguished: superficial squamous cells, an intermediate layer, a parabasal layer, and a basal layer. The topmost squamous cell layer *exfoliates* (sheds) and is seen in a vaginal smear. A vaginal smear during the *proliferative phase* of the uterine cycle (corresponding to the follicular phase of the ovaries, fig. 11.3), when the estrogen level is high, consists almost entirely of superficial squamous cells. Under the influence of progesterone, during the *secretory phase* of the cycle (corresponding to the luteal stage of the ovaries, fig. 11.3) and during pregnancy, the top layer of cells in the vaginal epithelium is the intermediate layer. These cells are seen in a vaginal smear taken during the second half of the cycle and during pregnancy. In the absence of all sex steroids, such as prior to the onset of puberty in females, or *menarche*, the vaginal epithelium is atrophic and parabasal cells are seen in a vaginal smear.

Vaginal smears can be used clinically to determine the stage of the menstrual cycle, the effectiveness of exogenous ("from outside") hormone treatments, and the stage and health of pregnancy. Vaginal smears are also helpful in the diagnosis of pathological states, such as *primary* or *secondary amenorrhea* (the absence of a menstrual period when one would normally be expected), inflammation, and cancer. Indeed, since malignant tissue exfoliates to a greater extent than normal tissue, this technique may discover cancers that are too small to detect by other means. The staining technique most often used clinically for vaginal smears was developed by an American physician, George Papanicolaou. Since the introduction of this technique in 1942, the **Pap smear** has become synonymous with the clinical vaginal smear.

PROCEDURE

1. Anesthetize the female rat with ether.
2. Moisten the cotton tip of a swab with isotonic saline and insert it into the vagina. Smear this on a clean microscope slide. (One swab is sufficient to make approximately six slides.)
3. Stain the slides as follows:
   (a) Dry the slide in air.
   (b) Immerse the slide in absolute methyl alcohol (5 seconds).
   (c) Air dry the slide.
   (d) Place it in Giemsa's stain (1:50) for 30 minutes.
   (e) Rinse it in tap water.
   (f) Dry the slide and observe it without a coverslip under the microscope.

Laboratory Report 11.1

REVIEW ACTIVITIES FOR EXERCISE 11.1

Test Your Knowledge of Terms and Facts

1. The stage of the estrous cycle characterized by the appearance of cornified epithelial cells is called the _____ phase.
2. In the stage described in question 1, which ovarian hormone is secreted in high amounts? _____
3. The phase of the human menstrual cycle which corresponds to the phase named in question 1 is the _____ phase.
4. The stage of the estrous cycle in which estrogen and progesterone secretion declines is called the _____ phase.
5. The phase of the human menstrual cycle that corresponds to the phase named in question 4 is the _____ phase.
6. The proliferative phase of the endometrium occurs during the _____ phase of the ovarian cycle.
7. The secretory phase of the endometrium occurs during the _____ phase of the ovarian cycle.

Test Your Understanding of Concepts

8. Describe the events that occur in the ovaries as a rat goes through the estrus cycle. Relate these changes to the cyclic changes of the vaginal epithelium.

9. Describe the changes that occur in the human endometrium during the menstrual cycle. Relate these changes to the events that occur in the human ovaries during the menstrual cycle.

10. What would a vaginal smear performed on an ovariectomized rat (one with its ovaries removed) look like? Explain.

11. The contraceptive pill is composed of estrogen and progesterone. Suppose your laboratory rat, with an estrous cycle of 4 days, were to take the pill on a regimen of "3 days on, 1 day off." What would the vaginal smears taken on each of the 4 days look like? Explain.

Human Chorionic Gonadotropin and the Pregnancy Test

MATERIALS

1. Urine collection cup
2. Pregnancy kit (DAP test kit—Wampole; Qupid test—Stanbio; ICON II—Hybritech; or similar kit)

Shortly after fertilization occurs, cells that are going to become part of the placenta secrete a hormone called human chorionic gonadotropin (hCG). Pregnancy is commonly tested by an assay for hCG in the plasma or urine.

OBJECTIVES

1. Describe the fate of the corpus luteum at the end of a nonfertile cycle.
2. Describe the source of hCG and the physiological role of this hormone in pregnancy.
3. Demonstrate a pregnancy test and explain how this procedure determines pregnancy.

Textbook Correlations

Before performing this exercise, you may want to consult the following references in *Human Physiology*, seventh edition, by Stuart I. Fox:

- *Implantation of the Blastocyst and Formation of the Placenta.* Chapter 20, pp. 672–675.

Those using different physiology textbooks may want to consult the corresponding information in those books.

About a week after fertilization occurs, the developing embryo implants into the wall of the endometrium (a process called *implantation*). Near the end of the cycle, the usual fall in the secretion of ovarian hormones with the resultant shedding of the upper two-thirds of the endometrium would result in an abortion. Therefore, the hormonal secretions of the ovary after implantation must be freed from regulatory control by pituitary hormones.

The tiny implanted embryo saves itself from being aborted by secreting a hormone that indirectly prevents menstruation. Cells of the embryo, which will later become part of the **placenta,** secrete **chorionic gonadotropin,** or **hCG** (the **h** stands for "human"), which is identical to the anterior pituitary luteinizing hormone (LH) in its effects. Like LH, therefore, hCG stimulates the corpus luteum of the ovary to continue secreting large amounts of estrogen and progesterone for the first 10 weeks of pregnancy. At this time, the growing placenta becomes the primary source of both estrogen and progesterone and takes over support of the pregnancy until birth.

Large amounts of hCG are present in the plasma and excreted in the urine of pregnant women, whereas no hCG is present in the urine of nonpregnant women. An assay for the presence of hCG in urine is thus an accurate test for pregnancy.

In this exercise, the presence of hCG will be assayed by means of an antigen-antibody bonding reaction. An *antigen* is a large molecule (usually a foreign protein) that is capable of stimulating lymphocytes to produce antibodies. *Antibodies* are proteins that are capable of bonding to specific antigens. The antigen-antibody reaction in the body is part of the *immune response,* and an assay that makes use of this reaction is termed an **immunoassay.**

In this immunoassay for hCG, antibodies are stuck onto tiny white latex particles to make them visible to the naked eye. If urine containing hCG (the antigen) is added, chemical bonds form between the antigens and antibodies, causing the latex particles to clump together (*agglutinate*). Since antibodies are relatively specific in their reaction with antigens, no agglutination reaction will occur if hCG is absent from the urine. (See fig. 11.5 for examples of positive and negative agglutination reactions.)

(a)

(b)

Figure 11.5 A pregnancy test. (a) Negative—no agglutination reaction occurs when urine is added to a control solution containing white latex particles with rabbit gamma globulin (from a rabbit not sensitized to hCG). (b) Positive— agglutination of latex particles occurs when urine from a pregnant woman is added to latex coated with antibodies from a rabbit sensitized to hCG.

The immunoassay of hCG in urine is a very accurate test for pregnancy and, in one form or another, is now the most widely used type of pregnancy test. If this test is performed too soon after conception, however, the hCG level in the urine may be below the sensitivity of the assay and may produce false negative results. Therefore, the test should not be performed within the first 2 weeks following a missed period. False positive tests are less common and may be due to tumors that secrete hCG (such as choriocarcinomas).

Extremely sensitive techniques utilize monoclonal antibodies against a specific antigenic subunit of hCG and radioimmunoassay of hCG levels in the blood. Such techniques make it possible to detect pregnancy in a clinical laboratory as early as 7 to 10 days after conception.

PROCEDURE

1. Allow the urine sample to reach room temperature.
2. Fill the plastic reservoir provided in the pregnancy kit with urine and insert the filtering attachment.
3. Expel the urine onto two circles on the disposable slide provided by gently squeezing the reservoir.
4. Shake the latex control (latex particles with gamma globulin antibody) and add 1 drop to the first circle. Mix the urine and reagent with an applicator stick by spreading the mixture over the entire circle.
5. Shake the bottle of reagent (latex particles with antibodies against hCG) and add 1 drop to the second circle. Mix as before.
6. Rock the slide gently for 1 minute; then look for agglutination. If *negative:* the solution will remain milky (fig. 11.5*a*). If *positive:* the solution will appear grainy (fig. 11.5*b*).

Laboratory Report 11.2

Name _____

Date _____

Section _____

REVIEW ACTIVITIES FOR EXERCISE 11.2

Test Your Knowledge of Terms and Facts

1. Menstruation is caused by a(n) _____ (increase/decrease) in the secretion of estrogen and progesterone.
2. The structure that secretes estrogen and progesterone for the first 10 weeks of pregnancy is the _____
   _____ .
3. The hormone tested for in a pregnancy test is _____ .
4. The hormone named in question 3 is produced by the _____ .
5. The hormone named in question 3 has an action similar to which pituitary hormone? _____ .

Test Your Understanding of Concepts

6. Describe the formation, function, and fate of the corpus luteum during an unfertile menstrual cycle. What happens to the corpus luteum if fertilization occurs?

7. Why is this pregnancy test called an immunoassay? Explain how this test works to detect pregnancy.

8. Why are most pregnancy tests not valid unless performed at least a couple of weeks after a missed period?

9. Suppose a man has a tumor that secretes hCG. Would he give a positive result on a pregnancy test? Would hCG have any physiological effect on him? Explain

Patterns of Heredity

MATERIALS

1. Phenylthiocarbamide (PTC) paper (VWR Scientific Products, Ward's)
2. Sickle hemoglobin test (Sickle-Sol tube test—Dade; SickleScreen Assay—Pacific Hemostasis); prepared slides of sickle cell anemia and normal blood.
3. Ishihara color-blindness charts

The ways in which many aspects of body structure and function are inherited can be understood by applying relatively simple concepts. The patterns of heredity are important in anatomy and physiology because of the numerous developmental and functional disorders that have a genetic basis. The knowledge of which disorders and diseases are inherited finds practical application in the genetic counseling of prospective parents.

OBJECTIVES

1. Define the terms dominant, recessive, homozygous, and heterozygous.
2. Distinguish between autosomal and sex-linked inheritance.
3. Explain the nature of sickle-cell anemia and describe how it is inherited.
4. Describe how hemophilia and color blindness are inherited.

Textbook Correlations

Before performing this exercise, you may want to consult the following references in *Human Physiology*, seventh edition, by Stuart I. Fox:

- *DNA Synthesis and Cell Division.* Chapter 3, pp. 69–77.
- *Cones and Color Vision.* Chapter 10, p. 271.
- *Inherited Defects in Hemoglobin Structure and Function.* Chapter 16, p. 511.

Those using different physiology textbooks may want to consult the corresponding information in those books.

A person inherits two sets of genes controlling every trait: one from the mother and one from the father (if these genes are *autosomal*—that is, not located on the sex chromosomes). If both genes are identical, the person is said to be **homozygous** for that trait. A person who is homozygous for normal adult hemoglobin A, for example, has the **genotype** AA; a person who is homozygous for the sickled hemoglobin S has the genotype SS.

If a person inherits the gene for hemoglobin A from one parent and the gene for hemoglobin S from the other parent, this person is said to be **heterozygous** for that trait and has the genotype AS. This person is a carrier and has the sickle-cell *trait* but does not have sickle-cell *disease*. The **phenotype** (in this case the absence of sickle cell disease) is the same for the heterozygote as it is for the person who is homozygous normal. Thus, the gene for hemoglobin A is **dominant** to the gene for hemoglobin S (or, stated another way, the gene for hemoglobin S is **recessive** to the gene for hemoglobin A).

Although the heterozygote does not display the phenotype of sickle cell disease, this person is a carrier of the sickle cell trait since one-half of the gametes will contain the gene for hemoglobin A and one-half will contain the gene for hemoglobin S. (In the process of gamete formation, known as *meiosis*, the chromosome number is halved). If this individual mates with one who is homozygous AA, the probability is half the progeny will be homozygous AA, and half will be heterozygous AS.

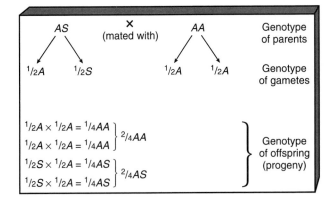

If two individuals who are both heterozygous AS mate, one-fourth of the progeny will have the genotype AA, one-fourth will have the genotype SS, and one-half will have the genotype AS. Although individuals with the homozygous genotype AA and the heterozygous genotype AS are healthy, there is a one-in-four (25%) probability that a child from this mating will have the phenotype of sickle cell disease (genotype SS).

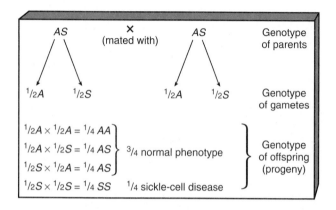

Most of the concepts of heredity discussed in this exercise were discovered in the 1860s by an Austrian monk named Gregor Mendel; consequently, these patterns of heredity are often called *simple Mendelian heredity*. A proper knowledge of these patterns is obviously needed for genetic counseling of carriers of genetic diseases. If both parents are carriers of such diseases as sickle cell anemia, Tay-Sachs disease, phenylketonuria (PKU), and others that are inherited as *autosomal recessive* traits, they should be aware that there is a 25% chance that their children will get the disease. If only one parent is a carrier, they should know that there is no chance of their children getting the disease. Further, couples should be informed that whether they have no children or a dozen, the probability that their next child will get the disease will always remain the same.

A. Sickle-Cell Anemia

Sickle-cell anemia is an autosomal recessive disease affecting 8–11% of the African-American population. In this disease, a single base change in the DNA, through the mechanisms of transcription and translation, results in the production of an abnormal hemoglobin (hemoglobin S). Hemoglobin S differs from the normal adult hemoglobin (hemoglobin A) by the substitution of one amino acid for another (valine for glutamic acid) in one position of the protein. A quick test for sickle-cell anemia is based on the fact that, under conditions of reduced oxygen tension, hemoglobin S is less soluble than hemoglobin A and tends to make a solution turbid, or cloudy (fig. 11.6a).

PROCEDURE

1. Fill a calibrated capillary tube with blood up to the line. Then, expel the blood into a test tube containing 2.0 mL of test reagent (contains sodium dithionite, which produces low oxygen tension).
2. If the solution does not become cloudy within 5 minutes the test is negative.
3. Place a drop of solution on a slide, add a coverslip and compare your cells with those in figure 11.6b. If available, compare your sample to that of a prepared slide of sickle cell anemia.
4. Record your data in the laboratory report.

B. Inheritance of PTC Taste

The ability to taste PTC paper (phenylthiocarbamide) is inherited as an autosomal dominant trait. Therefore, if T is a taster and t is a nontaster, tasters have the genotype TT or Tt and nontasters have the genotype tt.

PROCEDURE

1. Taste the PTC paper by leaving a strip of it on the tongue for a minute or so. If the paper has an unpleasantly bitter taste, you are a taster.
2. Determine the number of tasters and nontasters, calculate the proportion of each in the class, and enter this data in your laboratory report.

C. Sex-Linked Traits: Inheritance of Color Blindness

The sex of an individual is determined by one pair of the twenty-three pairs of chromosomes inherited from the parents. These are the sex chromosomes, X and Y. The female has the genotype XX and the male has the genotype XY. Traits that are determined by genes located on the X sex chromosome (as opposed to the other, autosomal chromosomes) are called **sex-linked traits** (the Y chromosome apparently carries very few genes).

Unlike the patterns of heredity previously considered, where the genes are carried on autosomal chromosomes, the inheritance of genes carried on the X chromosome follows a different pattern for males than for females. This is because the male inherits only one X chromosome (and only one set of sex-linked traits) from his mother, whereas the female inherits an X chromosome from both parents.

The genes for color vision and for some of the blood-clotting factors are carried on the X chromosome, where the phenotypes for **color blindness** and **hemophilia** are recessive to the normal phenotypes. A normal female

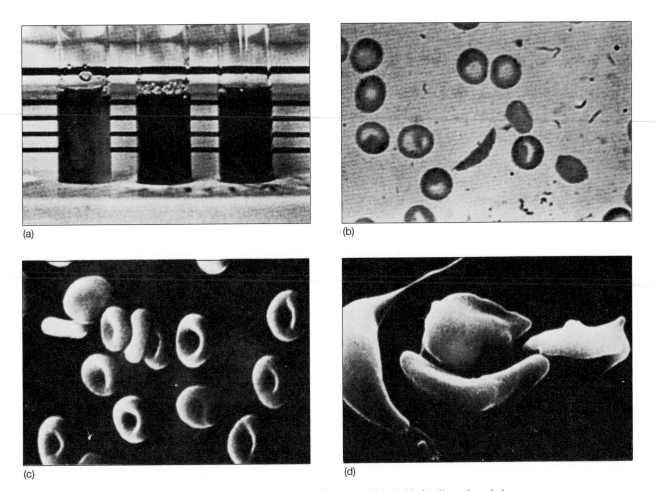

Figure 11.6   Sickle-cell anemia. (a) A turbidity test for sickle-cell anemia. (b) Sickled cells under a light microscope. (c) Normal red blood cells under a scanning electron microscope. (d) Sickled red blood cells under a scanning electron microscope.

may have either the homozygous or the heterozygous ("carrier") genotypes, whereas a male must have either the normal or the affected phenotypes.

Let's suppose a man with the normal phenotype mates with a woman who is a carrier for color blindness (C is normal, c is color blind).

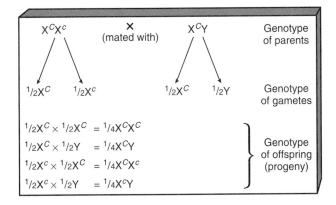

The probability that a child formed from this union will be color blind is one in four (25%). And in the event of a color-blind offspring, that offspring is certain to be male (100% probability). All female children formed from this union will of course have the normal phenotype, but the probability that a given female child will be a carrier for color blindness is one in two (50%).

The perception of color is due to the action of certain photoreceptor cells, known as **cones,** in the retina of the eye. According to the *Young-Helmholtz theory* of color vision, the perception of all the colors of the visible spectrum is due to the stimulation of only three types of cones—*blue, green,* and *red.* Their names refer to the regions of the wavelength spectrum at which each type of cone is maximally stimulated. When one of these three types of cones is defective owing to the inheritance of a sex-linked recessive trait, the ability to distinguish certain colors is diminished.

PROCEDURE

In the *Ishihara test,* colored dots are arranged in a series of circles in such a way that a person with normal vision can see a number embedded within each circle. By contrast, a color-blind person will see only an apparently random array of colored dots.

Laboratory Report 11.3

Name _____

Date _____

Section _____

DATA FROM EXERCISE 11.3

A. Sickle-Cell Anemia

Was your test positive or negative? _____

Describe the appearance of your red blood cells in the microscope; compare to those of a prepared slide of sickle-cell disease.

B. Inheritance of PTC Taste

1. Are you a taster? _____
2. Enter the number of tasters and nontasters in your class in the table below.
3. Calculate the proportion of tasters (the number of tasters divided by the total number of students). Enter this value in the table below.

Phenotype	Number in Class	Proportion of Tasters
Tasters		
Nontasters		

C. Sex-Linked Traits: Inheritance of Color Blindness

Are you color-blind? YES or NO (circle one). If YES, what type of color blindness do you have? _____

REVIEW ACTIVITIES FOR EXERCISE 11.3

Test Your Knowledge of Terms and Facts

1. Chromosomes other than the sex chromosomes are called _____ chromosomes.
2. If a person has two identical genes for a trait, the person is said to be _____ for that trait.
3. If a person inherits a different gene from one parent than the other for a trait, the person is said to be _____ for that trait.
4. Genes inherited on the X chromosomes code for _____ traits.
5. The physical manifestation of a genotype is called a _____ .

6. Describe the inheritance of sickle-cell disease. Include the terms genotype, phenotype, homozygous, heterozygous, dominant, and recessive in your description.

7. Describe the inheritance of color blindness, and explain why color blindness is much more common in men than women.

Test Your Ability to Analyze and Apply Your Knowledge

8. If a man with sickle-cell disease marries a woman with sickle-cell trait, what is the probability that their children will have (a) sickle-cell trait, and (b) sickle-cell disease?

9. A man with normal blood clotting marries a woman who is a carrier for hemophilia, which is a sex-linked trait. What is the probability that their first child will have hemophilia? If this child is hemophilic, what is its sex? What is the probability that this couple's next child will have hemophilia? Explain.

Appendix 1

Basic Chemistry

Atoms

All of the matter on Earth, living as well as nonliving, is composed of about one hundred different types of atoms. Each atom consists of a central positively charged *nucleus* surrounded by a region containing one or more swiftly moving, negatively charged *electrons*. A given electron can occupy any position in a certain volume of space surrounding the nucleus. The outer boundary of this volume of space is called the *orbital* of the electron. Orbitals are like *energy shells*, or barriers, beyond which the electron usually does not pass; they are often represented as a series of concentric circles around the nucleus. The nucleus makes up most of the mass of the atom but accounts for only a tiny fraction of its volume. It consists of positively charged *protons* and, with one exception (H^1), noncharged particles known as *neutrons*.

Element	Symbol	Atomic Number	Atomic Weight
Hydrogen	H	1	1.01
Carbon	C	6	12.01
Nitrogen	N	7	14.01
Oxygen	O	8	16.00
Sodium	Na	11	23.00
Magnesium	Mg	12	24.31
Phosphorous	P	15	30.97
Sulfur	S	16	32.06
Chlorine	Cl	17	35.45
Potassium	K	19	39.10
Calcium	Ca	20	40.08
Iron	Fe	26	55.85
Copper	Cu	29	63.54
Iodine	I	53	126.90

H^1	H^2 (Deuterium)	H^3 (Tritium)
1 proton	1 proton	1 proton
1 electron	1 electron	1 electron
	1 neutron	2 neutrons

A given element may exist in different forms (**isotopes**) because of the presence of different numbers of neutrons in the nucleus.

The superscript above the symbol of the element is known as the *mass number* and indicates the total number of protons and neutrons in the isotope. Although some isotopes are stable, others (e.g., H^3, or tritium) are unstable and undergo radioactive decomposition, emitting gamma rays (very high energy light) or beta particles (high-energy electrons).

The *atomic number* of an element refers to the number of its protons. The *atomic weight* of an element refers to its weight relative to that of carbon, which is given as 12, and is approximately equal to the number of protons and neutrons in the element.

Chemical Bonds

Each electron shell surrounding the nucleus of an atom can accommodate only a limited number of electrons. From the inner shell outward, the maximum number of electrons is 2, 8, 18, 32, and so forth. The chemical properties of the atoms are determined by the number of electrons in the outer shell. If this number is less than the maximum, the difference can be made up by sharing electrons with another atom. Bonds formed by the mutual sharing of electrons are very strong and are called **covalent bonds.**

Hydrogen, for example, has only one electron and needs one more electron to complete its outer shell. Oxygen has eight electrons, two in its inner shell and six in its outer shell; it needs two more electrons to complete its outer shell. This requirement can be met by sharing electrons with two hydrogen atoms, forming a molecule of water.

$$H\cdot + H\cdot + \cdot \ddot{\underset{..}{O}}\cdot \longrightarrow H:\ddot{\underset{..}{O}}:H, \text{ or } H-O-H, \text{ or } H_2O$$

Oxygen gas is composed of oxygen molecules formed by the covalent bonding of two oxygen atoms. In this case, two pairs of electrons are shared by the two atoms, forming a double bond between them.

$$:\ddot{O}: + :\ddot{O}: \longrightarrow :\ddot{O}::\ddot{O}:, \quad \text{or} \quad O=O, \quad \text{or} \quad O_2$$

An atom of nitrogen has seven electrons, two in its inner shell and five in its outer shell. It requires three electrons to complete its outer shell. This requirement may be met by sharing electrons with three hydrogen atoms, forming a molecule of ammonia, or by sharing three pairs of electrons with another atom of nitrogen, forming a molecule of nitrogen gas.

$$:\dot{N}\cdot + 3H\cdot \longrightarrow \overset{\displaystyle H}{\underset{\displaystyle H}{:\dot{N}:H,}} \quad \text{or} \quad \overset{\displaystyle H}{\underset{\displaystyle H}{N-H,}} \quad \text{or} \quad NH_3$$

$$\cdot\dot{N}: + :\dot{N}\cdot \longrightarrow \dot{N}:::\dot{N}, \quad \text{or} \quad N\equiv N, \quad \text{or} \quad N_2$$

When the electrons are not shared equally, but instead are held by only one of the two nuclei, the atom that captures the electron has a negative charge and the atom that loses the electron has a positive charge. These charged atoms (called *ions*) may be held together by a weak electronic attraction known as an **ionic bond.**

Ions and Electrolytes

When a compound that is held together by weak ionic bonds is dissolved in water, it dissociates into positively charged ions (*cations*) and negatively charged ions (*anions*). These ions can conduct electricity, and hence the original ionic compound is called an **electrolyte.** The most ubiquitous electrolyte is common table salt (NaCl).

$$\underset{\text{Ionic compound}}{NaCl} \longrightarrow \underset{\text{Cation}}{Na^+} + \underset{\text{Anion}}{Cl^-}$$

Some atoms form ionic bonds as a group with other atoms and remain grouped when the ionic compound dissociates. These groups are called *radicals*. Examples of radicals include sulfate (SO_4^{2-}), phosphate (PO_4^{3-}), ammonium (NH_4^+), and hydroxyl (OH^-).

$$\underset{\text{Ammonium sulfate}}{(NH_4)_2SO_4} \longrightarrow 2\,\underset{\text{Ammonium}}{NH_4^+} + \underset{\text{Sulfate}}{SO_4^{2-}}$$

Notice that the sulfate radical has two negative charges and that two ammonium radicals are needed to retain electrical neutrality.

Cation	Symbol	Anion	Symbol
Sodium	Na^+	Chloride	Cl^-
Potassium	K^+	Sulfate	SO_4^{2-}
Calcium	Ca^{2+}	Bicarbonate	HCO_3^-
Magnesium	Mg^{2+}	Phosphate	PO_4^{3-}
Hydrogen	H^+	Hydroxyl	OH^-
Ammonium	NH_4^+	Carbonate	CO_3^{2-}

pH and Buffers

The hydrogen ion concentration of a solution can vary between 10^{-14} molar and zero (see exercise 2.6 for a discussion of molarity). Pure water, which has a hydrogen ion concentration of 10^{-7} molar, is considered neutral.

$$H-O-H \longrightarrow H^+ + OH^-$$

Any substance that increases the H^+ concentration is called an **acid,** and any substance that decreases the H^+ concentration is called a **base.** Bases decrease the H^+ concentration by adding OH^- to the solution. The OH^- can combine with free hydrogen ions to form water.

$$\underset{\text{Hydrochloric acid}}{HCl} \longrightarrow H^+ + Cl^-$$

$$\underset{\substack{\text{Sodium hydroxide}\\ \text{(a base)}}}{NaOH} \longrightarrow Na^+ + OH^-$$

When equal amounts of hydrogen cation and hydroxyl anion are added to a solution, the acid and base neutralize each other, forming water and a **salt.**

$$\underset{\text{Acid}}{HCl} + \underset{\text{Base}}{NaOH} \longrightarrow \underset{\text{Salt}}{NaCl} + \underset{\text{Water}}{H_2O}$$

A convenient way of expressing the hydrogen ion concentration of a solution is by means of the symbol **pH,** which is the negative logarithm of the hydrogen ion concentration.

$$pH = \log \frac{1}{[H^+]}$$

Thus, pure water, with 10^{-7} moles of hydrogen ions/L, has a pH of 7.000. Since the pH is an inverse function of the H^+ concentration, an increase in the hydrogen concentration above that of water (i.e., an *acidic* solution) is indicated by a pH of less than 7.000, whereas a decrease in the H^+ concentration (i.e., a *basic* solution) has a pH between 7.000 and 14. A solution that has 10^{-2} moles of hydrogen ions/L (pH 2) is acidic, whereas one that has 10^{-12} moles of hydrogen ions/L (pH 12) is basic.

Acid	Symbol	Base	Symbol
Hydrochloric acid	HCl	Sodium hydroxide	NaOH
Phosphoric acid	H_3PO_4	Potassium hydroxide	KOH
Nitric acid	HNO_3	Calcium hydroxide	$Ca(OH)_2$
Sulfuric acid	H_2SO_4	Ammonium hydroxide	NH_4OH

A **buffer** is a compound that serves to prevent drastic pH changes when acids or bases are added to a solution. It does this by replacing a strong acid or base (one that ionizes completely) with a weak acid or base (one that does not completely ionize).

$$NaHCO_3 + HCl \longrightarrow H_2CO_3 + NaCl$$

Sodium bicarbonate buffer   Hydrochloric acid   Carbonic acid   Sodium chloride

Notice that the strong hydrochloric acid was replaced by the weaker carbonic acid, thus minimizing the change in pH that would have been induced had HCl been added to the solution in the absence of buffer. The carbonic acid/bicarbonate *buffer system* also minimizes the effect of added base on the pH of the solution.

$$H_2CO_3 + NaOH \longrightarrow NaHCO_3 + H_2O$$

Carbonic acid   Sodium hydroxide   Sodium bicarbonate   Water

Organic Chemistry

The chemistry of organic compounds is based on the ability of *carbon atoms* to form chains and rings with other carbon atoms. Carbon, which has six electrons (two in the first shell and four in the second shell), requires four more electrons to complete its outer shell; hence, it is said to have four *bonding sites*.

Hydrogen (one bonding site)   H:H, or H—H, or H_2

Oxygen (two bonding sites)   H:O:H, or H—O—H, or H_2O

Nitrogen (three bonding sites)   H:N:H, or H—N—H, or NH_3

Carbon (four bonding sites)   H:C:H, or H—C—H, or CH_4

Carbon atoms can be covalently bonded to each other by sharing one pair of electrons (single bond) or by sharing two pairs of electrons (double bonds). Carbon-carbon double bonds are called sites of *unsaturation*, since they do not have the maximum number of hydrogen atoms.

H—C—C—H or CH_3CH_3

$C=C$ or CH_2CH_2

Carbon atoms can be covalently bonded together to form long chains or rings.

or

$$CH_3 - CH_2 - CH_2 - CH_2 - CH_2 - CH_3$$

or

C_6H_{14} (hexane)

or

C_6H_{12} (cyclohexane)

Notice that in the shorthand structural formulas for cyclic carbon compounds, the carbon atoms are represented by the corners of the figure and hydrogen atoms are not indicated.

Cyclic carbon compounds based on the structure of benzene are known as *aromatic* compounds. The common feature of their structural formula is the presence of three alternating double bonds in a six-carbon ring. This structural formula is in a sense misleading, since all the carbons in the aromatic ring are equivalent; hence, double bonds can be indicated between any two carbons in the ring.

or C_6H_6 (benzene)

FUNCTIONAL GROUPS

When carbon atoms are bonded together to form chains or rings, the remaining free bonding sites are available to combine with hydrogen atoms or with other compounds known as *functional groups*. These functional groups are generally more chemically reactive than the hydrocarbon backbone.

Some classes of organic compounds are named according to their functional groups. *Ketones*, for example, have a carbonyl group within the carbon chain, whereas *aldehydes* have a carbonyl group at one end of the chain. *Alcohols* have a hydroxyl group at one end of the chain, whereas *acids* have a carboxyl group at one end of the carbon chain (see p. 402).

Molecules that are identical in terms of the type and arrangement of their atoms but which differ with respect to the spatial orientation of key functional groups are called **optical isomers.** This name derives from the fact that these isomers can rotate plane-polarized light to the right or to the left, depending on the orientation of

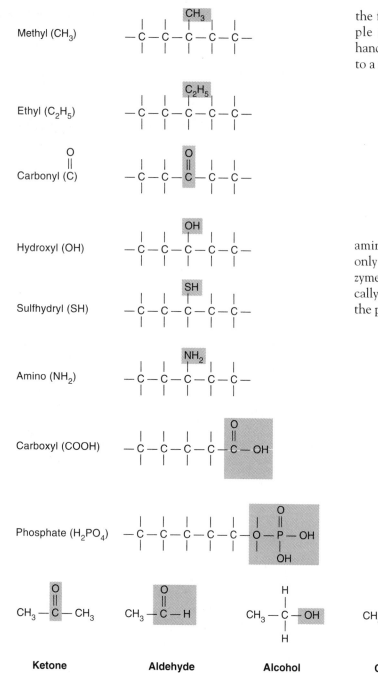

Methyl (CH₃)

Ethyl (C₂H₅)

Carbonyl (C)

Hydroxyl (OH)

Sulfhydryl (SH)

Amino (NH₂)

Carboxyl (COOH)

Phosphate (H₂PO₄)

Ketone **Aldehyde** **Alcohol** **Organic acid**

the functional group. The two optical isomers of the simple sugars and the amino acids are named *D* (right-handed) or *L* (left-handed), according to their similarity to a reference molecule.

L-glyceraldehyde   **D-glyceraldehyde**

Even though a synthetic mixture of simple sugars or amino acids will contain equal amounts of both forms, only one of these two optical isomers can be utilized by enzymes in cellular metabolism. Thus, all of the physiologically significant simple sugars are D-isomers, whereas all of the physiologically significant amino acids are L-isomers.

Sources of Equipment and Solutions

Note: Many of the following vendors are undergoing changes in addresses or telephone numbers, or closure of satellite offices. Many of these companies have toll-free telephone and/or FAX numbers for your convenience. Many larger suppliers have internet home pages, complete with catalogs and help features for ordering supplies through websites as shown, where available, in the following list.

American Scientific Products, Division of American Hospital Supply Corporation

Atlanta
American Scientific Products
1750 Stoneridge Drive
Stone Mountain, GA 30083
(404)943–4070
(800)232–3550 (GA)
(800)241–6640 (Out of state)

Boston
American Scientific Products
20 Wiggins Avenue
Bedford, MA 01730
(617)275–1100
(800)842–1208 (MA)
(800)225–1642 (Out of state)

Charlotte
American Scientific Products
8350 Arrowridge Boulevard
Charlotte, NC 28210
(704)525–1021
(800)432–6997 (NC)
(800)438–1234 (Out of state)

Chicago
American Scientific Products
1210 Waukegan Road
McGaw Park, IL 60085
(312)689–8410
(800)942–4591 (IL)
(800)323–4515 (Out of state)

Cleveland
American Scientific Products
3201 East Royalton Road
Broadview Heights, OH 44147
(216)526–2430
(800)362–9111 (OH)

Columbus
American Scientific Products
2340 McGaw Road
Obetz, OH 43207
(614)491–0050
(800)848–9670 (OH)
(800)282–9640 (Out of state)

Dallas
American Scientific Products
210 Great Southwest Parkway
Grand Prairie, TX 75050
(214)647–2000
(800)492–4820 (TX)
(800)527–6230 (Out of state)

Denver
American Scientific Products
4910 Moline Street
Denver, CO 80239
(303)371–0565
(800)332–1241 (CO)
(800)525–1251 (Out of state)

Detroit
American Scientific Products
30500 Cypress
Romulus, MI 48174
(313)729–6000
(800)482–3740 (MI)
(800)521–0757 (Out of state)

Honolulu
American Scientific Products
274 Puuhale Road
Honolulu, HI 98619
(808)847–1585

Houston
American Scientific Products
4660 Pine Timbers
Houston, TX 77041
(713)462–8000
(800)392–2054 (TX)

Kansas City
American Scientific Products
1118 Clay Street
North Kansas City, MO 64116
(816)221–2533
(800)892–2433 (MO)
(800)821–2006 (Out of state)

Los Angeles
American Scientific Products
17111 Red Hill Avenue
P.O. Box C19505
Irvine, CA 92713
(714)540–5320
(800)432–7141 (CA)

Miami
American Scientific Products
1900 N.W. 97th Avenue
Miami, FL 33152
(305)592–4620

Minneapolis
American Scientific Products
13505 Industrial Park Boulevard
Minneapolis, MN 55441
(612)553–1171
(800)642–3220 (MN)
(800)328–7195 (Out of state)

New Orleans
American Scientific Products
155 Brookhollow Esplanade
P.O. Box 23628
Harahan, LA 70183
(504)733–7571
(800)452–8738 (LA)
(800)535–7333 (Out of state)

New York City
American Scientific Products
100 Raritan Center Parkway
Edison, NJ 08817
(201)494–4000
(201)964–3500 (New York City)
(800)526–7510 (Out of state)

Ocala
American Scientific Products
601 S.W. 33rd Avenue
Ocala, FL 32670
(904)732–3480
(800)342–0191 (FL)

Philadelphia
American Scientific Products
2550 Boulevard of Generals
Valley Forge, PA 19482
(215)631–9300

Phoenix
American Scientific Products
602 West 22nd Street
Tempe, AZ 85282
(602)968–3151
(800)352–1431 (AZ)
(800)528–4471 (Out of state)

Puerto Rico
American Scientific Products
G.P.O. 2796
San Juan, PR 00936
(809)788–1200

Rochester
American Scientific Products
2 Town Line Circle
Rochester, NY 14623
(716)475–1470
(716)856–0114 (Buffalo)
(315)242–0747 (Syracuse)
(800)462–5673 (New York State)

St. Louis
American Scientific Products
10888 Metro Court
Maryland Heights, MO 63043
(314)569–2960
(800)392–4234 (MO)
(800)325–4520 (Out of state)

Salt Lake City
American Scientific Products
P.O. Box 27568
Salt Lake City, UT 84125
(801)972–3032
(800)453–4690, 4691

San Francisco
American Scientific Products
255 Caspian Drive
Sunnyvale, CA 94086
(408)743–3100
(800)538–1670 (northern CA)
(800)672–8610 (southern CA)

Seattle (office handling Alaska)
American Scientific Products
3660 148th Avenue, N.E.
Redmond, WA 98052
(206)885–4131
(800)562–8060 (WA)
(800)426–2950 (ID, MT, OR)
(800)426–6360 (AK)

Washington, D.C.
American Scientific Products
8855 McGaw Road
Columbia, MD 21045
(301)992–0800
(800)638–2813

Bio-Analytic Laboratories, Inc.

3473 Palm City School Road
P.O. Box 388
Palm City, FL 34990
(407)287–3340
(800)327–8282

Carolina Biological Supply Company

Main Office
Burlington, NC 27215
(800)334–5551
West Coast Customers
Powell Laboratory Division
Gladstone, OR 97027
(800)547–1733

Curtin Matheson Scientific, Inc.

General Offices
P.O. Box 1546
Houston, TX 77251–1546
(713) 820–9898
(713) 878–2221 FAX
International Division
9999 Veterans Memorial Drive
Houston, TX 77038–2499
(713) 820–9898
International Sales Only
(713) 878–2444 FAX

Atlanta
955 Cobb Place Boulevard
Kennesaw, GA 30144–6802
(404)590–9006

Boston
17 Jonspin Road
Wilmington, MA 01887–1020
(508)657–5970

Chicago
1225 N. Michael Drive
Wood Dale, IL 60191
(708)766–5670

Cincinnati
7383 Empire Drive, #A
Florence, KY 41042–2925
(606)371–1311

Cleveland
9199 Market Place
Broadview Heights, OH 44147–2835
(216)526–7640

Dallas
2110 Hutton Drive, #1000
Carrollton, TX 75006–6807
(214)243–7500

Denver
20100 E. 35th Drive
Aurora, CO 80011–8154
(303)375–8989

Honolulu
99–1169 Iwaena Street
Aiea, HI 96701–3259
(808)487–7220

Houston
9999 Veterans Memorial Drive
Houston, TX 77038–2499
(713)878–3500

Los Angeles
22745 Savi Ranch Parkway
Yorba Linda, CA 92687–7904
(714)637–2556

Minneapolis
7677 Equitable Drive
Eden Prairie, MN 55344–3676
(612)934–1793

New York
500 American Road
Morris Plains, NJ 07950–2452
(201)644–9500

Orlando
6301 Hazeltine National Drive, #100
Orlando, FL 32822–5119
(407)859–0300

San Francisco
48201 Fremont Boulevard
Fremont, CA 94538–6510
(510)683–6733

Seattle
822 S. 333rd Street
Federal Way, WA 98003–6343
(206)874–4400

Washington, D.C.
8291 Patuxent Range Road, #F
Jessup, MD 20794–8067
(301)498–5210

Fisher General Scientific (SEA)

USA: 1–800–766–7000
FAX: 1–800–926–1166
www.fisher1.com
Canada: 1–800–2FISHER, or
1–800–234–7437
www.fishersci.ca

Medical Analysis Systems, Inc.

Lincoln Technology Park
542 Flynn Road
Camarillo, CA 93012
(800)582–3095
(805)987–7891

Sigma Chemical Company

P.O. Box 14508
St. Louis, MO 63178
(800)325–3010
E-mail: custserv@sial.com
www.sigma.sial.com

Stanbio Laboratory, Inc.

2930 East Houston Street
San Antonio, TX 78202
(800)531–5535
(512)222–2108

VWR Scientific Products, Sargent Welch

P.O. Box 5229
Buffalo Grove, IL 60089–5229
1–800–932–5000
FAX: 1–800–477–4897
E-mail: sarwel@sargentwelch.com
www.sargentwelch.com
www.vwrsp.com

Ward's Biology

P.O. Box 92912
Rochester, NY 14692–9012
1–800–962–2660
FAX: 1–800–635–8439
www.wardsci.com

Warren E. Collins, Inc.

220 Wood Road
Braintree, MA 02184
(617)843–0610

Appendix 3

Multimedia Correlations

This appendix links the physiology exercises in this laboratory manual to important structural and functional concepts provided in the following multimedia products: (1) five CD-ROM modules entitled *InterActive PHYSIOLOGY* by A.D.A.M. and Benjamin/Cummings; (2) 10 laboratory exercises featured in the *Virtual Physiology Lab* CD-ROM by McGraw-Hill and Cypris Publishing; (3) Intelitool *Physiology Laboratory Exercises; and* (4) *Biopac* laboratory exercises. From the list below, find the number of the laboratory exercise currently being studied from the *Laboratory Guide to Human Physiology*, ninth edition, by Stuart Fox. Then determine the appropriate A.D.A.M. *InterActive PHYSIOLOGY* module and topic, the appropriate *Virtual Physiology Lab* exercise, the Intelitool *Physiology Laboratory Exercise* or the *Biopac* laboratory exercise for application, supplementation, or review. Be sure to read the computer system requirements before purchasing any software.

Correlations to A.D.A.M *InterActive PHYSIOLOGY*

Section 3 The Nervous System and Sensory Physiology
Nervous System
Exercise 3.1 Recording the Nerve Action Potential
A.D.A.M. *Topics: The Neuron: The Action Potential*
Orientation
Anatomy Review
Exercise 3.2 Electroencephalogram (EEG)
A.D.A.M. *Topics: Ion Channels*
The Membrane Potential

Section 5 Skeletal Muscles
Muscular System
Exercise 5.1 Neural Control of Muscle Contraction
A.D.A.M. *Topics: Anatomy Review—Skeletal Muscle Tissue*
The Neuromuscular Junction
Sliding Filament Theory
Contraction of Motor Units
Contraction of Whole Muscle

Exercise 5.2 Summation, Tetanus, and Fatigue
A.D.A.M. *Topics: Muscle Metabolism*
Contraction of Motor Units
Contraction of Whole Muscle
Exercise 5.3 Electromyogram (EMG)
A.D.A.M. *Topics: Anatomy Review—Skeletal Muscle Tissue*
The Neuromuscular Junction
Sliding Filament Theory
Contraction of Motor Units
Contraction of Whole Muscle

Section 7 The Cardiovascular System
Cardiovascular System
Exercise 7.1 Effects of Drugs on the Frog Heart
A.D.A.M. *Topics: Heart Physiology—Anatomy Review of the Heart*
Cardiac Cycle
Exercise 7.2 Electrocardiogram (ECG)
A.D.A.M. *Topics: Intrinsic Conduction System*
Cardiac Action Potential
Exercise 7.3 Effects of Exercise on the Electrocardiogram
A.D.A.M. *Topics: Cardiac Output*
Exercise 7.5 Heart Sounds
A.D.A.M. *Topics: Heart Physiology—Anatomy Review of the Heart*
Anatomy Review of Blood Vessel Structure and Function
Cardiac Cycle
Exercise 7.6 Measurements of Blood Pressure
A.D.A.M. *Topics: Heart Physiology—Anatomy Review of the Heart*
Anatomy Review of Blood Vessel Structure and Function
Measuring Blood Pressure
Factors That Affect Blood Pressure
Blood Pressure Regulation
Autoregulation and Capillary Dynamics

Section 8 Respiration and Metabolism
Respiratory System
Exercise 8.1 Measurements of Pulmonary Function
A.D.A.M. *Topics: Anatomy Review— Respiratory Structures*
Pulmonary Ventilation
Gas Exchange

Correlations to *Virtual Physiology Lab* CD-ROM

Correlations to the *Intelitool Physiology Laboratory Exercises*

Correlations to the *Biopac Student Lab Exercises*

Line Art

Fig 2.5: Courtesy of Gelman Sciences, Inc.; Fig 6.9: Adapted from A. Marchand, "Case of the Month, Circulating Anticoagulants: Chasing the Diagnosis" in Diagnostic Medicine, June 1983, page 14. Copyright © 1984 Medical Economics Company, Inc., Oradell, NJ; Fig 7.17: Courtesy of Hewlett-Packard Company. Text Art 7.2 Courtesy of Hewlett-Packard Company; Fig 7.22: Modified from L. A. Geddes, The Direct and Indirect Measurement of Blood. Copyright © 1970 Year Book Medical Publishers. Used by permission of Mosby-Year Book, Inc., Chicago, IL; Fig 8.10: Courtesy of Warren E. Collins, Inc., Braintree, MA.

Photos

Chapter 1
1.1: Courtesy of Reichert Scientific Instruments; 1.4a-1: © Ray Simons/Photo Researchers, Inc.; 1.4 b-1and 1.4c-1: © Edwin A. Reschke; 1.5 (all): © Edwin A. Reschke;1.6a-1: © Biology Media/Photo Researchers, Inc.; 1.6b-1 to 1.6c-1: © Edwin A. Reschke; 1.6d-1 © Edwin A. Reschke/Peter Arnold, Inc.; 1.7a-1: © Edwin A. Reschke; 1.7b-1 and c-1: Edwin Reschke/Peter Arnold, Inc.; 1.8a-1, 1.9a-1 to 1.9c-1: © Edwin A. Reschke;1.10a: © John D. Cunningham/Visuals Unlimited; 1.10b: © Martin M. Rotker/Photo Researchers, Inc.

Chapter 2
2.1ab: © Provided by Stuart I. Fox, Ph.D.; 2.2: Courtesy of Bausch & Lomb Analytical Systems Division; 2.4: © Provided by Stuart I. Fox, Ph.D.; 2.6a-h: Courtesy of Pall Gelman Sciences; 2.11: Copyright © R.G. Kessel and R. Kardon, Tissues and Organs: A Text Atlas of Scanning Electron Microscopy. Copyright © 1979 W.H. Freeman and Company.; 3.12a-e: Courtesy of Dr. Sheril Burton assisted by Dr. Douglas W. Hacking; 3.19: From D. Vaugh and T. Asbury, General Opthmology, 7th edition, 1974. Copyright © Lange Medical Publications, Los Angeles, CA

Chapter 4
4.2ab: and 4.4: © Edwin A. Reschke; 4.6: © Martin M. Rotker; 4.9a: © R. Kessel/Visuals Unlimited; 4.9b: © Provided by Stuart I. Fox, Ph.D.; 5.1b: By permission of Dr. H.E. Huxley

Chapter 5
5.2 and 5.3: Courtesy of Narco Bio-Systems, John F. Pritchett, photographer; 5.4ab, 5.5ab, 5.6a-c, 5.9, 5.10: © Provided by Stuart I. Fox, Ph.D.; 5.13ab: Courtesy Kent M. Van De Graaff; 5.14, 5.15ab: © Provided by Stuart I. Fox, Ph.D.

Chapter 6
6.2, 6.3ab, 6.4, 6.6: © Provided by Stuart I. Fox, Ph.D
Plate 2: © Provided by Stuart I. Fox, Ph.D.

Chapter 7
7.2ab, 7.4ab, 7.5, 7.20a: © Provided by Stuart I. Fox, Ph.D.

Chatper 8
8.1: Courtesy of Warren E. Collins, Inc.; 8.2: Courtesy of Phipps & Bird, Inc./Intelitool; 8.6-8.9, 8.11-8.12: © Provided by Stuart I. Fox, Ph.D

Chatper 9
9.1b: © Biology Media/Photo Researchers, Inc.; 9.3ab: © Provided by Stuart I. Fox, Ph.D.; 9.6 (all): Courtesy of Abbott Laboratories; 9.8 (all): Courtesy of Hycor Biomedical, Inc.

Chapter 10
10.3 and 10.4: © Edwin A. Reschke; 10.6: From Mariano S.H. DiFore, Atlas of Human Histology, 1981. Copyright © Williams & Wilkins, a Waverly Company; 10.7a: © Edwin A. Reschke; 10.7b: Copyright © R.G. Kessel and R. Kardon, Tissues and Organs: A Text Atlas of Scanning Electron Microscopy. Copyright © 1979 W.H. Freeman and Company.; 10.8c: © Victor B. Eichler, Ph.D.; 10.9a: Copyright © R.G. Kessel and R. Kardon, Tissues and Organs: A Text Atlas of Scanning Electron Microscopy. Copyright © 1979 W.H. Freeman and Company.; 10.11: © Edwin A. Reschke

Chapter 11
11.4a-c: © Provided by Stuart I. Fox, Ph.D.;11.5ab: © Provided by Stuart I. Fox, Ph.D.; 11.6a-d: From P.R. McCurdy, Sickle Cell Disease, © 1973 Medcom, Inc.

Index

Note: Page references followed by the letters *f* and *t* indicate figures and tables, respectively.

Sympathomimetic drugs, 235
Symphysis pubis, 382*f*
Synapse, 93
Synaptic cleft, 93, 94*f*
Synaptic integration, 93, 95*f*
Synaptic transmission, 93, 94*f*
Synaptic vesicle, 93, 94*f*
Systole, 241, 242, 251, 261
Systolic pressure, 271

T

T_3. *See* Triiodothyronine
T_4. *See* Tetraiodothyronine
Tachycardia
    paroxysmal atrial, 246
    supraventricular, 246
    ventricular, 246, 246*f*
Tallquist paper blot method, 203
Target organs, 148
Taste buds, 139, 140*f*, 141*f*
Taste perception, 139–144, 141*f*
    modalities of, 139
    olfaction and, 140
    of phenylthiocarbamide, inheritance
        of, 394
Taste pore, 140*f*
Tay-Sachs disease, 69
Tectorial membrane, 127, 129*f*
Teeth, 344*f*
Telophase, 7*t*, 8*f*
Telophase I, 9*f*
Telophase II, 9*f*
Temperature optimum, 60, 60*f*
Temperature receptors
    adaptation of, 110
    mapping of, 107–108
Temporal bone, 128*f*
Temporal lobe, inferior, 124
Testis, 6, 382*f*
    endocrine function of, 147*t*, 150, 160
    histology of, 150*f*, 150–151
    regulation of, 381
Testosterone, 147*t*, 150, 160, 160*f*, 162
Tetanus, 179–185, 180*f*
    in frog gastrocnemius muscle, 180
    in human muscle, 180–182
Tetany, 179, 180
Tetraiodothyronine (T_4), 147*t*,
    152–153, 372
Theca interna, 149*f*
Theophylline, 239
Theta waves, 94, 96*f*
Thick filaments, 169
Thin filaments, 169
Thin-layer chromatography
    of amino acids, 49–52
    of steroids, 161–162
Thin-layer plate, 49, 161
Thistle tube setup, 76, 76*f*
Threshold potential, 85
Threshold stimulation, 174
Thrombin, 223, 224*f*
Thrombocytes, Color Plate 1, 17, 18*f*, 197
Thromboplastin, tissue, 223, 224*f*
Thromboplastin time, activated partial,
    224, 224*t*, 225
Thrombosis, coronary, 252
Thrombus, 38
Thymine, 68
Thymopoietin, 147*t*
Thymus
    endocrine function of, 147*t*
    in lymphocyte production, 210
Thyrocalcitonin. *See* Calcitonin
Thyroid follicles, 152, 153*f*
Thyroid gland
    endocrine function of, 147*t*, 152–153
    histology of, 152–153, 153*f*

Thyroid stimulating-hormone (TSH), 153
Thyroxine. *See* Tetraiodothyronine
Tickle, sensation of, 107
Tidal volume (TV), 284
    calculation of, 286
    in spirometry recording, 284*f*,
        287*f*, 288*f*
Tissue(s), 13–24. *See also specific types*
    definition of, 1, 13
    microscopic examination of, 13–20
    in skin, 20
    types of, 13
Tissue macrophages, 210
Tissue thromboplastin, 223, 224*f*
TLC. *See* Total lung capacity
T lymphocytes, 210
Tongue, 139, 140*f*, 141*f*, 344*f*
Tonicity, 76–78
Tonometer, 120
Tonsils, 140*f*
Total lung capacity (TLC), 283, 284*f*,
    289, 290*t*
Touch sensation
    receptors for
        distribution of, 107, 109
        mapping of, 107–108
        two-point threshold in, 109–110
Trace elements, 372
Training cardiac rate, 275, 275*t*
Transcription, 68, 68*f*
Transducer
    in physiograph, 171, 172*f*
    sensory, 110
Transfer RNA (tRNA), 68
Transitional epithelium, 14, 15*f*
Translation, 68, 68*f*
Transmittance, percent, 35, 36
Transport
    active, 74
    of oxygen, 199–200, 281, 305
    passive, 73–74
Transport maximum, for glucose, 336
Transversus abdominis, 285*f*
Tremor, flapping, 190
Triceps brachii muscle, 189
Triceps-jerk reflex, 103, 104*f*
Tricuspid valve, 230*f*
Trigeminal ganglion, 141*f*
Trigeminal nerve, 141*f*
Triglycerides, 38, 371
Triiodothyronine (T_3), 147*t*, 152
Trimers, 35
tRNA. *See* Transfer RNA
Trochlea, 121*f*
Trophic hormones, 147*t*, 154*f*
Tropomyosin, 169
Troponin, 169
Trypsin, 59, 351*t*, 354, 361
TSH. *See* Thyroid stimulating-hormone
Tunics, 347
Turbulent flow, 269
TV. *See* Tidal volume
T wave, 242
Twitch, 179
    in frog gastrocnemius muscle, 180
    in human muscle, 180–182
Two-point threshold test, 109–110
Tympanic cavity, 128*f*
Tympanic membrane, 127, 128*f*
Tympanitis, 128
Tyrosine, 68
    in urine, 339*f*, 340*f*

U

Ulcers
    duodenal, 361
    gastric, 110, 360, 361
Units of activity, 60

Unmyelinated axons, 19
    conduction of action potentials in,
        87, 87*f*
Unopette system, 200, 200*f*, 201*f*
Unsaturated fatty acids, 38
Unsaturation, sites of, 38, 401
Uracil, 68
Urea
    measurement of, 331–332
    normal range of, 332
    production of, 355
    renal plasma clearance of,
        329–334, 330*f*
Urea nitrogen, blood, 330
Ureter, 321, 322*f*, 382*f*
Urethra, 338*f*, 381, 382*f*
Uric acid crystals, 337, 339*f*, 340*f*
Urinary bladder, 382*f*
Urine
    abnormal appearance of, 335*t*
    chloride concentration of, 324–326
    chorionic gonadotropin in,
        389–390, 390*f*
    clinical examination of, 335–343
    formation of, 321
    pH of, 325
    sediment, 337, 338*f*, 339*f*, 340*f*
    specific gravity of, 325, 325*f*
    volume of, 325
Urinometer float, 325, 325*f*
Uterine tubes, 381, 382*f*
Uterus, 381, 382*f*
Utricle, 133, 134*f*
Uvula, 141*f*
U wave, 242

V

Vacuole, 7*t*
Vagina, 382*f*
Vaginal fornix, 382*f*
Vaginal orifice, 382*f*
Vaginal smear, 383–388
Vagus nerve, 139, 141*f*
Vallate papillae, 140*f*
Valves, heart, 229, 230*f*
Vas (ductus) deferens, 150, 381, 382*f*
Vasa recta, 322*f*
Vascular tree, 229
Vasoconstriction, 223
Vasopressin. *See* Antidiuretic
    hormone
VC. *See* Vital capacity
Veins
    arcuate, 322*f*
    central, 118*f*, 351, 352*f*, 353*f*
    hepatic, 351, 353*f*
    hepatic portal, 351, 352*f*
    interlobular, 322*f*
    pulmonary, 230*f*
    renal, 322*f*
    structure of, 229
Vena cava
    inferior, 230*f*
    superior, 230*f*
Venous sinus, 120
Ventilation, 281
    average frequency of, 301
    functions of, 313
    regulation of, carbon dioxide in, 316
Ventral root, 101
Ventricles
    left, 229, 230*f*
    mean electrical axis of, 255–260
    right, 229, 230*f*
Ventricular contractions, premature,
    246, 246*f*
Ventricular fibrillation, 246, 246*f*
Ventricular tachycardia, 246, 246*f*

Venules
    in skin, 21*f*
    structure of, 229
Vertigo, 134
Vesicles, 6*f*
    seminal, 150, 382*f*
    synaptic, 93, 94*f*
Vestibular apparatus, 133–138, 134*f*
Vestibular membrane, 129*f*
Vestibular nerve, 128*f*, 134*f*
Vestibular nystagmus, 133
Vestibulocochlear nerve, 129*f*, 133
Villi, 345, 348*f*, 350, 350*f*
Virtual Physiology Lab CD-ROM, x–xi
Vision, 117–126
    color, 124, 394–395
    near point of, 120
    photopic, 123
    scotopic, 123
Visual acuity, 119
    test for, 119
Visual cortex, 124
Vital capacity (VC), 283
    calculation of, 289–290
    predicted values for, 289–290
        for females, 293*t*
        for males, 294*t*
    in spirometry recording, 284*f*, 287*f*
    timed, 290–292, 291*f*
Vitamin(s)
    fat-soluble, 371, 372
    water-soluble, 372
Vitamin A_1, 124
Vitamin D_1, 147*t*
Vitamin K deficiency, 224
Vitreous humor, 117, 118*f*
Voltage. *See* Potential difference

W

Waist-to-hip ratio, 371
Warmth, 107
Water, as essential nutrient, 372
Water bath, constant-temperature,
    negative feedback in, 25, 26
Water-soluble vitamins, 372
Waxy cast, 338*f*
Weber's test, 127–128
Weight
    atomic, 399
    equivalent, 324–325
    ideal body, 373
Wenckebach's phenomenon, 245
White blood cell(s). *See* Leukocytes
White blood cell count
    differential, Color Plate 1, 211–212
    total, 210
Wright's stain, 211–212

Y

Young-Helmholtz theory of color vision,
    124, 395

Z

Z lines, 170*f*
Zona fasciculata, 152, 152*f*
Zona glomerulosa, 151, 152*f*
Zona pellucida, 149*f*, 150
Zona reticularis, 152, 152*f*
Zonular fibers, 118*f*
Zwitterion formula, 53
Zygote, 381